O. N. Hartmann · P. Bühler · J. Marton · K. Suzuki ·
E. Widmann · J. Zmeskal
Editors

EXA 2011

Proceedings of the International Conference on Exotic Atoms and Related Topics (EXA 2011) held in Vienna, Austria, 5-9 September 2011

Previously published in *Hyperfine Interactions*
Volumes 209 to 211, 2012

Springer

Editors
Olaf N. Hartmann
Austrian Academy of Sciences
Stefan-Meyer-Institute for Subatomic Physics
Boltzmanngasse 3, 1090 Vienna, Austria
olaf.hartmann@oeaw.ac.at

Paul Bühler
Austrian Academy of Sciences
Stefan-Meyer-Institute for Subatomic Physics
Boltzmanngasse 3, 1090 Vienna, Austria
paul.buehler@oeaw.ac.at

Johann Marton
Austrian Academy of Sciences
Stefan-Meyer-Institute for Subatomic Physics
Boltzmanngasse 3, 1090 Vienna, Austria
johann.marton@oeaw.ac.at

Ken Suzuki
Austrian Academy of Sciences
Stefan-Meyer-Institute for Subatomic Physics
Boltzmanngasse 3, 1090 Vienna, Austria
ken.suzuki@oeaw.ac.at

Eberhard Widmann
Austrian Academy of Sciences
Stefan-Meyer-Institute for Subatomic Physics
Boltzmanngasse 3, 1090 Vienna, Austria
eberhard.widmann@oeaw.ac.at

Johann Zmeskal
Austrian Academy of Sciences
Stefan-Meyer-Institute for Subatomic Physics
Boltzmanngasse 3, 1090 Vienna, Austria
johann.zmeskal@oeaw.ac.at

ISBN 978-94-007-4889-7
Springer Dordrecht Heidelberg New York London

Library of Congress Control Number: 2012941295

Printed on acid-free paper

Springer is part of Springer Science+Business Media (www.springer.com)

Table of Contents

Exotic Atoms

Hadron Physics

Kaonic Atoms

Precision Experiments

Strangeness in Matter

Facilities, Experiment, Instrumentation

Symposium

Hyperfine Interact (2012) 209:1–7
DOI 10.1007/s10751-012-0628-1

EDITORIAL

Foreword

O. Hartmann · P. Bühler · J. Marton · K. Suzuki · E. Widmann · J. Zmeskal

Published online: 20 April 2012

This proceedings volume contains the contributions of the 4th EXA Conference EXA 2011 which took place in Vienna, in the Austrian Academy of Sciences from September 5th to 9th, 2011. Like the previous EXA Conferences it was organized by SMI (Co-Chairmen: J. Marton, E. Widmann and J. Zmeskal). EXA 2011 attracted more than 120 scientists working in the broad research field covered by this conference on exotic atoms and related topics. The Conference was opened by Eberhard Widmann and overall 17 sessions devoted to the broad scientific scope followed. In the first session the extremely successful two decades of research on antiprotonic helium was celebrated in an overview talk by Ryu Hayano followed by presentations about the current research and results in theory and experiment. The recent advances in antihydrogen studies and CPT and gravity tests with antimatter were discussed by distinguished speakers from the collaborations working at the AD of CERN. Another session was devoted to other exotic atoms where an exciting new result on the size of the proton was one of the highlights.

Hot topics in precision experiments - using different probes like neutrons and muons - were discussed. The topic strong interaction with strangeness in the low energy regime was presented in talks on kaonic atoms and nuclei. The new result of the SIDDHARTA experiment at DAFNE/LNF on the strong interaction observables from kaonic hydrogen - representing another highlight of EXA 2011- was discussed by M. Cargnelli/SMI. On the occasion of the 80th birthday of P. Kienle a symposium „Creativity-Innovation-the Seed for Frontier Science“ took place in the framework of EXA 2011. The talks were given by distinguished scientists and summarized the scientific topics in which P. Kienle delivered and delivers new ideas with crucial impact.

Many young scientists attended the Conference and gave very interesting talks respectively poster presentations about their work. This fact is an extremely good

O. Hartmann · P. Bühler · J. Marton · K. Suzuki · E. Widmann (✉) · J. Zmeskal
Stefan Meyer Institute for Subatomic Physics, Austrian Academy of Sciences, Boltzmanngasse 3, 1090 Vienna, Austria
e-mail: eberhard.widmann@oeaw.ac.at

sign for the future research in the broad scientific field covered by the EXA Conference series.

The organizers are very grateful to the members of the International Advisory Committee of EXA 2011: S. Bertolucci, T. Ericson, A. Gal, C. Guaraldo, R. Hayano, P. Indelicato, K. Jungmann, P. Kienle, H. Koch, J. Marton, K. Nagamine, S. Nagamiya, S. Paul, H. Rauch, D. Schwalm, B. Sharkov, H. Stöcker, E. Widmann, U. Wiedner, T. Yamazaki and J. Zmeskal for their impact in establishing an exciting scientific programme. We also thank the Austrian Academy of Sciences for hosting the Conference in the Theatersaal of the Academy.

The next conference of the EXA series will be hosted and organized again by the Stefan Meyer Institute and will take place in Vienna.

DI Robert Alexander Herbst

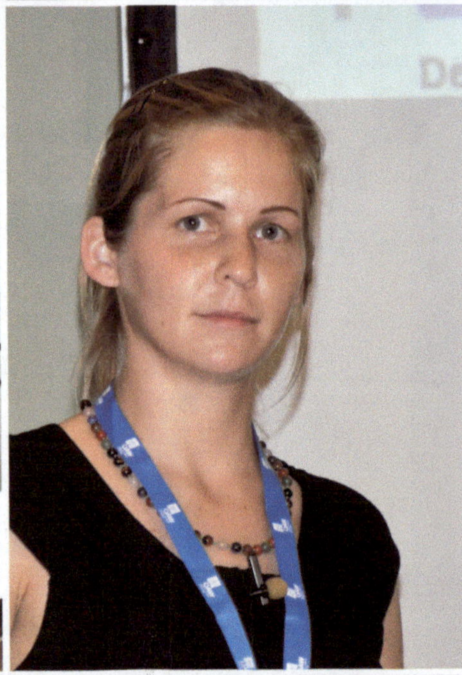

SALON
MONTLEART

Hyperfine Interact (2012) 209:9–14
DOI 10.1007/s10751-011-0527-x

20 years of antiprotonic helium

Overview and history

Ryugo S. Hayano

Published online: 31 January 2012

Abstract Twenty years ago, we published a paper entitled "Discovery of antiproton trapping by long-lived metastable states in liquid helium". In retrospect, this was the discovery of antiprotonic helium atoms, the study of which is actively being done at CERN's antiproton decelerator. A brief overview of this interesting exotic atom is given, together with some historical background.

Keywords Antiprotonic helium atoms · Fundamental constants · Antiproton decelerator

1 Introduction

It was in late 80's at the 12-GeV proton synchrotron at KEK, Japan, when we first discovered an intriguing anomaly. Our aim then was to search for a Σ hypernucleus, a nucleus containing a Σ hyperon [1]. The experiment was done by stopping K^- in a liquid helium target, and by measuring the momenta of outgoing pions using a magnetic spectrometer. In this experiment we obtained a momentum spectrum of negatively-charged particles as shown in Fig. 1, and it is in this spectrum we found the aforementioned anomaly [2].

The peak at 255 MeV/c is due to the formation of ${}^{4}_{\Lambda}$He hypernuclei, and the bump around 170 MeV/c corresponds to the Σ hypernuclear formation. The dominant peak at 235 MeV/c and another at 205 MeV/c are due respectively to $K^- \to \mu^- \bar{\nu}_\mu$ ($K_{\mu 2}$) and $K^- \to \pi^- \pi^0$ ($K_{\pi 2}$) weak decays. The observation of these peaks was unexpected, because the K^- cascade time is known to be ~ps, much shorter than the kaon lifetime of 12 ns.

Supported in part by the Japanese MEXT grant-in-aid 20002003.

R. S. Hayano (✉)
Department of Physics, The University of Tokyo,
Tokyo, Japan
e-mail: hayano@phys.s.u-tokyo.ac.jp

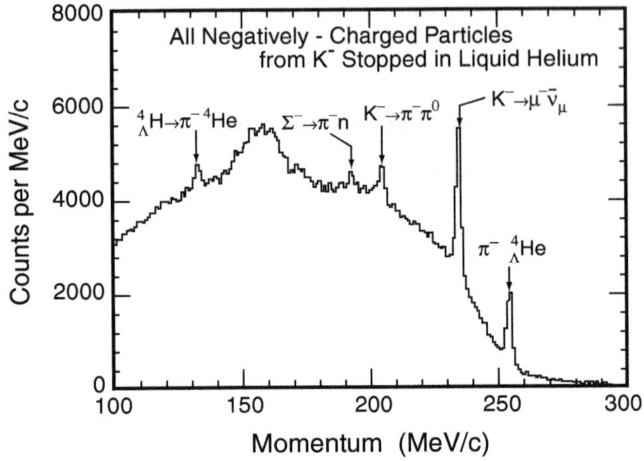

Fig. 1 Momentum spectrum of negatively charged particles emitted after K^- stopping in liquid helium. From [2]

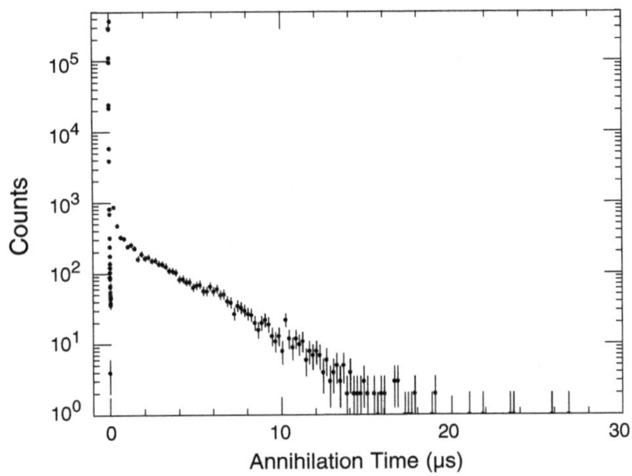

Fig. 2 The antiproton annihilation time spectrum measured by stopping antiprotons in liquid helium at KEK. From [3]

In a follow-up experiment we found a more spectacular effect of $\bar{p}$ longevity in helium [3]. In the $\bar{p}$ annihilation time spectrum shown in Fig. 2, the peak at $t = 0$ is due to the normal, prompt annihilation. This however is followed by an anomalous delayed component, indicating that about 3% of antiprotons trapped in liquid helium survive with a mean lifetime of 3 μs. This was the beginning of our long series of experiments on antiprotonic helium.

2 Metastability of antiprotonic helium

The anomalous longevity is now ascribed to the formation of the antiprotonic helium atoms (hereafter denoted $\bar{p}\text{He}^+$), which have the following remarkable features:

1. The metastability occurs when the antiproton occupies a near-circular orbit having a large $n(\sim 38)$ and also large $\ell(> 35)$.

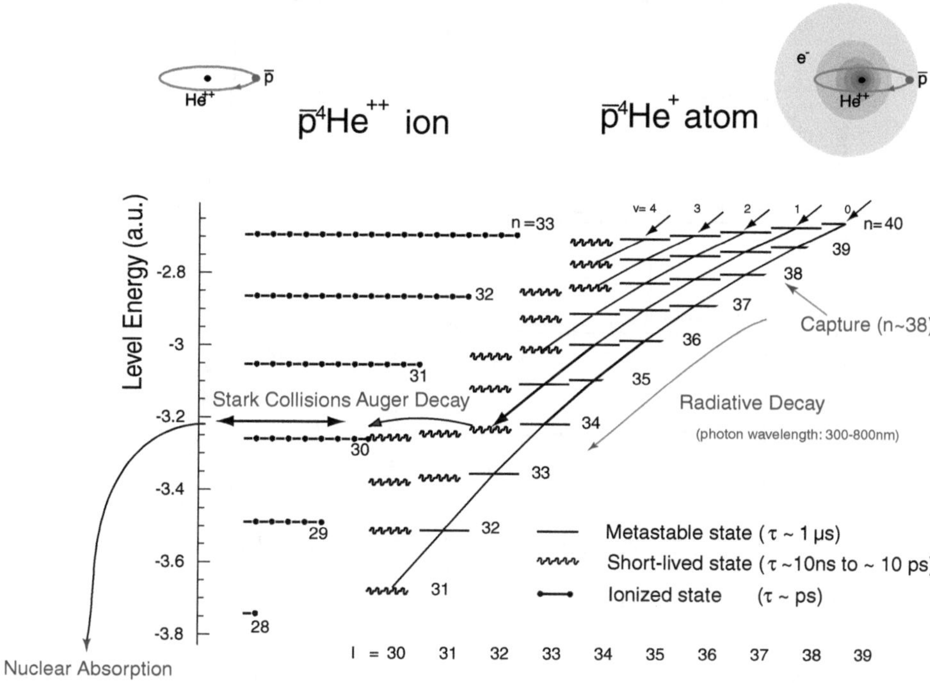

Fig. 3 The level diagram of $\bar{p}^4He^+$ in relation to that of $\bar{p}^4He^{++}$. The continuous and wavy bars stand for metastable and short-lived states, respectively, and the dotted lines are for ℓ−degenerate ionized states

2. We usually use low-temperature ($T \sim 10$ K) helium gas as the target. The $\bar{p}^4He^+$ atoms that are produced collide with the surrounding helium atoms and are thermalized, without being destroyed. Therefore, the antiprotonic helium atoms are already cold and are well suited for high-precision spectroscopy.

Figure 3 shows an energy level diagram of $\bar{p}^4He^+$. The levels indicated by the continuous lines have metastable (>1 μs) lifetimes and de-excite radiatively, while the levels shown by wavy lines are short lived (< 10 ns) and de-excite by Auger transitions to antiprotonic helium ion states (shown by dotted lines). Since the ionic states are hydrogenic, Stark collisions quickly induce antiproton annihilation on the helium nucleus, also indicated in Fig. 3.

3 Principle of $\bar{p}^4He^+$ laser spectroscopy

Laser spectroscopy of $\bar{p}^4He^+$ works as follows: As shown in Fig. 3, there is a boundary between metastable states and short-lived states. For example, $(n, \ell) = (35, 33)$ is metastable, while $(n, \ell) = (34, 32)$ is short lived. Thus, if we use a laser to induce a transition from (35,33) to (34,32), (and of course if an antiproton happens to be occupying the (35,33) level at the time of laser ignition), the antiproton is deexcited to the short-lived state, which then Auger decays to an ionic $(n_i, \ell_i) = (30, 29)$ state within < 10 ns. The ionic state is then promptly destroyed by Stark

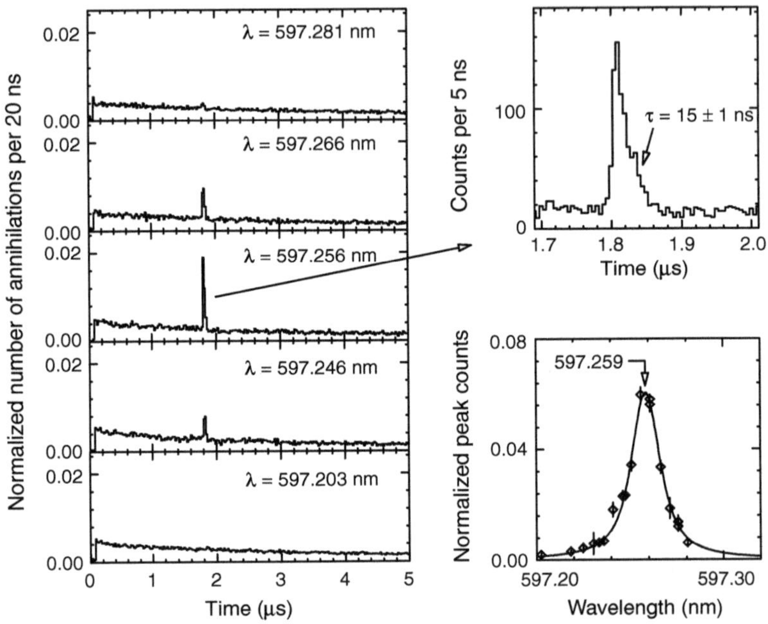

Fig. 4 Laser resonance of the (39, 35) → (38, 34) transition in $\bar{p}^4He^+$. *Left* Observed time spectra of delayed annihilation of antiprotons with laser irradiation of various wavelengths near 597.2 nm. *Upper right* Enlarged time profile of the resonance spike. *Lower right* Normalized peak count versus wavelength in the resonance region. From [4]

collisions, leading to the nuclear absorption or annihilation of the antiproton. As a result, we can observe a sharp increase in the $\bar{p}$ annihilation rate in coincidence with a resonant laser pulse, as shown in Fig. 4 [4]. We measure the intensity of the laser-induced annihilation spike as a function of laser detuning to obtain the transition frequency $\nu_{\rm exp}$. By comparing $\nu_{\rm exp}$ with $\nu_{\rm th}$ calculated by three-body QED theories [5, 6] (Korobov 1997, private communication), we can deduce $m_{\bar{p}}/m_e$.

4 Progress over the years

Figure 5 compares the progress of experiment and theory over the years [7]. At LEAR (left-most points), we reached a precision of $\delta\nu/\nu$ of $0.5 - 1 \times 10^{-6}$ [8]. In our first experiment at the CERN AD, we measured six transition frequencies of $\bar{p}^4He^+$ to $\delta\nu/\nu = 1 - 10 \times 10^{-7}$ [9]. The essential difference between the LEAR and the AD experiments is the time structure of the antiproton beam. At LEAR, antiprotons were slowly extracted from the LEAR ring, so that laser was fired for each $\bar{p}^4He^+$-candidate event which occurred randomly with a mean rate of some 300 Hz. In contrast, the AD provides a short pulse of ∼ 100-ns wide containing some $3 \times 10^7\,\bar{p}$s, repeated every ∼ 100 s. A single laser pulse in this case irradiates some 10^6 metastable atoms. The conventional event-by-event collection of antiproton-annihilation events used at LEAR is impossible with the pulsed beam at AD. We thus developed a new detection scheme based on analogue waveform recording of Čerenkov detectors viewed by gateable photomultipliers.

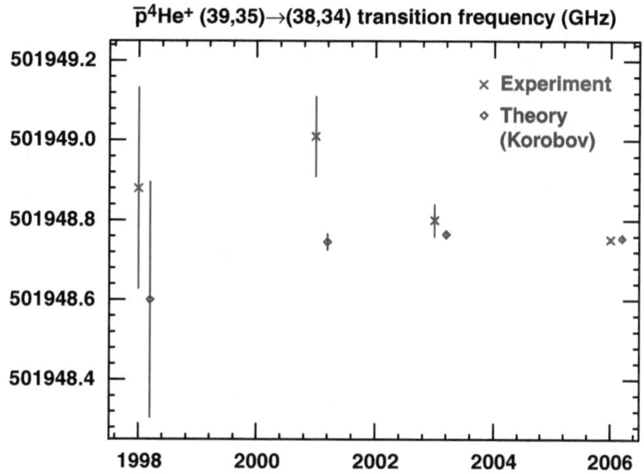

Fig. 5 Comparison of measured and calculated transition frequencies for the $(39, 35) \rightarrow (38, 34)$ transition of $\bar{p}^4$He. From [7]

In the second series at AD, in 2003, we determined seven transition frequencies of $\bar{p}^4He^+$ and 6 of $\bar{p}^3He^+$, with errors of $\delta\nu/\nu \sim 0.5 - 2 \times 10^{-7}$ [10]. This was made possible by constructing a radio-frequency quadrupole decelerator (RFQD), which decelerates the 5.3 MeV antiprotons ejected from the AD ring to < 100 keV with an efficiency of 20–25% [11], so that the antiprotons could be stopped in a much lower-density target having $\rho \sim 10^{17}$ cm^{-3}, removing the uncertainties due to collisional effects.

Having eliminated the collisional shift, the line width of the laser and its frequency calibration remained as the largest contributory factors to the errors on measured values of the resonant frequencies. Only pulsed lasers can provide the megawatt-scale intensities needed here to induce the $\bar{p}He^+$ transitions. However, fluctuations in their frequency and line width and the difficulty of calibrating the wide range of $\bar{p}He^+$ wavelengths (from infrared to ultraviolet) have limited our experimental precision. We circumvented these problems by basing our experiments on a continuous-wave (cw) laser whose frequency ν_{cw} could be stabilized with a precision $< 4 \times 10^{-10}$ against an optical comb. Its intensity was then amplified by a factor 10^6 to produce a pulsed laser beam of frequency $\nu_{pl} \sim \nu_{cw}$ with an accuracy and resolution $1 - 2$ orders of magnitude higher than before. This made it possible for us [12] to reach the precision indicated by the right-most point in Fig. 5.

5 Weighing the antiproton

The ν_{exp} values thus obtained were compared with theoretical values ν_{th} [6] (Korobov 1997, private communication), which included QED and nuclear-size ($\Delta_{nuc} = 2 -$ 4 MHz) effects, and used the 2002 CODATA recommended values for fundamental constants. Theory also provided coefficients for $d\nu_{th}/d(m_p/m_e)$. These we used to determine the antiproton-to-electron mass ratio, by minimizing the sum $\Sigma[\nu_{th}(m_{\bar{p}}/m_e) - \nu_{exp}]^2/\sigma_{exp}^2$ over the 12 transitions, to be 1,836.152674(5) [12]. This result was included in the evaluation of the CODATA 2006 fundamental physical constants [13].

Recently, we succeeded to achieve a higher frequency precision by performing a two-photon resonance (by using two counter-propagating beams), which canceled the first order Doppler width, and determined $m_{\bar{p}}/m_e$ to be 1,836.1526736(23) [14]. This result was included in the evaluation of the CODATA 2010 fundamental physical constants [15]. This, and future prospects, will be discussed in detail by Hori (experiment) [Hori, M., in preparation (2012)] and [Korobov, V.I., in preparation (2012)] in this session.

References

1. Hayano, R.S., et al.: Phys. Lett. B. **231**, 355 (1989)
2. Yamazaki, T., et al.: Phys. Rev. Lett. **63**, 1590 (1989)
3. Iwasaki, M., et al.: Phys. Rev. Lett. **67**, 1246 (1991)
4. Morita, N., et al.: Phys. Rev. Lett. **72**, 1180 (1994)
5. Korobov, V.I.: In: Proceedings of EXA05. Austrian Academy of Sciences Press, Vienna (2005)
6. Korobov, V.I.: Phys. Rev. A **73**, 022509 (2006)
7. Hayano, R.S., et al.: Rep. Prog. Phys. **70**, 1995 (2007)
8. Torii, H.A., et al.: Phys. Rev. A **59**, 223 (1999)
9. Hori, M., et al.: Phys. Rev. Lett. **87**, 093401 (2001)
10. Hori, M., et al.: Phys. Rev. Lett. **91**, 123401 (2003)
11. Lombardi, A.M., et al.: In: Proceedings of the 2001 Particle Accelerator Conference, Chicago, pp. 585. IEEE, Piscataway, NJ (2001)
12. Hori, M., et al.: Phys. Rev. Lett. **96**, 243401 (2006)
13. Mohr, P.J., Taylor, B.N.: Rev. Mod. Phys. **80**, 633 (2008)
14. Hori, M., et al.: Nature **475**, 484 (2011)
15. Mohr, P.J., Taylor, B.N., Newell, D.B.: The 2010 CODATA Recommended Values of the Fundamental Physical Constants (Web Version). http://physics.nist.gov/cuu/Constants/. National Institute of Standards and Technology, Gaithersburg, MD 20899

Hyperfine Interact (2012) 209:15–20
DOI 10.1007/s10751-012-0575-x

Collisional processes in exotic atoms

Grigory Ya. Korenman

Published online: 2 February 2012

Abstract Selected problems of collisional processes in exotic atoms are discussed. The emphasis is on the collisional effects in antiprotonic helium including formation of antiprotonic atoms, collisional quenching of hot and thermalized metastable antiprotonic helium, shift and broadening of E1 and M1 spectral lines.

Keywords Antiproton · Antiprotonic helium · Collisional transitions · Hyperfine splitting

1 Introduction

Experimental discovery of long-lived $\bar{p}$ states in liquid helium [1] at 1991 has opened up a new chapter in the study of exotic atoms and of antiproton properties. Initially, the long-lived states were studied by delayed annihilation time spectra (DATS) that brought an extensive data on this phenomenon in different helium phases involving targets with various admixtures (see [2, 3] and references therein). A possible existence of long-lived states in hadronic helium atoms was discussed a long time before [4, 5], however, the DATS experimental data have raised a number of extended theoretical problems in the physics of exotic atoms. At this stage it was important to clarify an overall dynamic picture of the involved processes, such as formation of antiprotonic atoms including fraction of metastable states, cascade of transition, mechanisms of collisional quenching of metastable states etc. New stage of the $\bar{p}He^+$ study is connected with the precision measurements of E1 transition from metastable to short-lived states by means of laser spectroscopy and, more

This work was supported by the joint Russian Foundation for Basic Research—Austrian Science Foundation grant No. 09-02-91000 and by the Russian Foundation for Basic Research grant No. 10-02-01096.

G. Ya. Korenman (✉)
Institute of Nuclear Physics, Moscow State University, 119991 Moscow, Russia
e-mail: korenman@nucl-th.sinp.msu.ru

recently, of M1 transitions between HFS substates by a triple laser-microwave-laser method. These methods open a possibility to investigate fundamental properties of antiproton and symmetries of antiparticles [3, 6]. It poses more refined challenges for the theory of antiprotonic helium. By now theoretical calculations of quantum 3-body system "antiproton—electron—helium nucleus" ($\bar{p}\,e\,\mathrm{He}^{++}$) involving fine and hyperfine interactions and other subtle effects have achieved a very high precision [7] comparable with the experimental accuracy. Experimental study of $\bar{p}\mathrm{He}^{+}$ by laser and microwave spectroscopy provides, along with the precision measurements of the spectral line, data on different collisional characteristics of individual $(\bar{p}\mathrm{He}^{+})^{*}_{nL}$ states (primary populations, collisional quenching, shift and broadening of spectral lines etc.). From theoretical point of view, collisional processes in different exotic atoms, especially at the first stage of their "life", have many common traits. But these processes in antiprotonic helium are of specific interest due to the most direct manifestation in the data on DATS and spectroscopy measurements. In this paper we discuss briefly some selected theoretical problems on collisional processes in exotic atoms, primarily connected with the antiprotonic helium studies.

2 Formation of exotic atoms and collisional quenching of hot $(\bar{p}\mathrm{He}^{+})^{*}$

When a heavy negative particle (say, for definiteness, $\bar{p}$) with a kinetic energy of several MeV enters the target it is initially slowed down due to elastic scattering, ionization and excitation of the target atoms or molecules. A competition between the energy losses and Coulomb capture into atomic or molecular orbits takes place below an energy of few tens eV depending on the target. Theoretical studies of the involved processes have a long history, beginning from the classical paper by Fermi and Teller [8]. A comprehensive review of different approaches to the problem and present status of the theory of exotic atom formation in various targets can be found in [9].

A general scheme of antiprotonic atom formation in monatomic gases can be described by a simple model of "black body with potential tale" [10, 11] that was developed on the base of semiclassical approximation for coupled two- and three-body channels [12, 13]. Let $V_0(R)$ and $V_1(R)$ be terms of $\bar{p}$ interaction with atom A and ion A^{+}, respectively, $I_0 = V_1(\infty) - V_0(\infty) = E_A^{+} - E_A$ be an atomic ionization potential, and R_0 be a crossing point of two terms, $V_0(R_0) = V_1(R_0)$. In the framework of semiclassical approximation, inelastic transitions $\bar{p} + A \to \bar{p} + A^{+} + e$ or $\bar{p} + A \to (\bar{p}A^{+}) + e$ can occur in the region $R \leqslant R_0$ with electron energy $\epsilon_e(R) = V_0(R) - V_1(R)$. The calculations for He show [12] that the most part of the transitions takes place nearly R_0 with $\epsilon_e \lesssim 2$ eV. If ϵ_e is neglected then inelastic process occurs at the distance R_0 with unit probability that corresponds to the "black body" model. Total inelastic cross section $\sigma_r(E) = \pi b_0^2(E)$ is defined by a maximum impact parameter $b_0(E)$ allowing $\bar{p}$ to reach a distance R_0, with $b_0^2(E) = R_0^2(1 + W_0/E)$, where $W_0 = -V_0(R_0)$. At very small kinetic energy ($E < E_p \sim 1$ eV), a "polarization capture" takes place with $b_0^2(E) = \sqrt{2\alpha/E}$, where α is an atomic polarizability. Because of the approximation $\epsilon_e = 0$ inelastic transitions with the cross section $\sigma_r(E)$ are attributed to atomic ionization at $E > I_0$ and to Coulomb capture at $E < I_0$. The parameters for He atom are $I_0 = 0.9036$ a.u. $= 24.59$ eV, $\alpha = 1.383$ a.u., $R_0 \simeq 1.0$ a.u., $W_0 = 0.184$ a.u., $E_p = 0.014$ a.u.

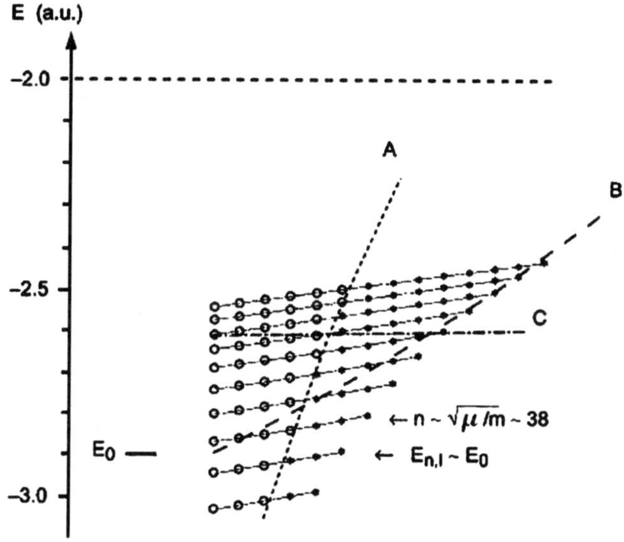

Fig. 1 Scheme of $\bar{p}\,^4He^+$ energy levels with boundaries of metastability (*curve A*), population (*curve B*) and hot collisional quenching (*line C*) areas

Primary kinetic characteristics were also obtained within the model. Neglecting energy losses due to elastic scattering we obtain a distribution of the $(\bar{p}A^+)^*$ system over inner energy ε and angular momentum L

$$p(\varepsilon, L) = (1/I_0)\theta(\varepsilon - E_A)\theta(E_A + I_0 - \varepsilon)\cdot\theta(L_m(\varepsilon) - L)\cdot(2L+1)/[L_m(\varepsilon)+1]^2, \tag{1}$$

where $L_m(\varepsilon) + 1/2 = \sqrt{2\mu E}b_0(E)$ and $E = \varepsilon - E_A$. A distribution over principal quantum number can be obtained from (1) as $p_{nL} = p(\varepsilon_{nL}, L)|d\varepsilon_{nL}/dn|$. A distribution over recoil kinetic energy T of the $(\bar{p}A^+)^*$ is uniform from 0 to $T_{\max} = I_0 \cdot m_{\bar{p}}/m_A$ with one-to-one correspondence between T and inner energy, $T_{nL} = (\varepsilon_{nL} - E_A)m_{\bar{p}}/m_A$. All distributions have sharp boundaries due to neglecting of ejected electron energy ϵ_e and of elastic scattering energy losses. More sophisticated theoretical approaches [9] refine detailed primary distributions with smoother and, more or less, shifted boundaries. Nevertheless, the simple model gives a reasonable general picture of exotic atom formation in helium and other inert gases.

Circular and nearly-circular states of $(\bar{p}He^+)^*_{nL}$ have slow radiative decay rates, $\lambda_\gamma \sim \mu Z^4_{eff}/(n^3L^2)\cdot 12\ \text{ns}^{-1} \sim 1\ \mu\text{s}^{-1}$. Auger transition rates are strongly dependent on the multipolarity $\Delta L_A = |L - L'|$ decreasing by three orders of magnitude as ΔL_A increases by one [14–16]. The condition $\lambda_A < \lambda_\gamma$ can be adopted as a metastability criterion. It is satisfied at $\Delta L_A \geq 4$ that leads to the boundary of metastability $\varepsilon_{nL} \leq -\mu Z^2/2(L-2)^2$. Using this condition and the distribution (1) we estimate a total population of metastable states about 20%, whereas experimental value is only 3%. But the main contribution to the estimation is found from very high antiprotonic states ($\varepsilon \gtrsim -2.6$ a.u., $n > 40$), which can be destroyed by collisions just after formation. It was shown [17] that collisional quenching cross sections are very sensitive to the Massey parameter $\eta = \Delta E \cdot R/\hbar \text{v}$ and, so, to the collision velocity. They are very small at $\text{v} < \text{v}_{cr}$, but are of order $\pi a_0^2/2$ at higher velocity. The critical velocity $\text{v}_{cr} = 3\cdot 10^{-3}\text{v}_a$ corresponds to $(\bar{p}He^+)^*$ recoil kinetic energy $T_{cr} \simeq 1$ eV.

The metastable antiprotonic helium with $T > 1$ eV in the ordinary (STP) medium will be destroyed by collisional transitions to short-lived states at the "hot" stage before thermalization. The survived part consists only 1/5 of the uniform primary distribution over T from 0 to $T_{max} \simeq 5$ eV for ^{4}He. So, collisional quenching at the "hot" stage reduces a total population of metastable states by, at least, factor 5. Primary populations of the individual metastable states extracted from experimental data [18] decrease to zero at $n > 40$ that also show effect of the hot collisional quenching. Figure 1 shows scheme of $(\bar{p}\,^4\mathrm{He}^+)^*$ energy levels with boundaries of populations, metastability and hot collisional quenching. It is seen that the area between three lines is a small part of initially populated metastable states lying between curves A and B.

3 Collisional processes in thermalized antiprotonic helium

Collisional quenching of individual states Collisional transitions of thermalized $\bar{p}\mathrm{He}^+$ are strongly suppressed as compare with hot atoms due to a large Massey parameter η. Experimental data [19, 20] show wide variations of the collisional quenching effect depending on the states. A decay rate for the state $(n, L) = (39, 35)$ is unsensitive to the target density up to liquid helium, $\lambda \sim 0.7\,\mu\mathrm{s}^{-1}$, whereas $\lambda(37, 34)$ increases nonlinearly from $0.8\,\mu\mathrm{s}^{-1}$ at the density $N = 10^{20}$ cm^{-3} to $8\,\mu\mathrm{s}^{-1}$ at $N = 3 \cdot 10^{21}$ cm^{-3}. Model calculations with free parameters of interatomic interaction [21] can not explain these data. More sophisticated calculations [22] using *ab initio* potential energy surface (PES) for the system $(\bar{p}\mathrm{He}^+ - \mathrm{He})$ give per-atom quenching rates $q(39, 35) = 1.6 \cdot 10^{-18}$ cm^3/s and $q(37, 34) = 3.2 \cdot 10^{-19}$ cm^3/s, whereas linear fit $\lambda = \lambda_0 + qN$ of the experimental data yields $q(39, 35) = (2.5 \pm 1.9) \cdot 10^{-18}$ and $q(37, 34) = (2.5 \pm 1.9) \cdot 10^{-15}$ cm^3/s. At present, there is no explanation of the strong qualitative distinction in the quenching of different states.

Shift and broadening of E1 spectral lines in antiprotonic helium Density shift and broadening of E1 spectral lines were measured for several laser-induced transitions in antiprotonic He [6, 23, 24]. The values $\Delta\nu$ and Γ are proportional to the density, with no significant dependence on the temperature. For all transitions the broadening is much smaller than the shift, contrary to ordinary atoms and molecules, where, as a rule, $\Delta\nu < \Gamma$. The reduced per-atom shifts $\Delta\nu/N$ are rather small as compare with that ones for ordinary atoms. Quantum theory of shift and broadening of spectral lines due to binary collisions gives complex shift $\Delta(\mathrm{v}) - i\gamma(\mathrm{v})$ in terms of S-matrix followed by average over thermal motion. In turn, to obtain S-matrix it is necessary to know an effective interaction for the $(\bar{p}\mathrm{He}^+ - \mathrm{He})$ system. Main qualitative features of the experimental data for $\Delta\nu$ and Γ were explained using the model interaction potential [25] of Van der Waals type, $V_0(R) = C_6(r_c^2 - R^2)/(R^2 + r_0^2)^4$. Quantitative agreement with the most part of the experimental data was achieved in [26] using *ab initio* PES and a semiclassical theory of shift and broadening.

Collisional effects on HFS transition in microwave field Transitions between hyperfine structure sublevels of antiprotonic helium are studied by the triple (laser-microwave-laser) resonance method for ^{4}He [27–29] and, recently, for ^{3}He [30]. In these experiments, collisional effects can be manifested by a relaxation of non-equilibrium populations HFS states (after first laser pulse) as well as by shift and broadening of M1 spectral lines. Theoretical consideration of these effects involving both the elementary ($\bar{p}He^+ - He$) collisions and kinetics of HFS transitions in the presence of resonance MW field was developed in the papers [31, 32]. Model interaction between $(\bar{p}He^+)^*$ and He atom contains scalar and tensor terms of Van der Waals type with asymptote strengths C_6 and $G_6 = -gC_6$. The tensor interaction includes an additional factor providing R^2 dependence at very small distance. The calculated values are proportional to G_6^2. In the initial calculations [31] we used the estimation $g = 0.37$, but subsequently [32, 33] we have reduced this value to $g = 0.125$ for fitting the results to the observed value [28] $p_{on}(F_+, t_d)/p_{off}(F_+, t_d)$, where t_d is a time delay of second laser pulse. General results for collisional HFS effects within the model can be summarized as follow. Rates of collisional HFS transitions in $\bar{p}^{3,4}He^+$ for the similar types of transition are of the same order, $\lambda \sim 10^6\ s^{-1}$ for single spin-flip transitions at $T = 6\,K$, $N = 3 \cdot 10^{20}\ cm^{-3}$ and decreases by three order of value with each additional spin-flip. Isotope ratio $^3He/^4He$ for similar transitions is about 1.5–2. At the same T and N, collisional shifts of the M1 spectral lines are very small, 1.6 kHz for ^4He, 4 kHz for ^3He, i.e., much less than the achieved experimental accuracy. Collisional broadening of the M1 spectral lines is about 0.3 MHz for ^4He, 0.4 MHz for ^3He. These values consist 12–18% of the total line widths calculated from a master equation for the density matrix of the system in the MW radiation. The obtained total widths are compatible with the observed values. Collisional transition rates, shifts and broadenings of M1 spectral lines decrease by several times with increasing temperature from 2 to 10 K due to the resonance scattering of antiprotonic helium on He atom at energy about 3–4 K for ^4He and 1 K for ^3He. Dependence of the ratio p_{on}/p_{off} on t_d for ^4He, as well as the value for ^3He at $t_d = 350\,ns$ are in agreement with the experimental results.

Acknowledgements Author would like to thank participants of ASACUSA collaboration, especially, T. Yamazaki, R. Hayano, E. Widmann and D. Horváth for numerous fruitful discussions and long-time contacts.

References

1. Iwasaki, M., Nakamura, S.N., Shigaki, K., Shimizu, Y., Tamura, H., Ishikawa, T., Hayano, R.S., Takada, E., Widmann, E., Outa, H., Aoki, M., Kitching, P., Yamazaki, T.: Phys. Rev. Lett. **67**, 1246–1249 (1991)
2. Yamazaki, T., Widmann, E., Hayano, R.S., Iwasaki, M., Nakamura, S.N., Shigaki, K., Hartmann, F.J., Daniel, H., von Egidy, T., Hofmann, P., Kim, Y.-S., Eades, J.: Nature **361**, 238–240 (1993)
3. Yamazaki, T., Morita, N., Hayano, R.S., Widmann, E., Eades, J.: Phys. Rep. **366**, 183–329 (2002)
4. Condo, G.T.: Phys. Lett. **9**, 65 (1964)
5. Russell, J.E., Phys. Rev. Lett. **23**, 63 (1969)
6. Hayano, R.S., Hori, M., Horváth, D., Widmann, E.: Rep. Prog. Phys. **70**(12), 1995–2065 (2007)
7. Korobov, V.I.: Phys. Rev. A **73**, 022509 (2006)
8. Fermi, E., Teller, E.: Phys. Rev. **72**, 399 (1947)
9. Cohen, J.S.: Rep. Prog. Phys. **67**, 1769 (2004)
10. Korenman, G.Ya.: Hyperfine Interact. **101/102**, 81–89 (1996)
11. Korenman, G.Ya.: Nucl. Phys. A **692**, 145c–152c (2001)
12. Dolinov, V.K., Korenman, G.Ya., Moskalenko I.V., Popov, V.P.: Muon Catal. Fusion **4**, 169 (1989)
13. Korenman, G.Ya.: Hyperfine Interact. **138**, 91–94 (2001)
14. Morita, N., Ohtsuki, K., Yamazaki, T.: Nucl. Instrum. Methods Phys. Res., Sect. A **330**, 439 (1993)
15. Korenman, G.Ya., Yudin, S.N.: Pisma v ZhETF (JETP Letters) **58**, 10 (1993)
16. Korobov, V.I., Shimamura, I.: Phys. Rev. A **56**, 4587 (1997)

17. Korenman, G.Ya.: Hyperfine Interact. **101/102**, 463 (1996)
18. Hori, M.: Nucl. Instrum. Methods Phys. Res. B **267**, 343–346 (2009)
19. Hori, M., Torii, H.A., Hayano, R.S., Ishikawa, T., Mass F.E., Tamura, H., Ketzer, B., Hartmann, F.J., Pohl, R., Maierl, C., Hasinoff, M., von Egidy, T., Kumakura, M., Morita, N., Sugai, I., Horváth, D., Widmann, E., Eades, J., Yamazaki, T.: Phys. Rev. A **57**, 1698 (1998); **58**, 1612 (1998)
20. Hori, M., Eades, J., Widmann, E., Yamazaki, T., Hayano, R.S., Ishikawa, T., Torii, H.A., von Egidy, T., Hartmann, F.J., Ketzer, B., Maierl, C., Pohl, R., Kumakura, M., Morita, N., Horváth, D., Sugai, I.: Phys. Rev. A **70**, 012504 (2004)
21. Russell, J.E.: Phys. Rev. A, **65**, 032509 (2002)
22. Obreshkov, B.D., Bakalov, D.D., Lepetit, B., Szalewicz, K.: Phys. Rev. A **69**, 042701 (2004)
23. Torii, H.A., Hayano, R.S., Hori, M., Ishikawa, T., Morita, N., Kumakura, M., Sugai, I., Yamazaki, T., Ketzer, B., Hartmann, F.J., von Egidy, T., Pohl, R., Maierl, C., Horváth, D., Eades, J., Widmann, E.: Phys. Rev. A **59**, 223 (1999)
24. Hori, M., Eades, J., Widmann, E., Yamaguchi, H., Sakaguchi, J., Ishikawa, T., Hayano, R.S., Torii, H.A., Juhász, B., Horváth, D., Yamazaki, T.: Phys. Rev. Lett. **87**, 093401 (2001)
25. Korenman, G.Ya.: Hyperfine Interact. **119**, 227–233 (1999)
26. Bakalov, D., Jeziorski, B., Korona, T., Szalewicz, K., Tchoukova, E.: Phys. Rev. Lett. **84**, 2350 (2000)
27. Widmann, E., Eades, J., Ishikawa, T., Sakaguchi, J., Tasaki, T., Yamaguchi, H., Hayano, R.S., Hori, M., Torii, H.A., Juhász, B., Horváth, D., Yamazaki, T.: Phys. Rev. Lett. **89**, 243402 (2002)
28. Pask, T., Barna, D., Dax, A., Hayano, R.S., Hori, M., Horváth, D., Juhász, B., Malbrunot, C., Marton, J., Ono, N., Suzuki, K., Zmeskal, J., Widmann, E.: J. Phys. B At. Mol. Opt. Phys. **41**, 081008 (2008)
29. Pask, T., Barna, D., Dax, A., Hayano, R.S., Hori, M., Horváth, D., Friedreich, S., Juhász, B., Massiczek, O., Ono, N., Sótér, A., Widmann, E.: Phys. Lett. B **678**, 55–59 (2009)
30. Friedreich, S., Barna, D., Caspers, F., Dax, A., Hayano, R.S., Hori, M., Horváth, D., Juhász, B., Kobayashi, T., Massiczek, O., Sótér, A., Todoroki, K., Widmann, E., Zmeskal, J.: Phys. Lett. B **700**, 1–6 (2011)
31. Korenman, G.Ya., Yudin, S.N.: J. Phys. B At. Mol. Opt. Phys. **39**, 1473–1484 (2006)
32. Korenman, G.Ya., Yudin, S.N.: Hyperfine Interact. **194**, 29–35 (2009)
33. Yudin, S.N., Korenman, G.Ya.: Hyperfine Interact. (2011). doi:10.1007/s10751-011-0525-z

Hyperfine Interact (2012) 209:21–24
DOI 10.1007/s10751-011-0525-z

Collisional transitions between HFS states of antiprotonic ^{3}He in the presence of microwave radiation

Sergey N. Yudin · Grigory Ya. Korenman

Published online: 21 December 2011

Abstract We consider collisional effects on the transitions between HFS states of ($\bar{p}\,^3He^+$) in the presence of microwave radiation. Cross sections and rates of collisional HFS transition, density shifts and broadening of M1 spectral lines are calculated within the approach developed previously for ($\bar{p}\,^4He^+$). The values averaged over thermal motion are used in the quantum master equation for the density matrix of HFS states in the microwave field. At the target density 3×10^{20} cm^{-3} and T = 6 K, partial rates of the single spin-flip transition in ($\bar{p}\,^3He^+$) are $\sim 1 \cdot 10^6$ s^{-1}. Effective relaxation rates by considering the kinetics are approximately twice more. At the same conditions collisional shifts of the M1 spectral lines are ≤ 4 kHz, collisional broadening of the lines is $\simeq 0.5$ MHz. The total width of the line $J^{--+} \rightarrow J^{+-+}$ at $t_d = 350$ ns is 2.4 MHz allowing for collisional and Fourier broadenings. Maximum MW resonance signal in ^{3}He is about 0.5 of that one in ^{4}He.

Keywords Antiproton · Antiprotonic helium · Hyperfine splitting · Collisional transitions

1 Introduction

Recently, the first experimental results for laser-microwave-laser spectroscopy of the hyperfine structure of ($\bar{p}\,^3He^+$) were published [1] complementing the analogous results for ($\bar{p}\,^4He^+$) [2–4]. Along with the measurements of hyperfine splitting of the levels and investigations of fundamental properties of antiproton, such experiments

This work was supported by the joint Russian Foundation for Basic Research - Austrian Science Foundation grant No. 09-02-91000 and by the Russian Foundation for Basic Research grant No. 10-02-01096.

S. N. Yudin · G. Ya. Korenman (✉)
Institute of Nuclear Physics, Moscow State University, 119991 Moscow, Russia
e-mail: korenman@nucl-th.sinp.msu.ru

provide an interesting information on collisional effects and interaction of antiprotonic He with a medium. Collisions between antiprotonic and ordinary He atoms lead to transitions between HFS states and to shifts and broadenings of the M1 spectral lines. A theory of these effects involves calculations of the elementary collision characteristics (cross sections, transition rates, collisional shifts and broadening of spectral lines *vs.* energy), an average of these values over thermal motion, and analysis of the time evolution of HFS density matrix with account for both the collisional effects and external microwave field. This program was realized in the papers [5, 6] for collisional HFS effects in antiprotonic ^{4}He using a model interaction between ($\bar{p}\,^4He^+$) and He atoms. In this paper we use the same model to consider collisional effects on HFS transitions in ($\bar{p}\,^3He^+$). The parameter of the strength of tensor interaction was taken from the fitting [6] of the results to the experimental data for ^{4}He.

2 Collisional effects on HFS states in antiprotonic ^{3}He

Antiprotonic ^{3}He at fixed n, L has 8 HFS states, which can be labeled [7] by quantum numbers F, G, J of angular momenta $\mathbf{F} = \mathbf{L} + \mathbf{s}_e$, $\mathbf{G} = \mathbf{F} + \mathbf{s}_n$, $\mathbf{J} = \mathbf{G} + \mathbf{s}_{\bar{p}}$, where s_e, s_n and $s_{\bar{p}}$ are spins of electron, nucleus and antiproton. For brevity, we enumerate the HFS states $|FGJ\rangle$ in order of increasing energy, $|1\rangle = |F^+ G^{++} J^{++-}\rangle$, $|2\rangle = |F^+ G^{++} J^{+++}\rangle$, $|3\rangle = |F^+ G^{+-} J^{+--}\rangle$, $|4\rangle = |F^+ G^{+-} J^{+-+}\rangle$, $|5\rangle = |F^- G^{--} J^{---}\rangle$, $|6\rangle = |F^- G^{--} J^{--+}\rangle$, $|7\rangle = |F^- G^{-+} J^{-+-}\rangle$, $|8\rangle = |F^- G^{-+} J^{-++}\rangle$, where the notations $F^{\pm}$, $G^{\pm\pm}$, $J^{\pm\pm\pm}$ correspond to [1]. We solve quantum close-coupling equations with the interaction from [6] in the basis $|FGJ, l : jm\rangle$, where l, j are orbital and total angular momenta of colliding atoms. With the obtained S-matrix we calculate elastic and inelastic ($i \leftrightarrow j$) cross sections as well as shifts and broadenings of M1 spectral lines depending on kinetic energy.

Elastic scattering cross sections for different HFS states are practically indistinguishable and show resonance-like behaviour at kinetic energy $E \sim 1$ K. Similar resonances for ^{4}He were found [5] at $E \sim 4$ K. After averaging over thermal motion we find transition rates λ_{ij}, shifts and broadenings Δ_{ij}, γ_{ij} of the lines depending on the target temperature. The HFS transitions can be classified by type of spin flip. Different groups of spin flip transitions involve several individual transitions (e.g., single electron spin flip transitions are $1 \leftrightarrow 7, 2 \leftrightarrow 8, 3 \leftrightarrow 5, 4 \leftrightarrow 6$). The mean values of λ, Δ and γ for different groups at T $= 6$ K and density $N = 3 \times 10^{20}$ cm^{-3} are given in Table 1. Variations around the mean value in the most important group of single s_e spin flip are within 10% for λ, 40% for Δ, and 14% for γ. In other spin flip groups the variations can be up to factor 2. Rates of direct ($i \to j$) and inverse ($j \to i$) transitions are connected by the detailed balance relationship

$$(2J_i + 1)\exp(-E_i/T)\lambda_{i\to j}(T) = (2J_j + 1)\exp(-E_j/T)\lambda_{j\to i}(T) \tag{1}$$

that we used to check an accuracy (0.1%) of averaging over thermal motion. With a rise of spin flip multiplicity the transition rates are decreased by 3 order of value, whereas shifts and broadenings of allowed M1 transitions remain of the same order of value.

Table 1 Collisional HFS transition rates, shifts and broadenings of allowed M1 transitions in $(\bar{p}\,^3He^+)_{36,34}$ at $N = 3 \times 10^{20}$ cm^{-3} and T = 6 K

	Single spin flip transitions			Double spin flip transitions	Triple spin flip transitions
	s_e	s_n	$s_{\bar{p}}$		
λ_{ij} (s^{-1})	$0.9 \cdot 10^6$	$1.2 \cdot 10^6$	$1.2 \cdot 10^6$	$0.9 \cdot 10^3$	0.8
Δ_{ij} (kHz)	3.5	1.4	0.2	2.2	4.0
γ_{ij} (MHz)	0.48	0.48	0.48	0.56	0.46

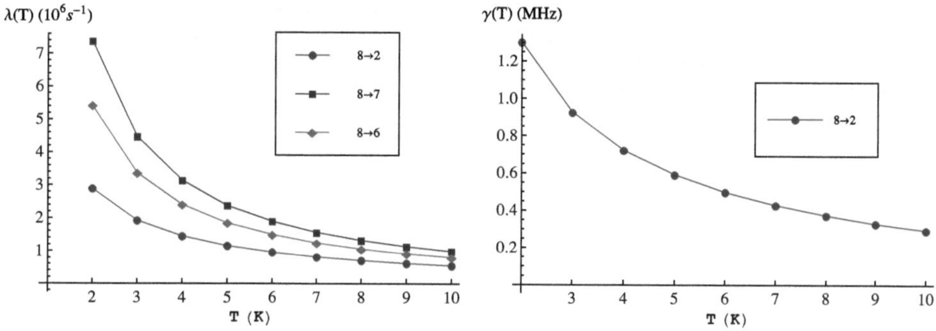

Fig. 1 Temperature dependence of HFS transition rates (*left*) and collisional broadening of M1 spectral line (*right*) in $(\bar{p}^3He^+)_{36;34}$

Figure 1 shows a typical temperature dependence of transition rates and collisional broadening of M1 spectral line. A relationship among single electron (8 → 2), nucleus (8 → 6) and antiproton (8 → 7) spin flip transition rates varies with temperature only slightly. Collisional broadenings of M1 spectral lines for different electron spin flip transitions in the figure scale are very close to the shown curve (8 → 2). Collisional shifts have a similar temperature dependence, but are less than the broadening by two orders of value. All transition rates, shifts and broadenings of the spectral lines fall with temperature up to $T = 10$ K due to a displacement of mean kinetic energy from the resonances at very low energy.

In order to consider density matrix of the eight HFS states depending on time and on the MW frequency detuning, we use the quantum master equation involving the relaxation matrix and interaction of antiprotonic He with external microwave field. The relaxation matrix is expressed in terms of the calculated elementary collision characteristics [6]. Time dependence of the density matrix was obtained by a numerical solution of the master equation within "rotating wave" approximation. Effective relaxation rate, i.e., a rate of decrease in the difference between populations of 4 upper and 4 lower levels, is found to be greater than the elementary electron spin flip transition rates, approximately, by factor 2. Total width of the line (6 → 4) at T = 350 ns is about 2.4 MHz allowing for collisional and Fourier broadenings that can be compared with experimental value 2.08(22) MHz. Maximum MW resonance signal $(p_{\text{on}}/p_{\text{off}} - 1)$ in ^{3}He is about 0.5 of that one in ^{4}He at the same delay time t_d between the two laser pulses.

Acknowledgements We are grateful to E. Widmann and S. Friedreich for fruitful discussions.

References

1. Friedreich, S., Barna, D., Caspers, F., Dax, A., Hayano, R.S., Hori, M., Horváth, D., Juhász, B., Kobayashi, T., Massiczek, O., Sótér, A., Todoroki, K., Widmann, E., Zmeskal, J.: Phys. Lett. B **700**, 1–6 (2011)
2. Widmann, E., Eades, J., Ishikawa, T., Sakaguchi, J., Tasaki, T., Yamaguchi, H., Hayano, R.S., Hori, M., Torii, H.A., Juhász, B., Horváth, D., Yamazaki, T.: Phys. Rev. Lett. **89**, 243402 (2002)
3. Pask, T., Barna, D., Dax, A., Hayano, R.S., Hori, M., Horváth, D., Juhász, B., Malbrunot, C., Marton, J., Ono, N., Suzuki, K., Zmeskal, J., Widmann, E.: J. Phys. B **41**, 081008 (2008)
4. Pask, T., Barna, D., Dax, A., Hayano, R.S., Hori, M., Horváth, D., Friedreich, S., Juhász, B., Massiczek, O., Ono, N., Sótér, A., Widmann, E.: Phys. Lett. B **678**, 55–59 (2009)
5. Korenman, G.Ya., Yudin, S.N.: J. Phys. B **39**, 1473–1484 (2006)
6. Korenman, G.Ya., Yudin, S.N.: Hyperfine Interact. **194**, 29–35 (2009)
7. Korobov, V.I.: Phys. Rev. A **73**, 022509 (2006)

Hyperfine Interact (2012) 209:25–28
DOI 10.1007/s10751-011-0526-y

Density shift and broadening of dipole transitions in antiprotonic helium

Dimitar Bakalov

Published online: 12 January 2012

Abstract We present the numerical values of the density shift and broadening slopes of laser-stimulated dipole transitions in antiprotonic helium atoms at temperature 1.5°K and 5.4°K, evaluated in the semiclassical approach with a realistic interatomic potential calculated *ab initio*. Compared to our 2000 paper, the present results cover the much wider set of transition lines that have been subject of precision spectroscopy since then.

Keywords Antiprotonic helium · Density shift and broadening · Laser spectroscopy

1 Introduction

At helium gas target pressure of the order of 1 bar the density shift and broadening of the spectral lines corresponding to dipole transitions in antiprotonic helium atoms have been shown [1] to be comparable with the leading relativistic corrections and, therefore, important to be taken into account in the interpretation of the precision laser spectroscopy experiments PS205 and AD3 performed at CERN [2, 3] and in the extraction of improved accuracy values of the fundamental characteristics of the antiproton. At such low target densities the shift of the resonance transition frequency $\delta\nu$ and its broadening $\delta\Gamma$ are, to a good approximation, linear functions of the number density n: $\delta\Gamma = n.\alpha$, $\delta\nu = n.\beta$. The slopes α and β have been evaluated in [1] for a few specific transition lines of experimental interest with a modified version of the semi-classical approach of Anderson [4] using the interatomic potential calculated *ab initio* in [1]. Table 1 compares the experimental results at temperatures

The work was partially supported by Grant DO 02-288 of the Bulgarian Scientific Fund.

D. Bakalov (✉)
Institute for Nuclear Research and Nuclear Energy, Bulgarian Academy of sciences,
Tsarigradsko ch. 72, Sofia 1784, Bulgaria
e-mail: dbakalov@inrne.bas.bg

Table 1 Slopes of the collisional shift β and of the broadening α (in 10^{-21} GHz.cm^3). A: experimental values; B: semiclassical results; C: quantum estimates from the elastic scattering phase shifts

Transition	A		B		C	
λ, nm	β	α	β	α	β	α
597	−4.05(7)	0.30(15)	−4.05	0.36	−3.94	0.36
470	−1.50(10)	<1.0	−1.47	0.08	−1.26	0.06
372	−0.4(1)		−0.22	0.03	−0.09	0.01
296	0.4(1)		0.35	0.02	0.44	0.01
726	−3.8(2)		−3.42	0.25	−4.07	0.38
617	3.8(5)		−2.07	0.11	−2.53	0.17
714	−3.3(8)		−3.75	0.30	−4.48	0.45

Table 2 Slopes α and β of the broadening and shift of favored transitions $(n, l) \to (n-1, l-1)$ in $\bar{p}^4$He, in the form $\beta(\alpha)$ (in units 10^{-21} GHz.cm^3), evaluated at T = 5.4°K or T = 1.5°K

n	T, °K	$l+1$	$l+2$	$l+3$	$l+4$	$l+5$
l						
31	5.4	0.46(0.02)				
	1.5	0.52(0.01)	0.34(0.01)			
32	5.4	0.35(0.02)	0.07(0.02)			
	1.5	0.40(0.01)	0.15(0.01)	−0.26(0.01)		
33	5.4	0.18(0.02)	−0.22(0.03)	−0.84(0.04)		
	1.5	0.26(0.01)	−0.11(0.01)	−0.68(0.02)	−1.51(0.07)	
34	5.4	−0.04(0.03)	−0.59(0.03)	−1.47(0.08)	−2.71(0.18)	
	1.5	0.05(0.01)	−0.46(0.02)	−1.25(0.05)	−2.38(0.16)	−3.95(0.40)
35	5.4	−0.37(0.03)	−1.16(0.06)	−2.35(0.15)	−4.05(0.36)	−6.50(0.87)
	1.5	−0.24(0.01)	−0.97(0.04)	−2.05(0.12)	−3.55(0.33)	−5.66(0.82)
36	5.4	−0.82(0.05)	−1.94(0.12)	−3.57(0.30)	−5.94(0.74)	
	1.5	−0.66(0.03)	−1.68(0.09)	−3.15(0.27)		
37	5.4	−1.50(0.09)	−3.07(0.23)	−5.35(0.61)		
38	5.4	−2.50(0.17)	−4.70(0.49)			
39	5.4	−3.96(0.37)				

about 5.4°K (column A) with the theoretical values, calculated in the semi-classical approach of [1] (column B) and from the quantum estimates of the elastic scattering phase shifts of [5] (column C).

The theoretical values were found to be in agreement with the experimental results except for the 617 nm line; the calculations of the collisional quenching rate for the initial state (36, 34) also failed. The possible explanation of this have been briefly discussed in [6].

Much progress has been achieved in the spectroscopy of antiprotonic helium since that time. Many more transitions lines have been measured, including two-photon [7] and magnetic hyperfine ones [8], and transitions in $\bar{p}^3$He. The experimental accuracy has been significantly improved by, among other, performing the measurements at lower helium gas target densities and temperatures, thus reducing some of the systematic uncertainties. Though at gas pressures of the order of a few mbar the density effects are suppressed, they still give a visible contribution as compared to the drastically reduced experimental uncertainties. This has urged the repeated calculation of the slopes α and β for a wider set of $E1$-transitions in both $\bar{p}^3$He and $\bar{p}^4$He and for different temperatures. In the present paper we give their numerical values, which could be used in the extrapolation of the experimental resonance

Table 3 Slopes α and β of the broadening and shift of unfavored transitions $(n, l) \rightarrow (n+1, l-1)$ in $\bar{p}^4$He, in the form $\beta(\alpha)$ (in units 10^{-21} GHz.cm^3), evaluated at T = 5.4°K or T = 1.5°K

n l	T, °K	$l+1$	$l+2$	$l+3$	$l+4$
31	5.4	−0.82(0.04)			
	1.5	−0.87(0.03)			
32	5.4	0.47(0.03)	0.31(0.03)		
	1.5	−0.52(0.02)	0.19(0.01)		
33	5.4	−0.05(0.03)	−1.06(0.06)	2.39(0.14)	
	1.5	−0.05(0.01)	0.88(0.03)	2.09(0.12)	
34	5.4	−0.76(0.05)	−2.07(0.11)	−3.75(0.30)	5.87 (0.68)
	1.5	0.60(0.02)	1.80(0.09)	3.32(0.28)	5.19(0.68)
35	5.4	−1.74(0.09)	−3.42(0.25)	−5.54(0.61)	−8.20(1.32)
36	5.4	−3.06(0.21)	−5.18(0.54)	−7.83(1.20)	
37	5.4	−4.82(0.47)	−7.50(1.10)		
38	5.4	−7.13(1.00)			

Table 4 Slopes α and β of the broadening and shift of favored transitions $(n, l) \rightarrow (n-1, l-1)$ in $\bar{p}^3$He (in units 10^{-21} GHz.cm^3, in the form $\beta(\alpha)$)

n l	T, °K	$l+1$	$l+2$	$l+3$	$l+4$	$l+5$
31	5.4	0.38(0.02)	0.07(0.02)			
	1.5	0.43(0.01)	0.16(0.01)			
32	5.4	0.20(0.02)	−0.23(0.02)	−0.91(0.04)		
	1.5	0.27(0.01)	−0.11(0.01)	−0.74(0.02)		
33	5.4	−0.05(0.02)	−0.66(0.04)	−1.60(0.08)	−2.97(0.19)	
	1.5	0.06(0.01)	−0.51(0.02)	−1.37(0.06)	−2.61(0.16)	
34	5.4	−0.39(0.03)	−1.27(0.06)	−2.57(0.15)	−4.45(0.38)	−7.21(0.93)
	1.5	−0.26(0.01)	−1.07(0.04)	−2.25(0.13)	−3.92(0.35)	−6.24(0.87)

Table 5 Slopes α and β of the broadening and shift of unfavoured transitions $(n, l) \rightarrow (n+1, l-1)$ in $\bar{p}^3$He (in units 10^{-21} GHz.cm^3, in the form $\beta(\alpha)$)

n l	T, °K	$l+1$	$l+2$	$l+3$	$l+4$
31	5.4	0.48(0.03)			
	1.5	−0.53(0.02)			
32	5.4	−0.09(0.03)	−1.18(0.05)		
	1.5	−0.02(0.01)	−0.98(0.03)		
33	5.4	−0.86(0.04)	−2.28(0.12)	−4.10(0.31)	
	1.5	0.69(0.02)	1.99(0.09)	3.64(0.29)	
34	5.4	−1.93(0.09)	−3.76(0.26)	−6.07(0.64)	−8.98(1.32)
	1.5	1.66(0.07)	3.32(0.25)	5.38(0.64)	7.91(1.39)

Table 6 Temperature dependence of the slopes β, α of the shift and broadening of the $(39, 35) \rightarrow (38, 34)$ transition line (in units 10^{-21} GHz.cm^3)

T, °K	1.5	2.0	2.5	3.5	4.5	5.4	6.0	7.0	8.0	10.0	12.0	15.0	20.0
β	−3.55	−3.68	−3.78	−3.91	−3.99	−4.05	−4.08	−4.15	−4.21	−4.33	−4.44	−4.58	−4.83
α	0.33	0.34	0.35	0.36	0.36	0.36	0.37	0.37	0.37	0.38	0.39	0.41	0.44

frequencies to zero target density. The density effects in two-photon and magnetic $M1$-transitions, which require a substantial extension of the semi-classical approach used here, will be considered in future papers.

2 Numerical results

The values of the slopes of the density shift, β, and density broadening, α, evaluated for a set of $E1$ transitions in $\bar{p}^4$He and $\bar{p}^3$He for temperatures T $= 5.4°K$ and $1.5°K$ using the method presented in [1], are given in Tables 2, 3, 4 and 5. We distinguish the cases of favored transitions with $\Delta n = \Delta l = \pm 1$ and unfavored transitions with $\Delta n = -\Delta l = \pm 1$ because of the different dependence on the principal and orbital momentum quantum numbers n, l of the initial metastable state of $\bar{p}$He.

The temperature dependence of the slopes α and β has been studied in some more details for the $(39, 35) \rightarrow (38, 34)$ line (see Table 6). The dependence is rather monotonous down up to $1.4°K$: both the shift and the width slopes decrease smoothly as the temperature is lowered. This dependence holds, of course, for very low densities only where critical phenomena do not take place.

Acknowledgements The author is grateful to the EXA2011 organizers for their support, and to his co-authors in Ref. [1] and to Dr. B. Obreshkov for their contribution at earlier stages of the work.

References

1. Bakalov, D., Jeziorski, B., Korona, T., Szalewicz, K., et al.: Phys. Rev. Lett. **84**, 2350 (2000)
2. Yamazaki, T., Morita, N., Hayano, R.S., Widmann, E., Eades, J.: Phys. Rep. **366**, 183 (2002)
3. Eades, J., Hartmann, F.J.: Rev. Mod. Phys. **71**, 373 (1999)
4. Anderson, P.W.: Phys. Rev. **76**, 647 (1949)
5. Obreshkov, B.: private communication
6. Obreshkov, B., Bakalov, D., Lepetit, B., Szalewicz, K.: Phys. Rev. A. **69**, 042701 (2004)
7. Hori, M., Soter, A., Barna, D., et al.: Nature **475**, 484 (2011)
8. Widmann, E., Eades, J., Ishikawa, T., et al.: Phys. Rev. Lett. **89**, 243402 (2002); Pask, T., Barna, D., Dax, A., et al.: J. Phys. B: AT. Mol. Opt. Phys. **41**, 081008 (2008)

Hyperfine Interact (2012) 209:29–33
DOI 10.1007/s10751-011-0534-y

CPT-symmetry studies with antihydrogen

Ralf Lehnert

Published online: 24 January 2012

Abstract Various approaches to physics beyond the Standard Model can lead to small violations of CPT invariance. Since CPT symmetry can be measured with ultra-high precision, CPT tests offer an interesting phenomenological avenue to search for underlying physics. We discuss this reasoning in more detail, comment on the connection between CPT and Lorentz invariance, and review how CPT breaking would affect the (anti)hydrogen spectrum.

Keywords CPT-symmtery violation · Antimatter · Gravity

1 Introduction

Present-day established physics is well described by two different, mutually incompatible frameworks: quantum theory, which governs nature at the microscopic level, and classical general relativity, which dominates physical phenomena at large distance scales. Although the description of the physical world by these two theories is tremendously successful phenomenologically, the apparent necessity for two *distinct* frameworks is somewhat unsatisfactory from a theoretical perspective. For example, certain situations, such as the description of the birth of our universe, are characterized by conditions that call for the simultaneous application of both quantum theory and general relativity.

For this reason, significant efforts are presently directed towards a more fundamental description of nature that contains both quantum theory and classical general relativity as limiting cases. While there are a number of theoretical approaches in this context, phenomenological progress is hampered, primarily by the expected Planck suppression of quantum-gravity effects. A common avenue to tackle this issue is to

R. Lehnert (✉)
Indiana University Center for Spacetime Symmetries, Bloomington, IN 47405, USA
e-mail: ralehner@indiana.edu

look for model prediction that are *not* Planck suppressed. Examples are large extra dimensions or novel particles.

Another path towards quantum-gravity phenomenology is a bottom-up approach. It involves the identification of physical relations or physics principles testable with present-day or near-future technology at sensitivity levels that can be interpreted as having Planck reach. A further condition on such a relation or principle is that it should hold *exactly* in established physics, so that even the smallest observed deviations would definitely imply new physics. The third desirable feature is that at least some theoretical approaches to quantum gravity should allow for departures from the established physics in questions, so as to provide motivation for research efforts along these lines.

Spacetime symmetries satisfy the above three requirements: they can be studied experimentally with ultra-high precision, they hold exactly in known physics, and various theoretical ideas in the field of quantum gravity can accommodate their breakdown [1]. Spacetime symmetries therefore represent an excellent area for research in this context.

Spacetime symmetries fall into two classes: continuous and discrete symmetries. The ten continuous ones are four translations, three rotations, and three Lorentz boosts. The discrete spacetime symmetry is CPT invariance. Note that these symmetries are closely intertwined. For example, in the context of axiomatic field theory with conventional quantum mechanics and local interactions, CPT violation goes hand in hand with Lorentz breaking [2].

In what follows, we will review the field-theory description of CPT violation at low energies and discuss its predictions for (anti)hydrogen spectroscopy as well as antihydrogen free-fall measurements.

2 Field theory for CPT violation

To test CPT invariance, it is desirable to employ a framework that allows for departures from this symmetry: such a framework places the identification and analysis of suitable CPT tests on a more solid footing. Various levels of CPT test models ranging from ad hoc assumptions of mass differences between particle and antiparticle to effective field theory (EFT) can in principle be considered. We proceed here at the level of EFT for the following reasons. EFT is a tremendously successful tool in various areas of physics including elementary-particle, nuclear, and condensed-matter physics. Note in particular that the latter area involves discrete backgrounds and non-relativistic dynamical aspects, features that may have to be modeled also in quantum-gravity phenomenology. Moreover, EFT can implement CPT violation at the most fundamental level in established physics, which can guarantee internal theoretical consistency at all levels of presently known physics. In particular, an EFT model can include the usual Standard Model and general relativity, so that essentially all presently feasible CPT tests can be studied.

Such an EFT framework for CPT violation—known as the Standard-Model Extension (SME)—is indeed available [3]. As per the aforementioned "Anti-CPT Theorem" [2], CPT- and Lorentz-breaking are closely linked in the SME. Note, however, that even though all operators for departures from CPT invariance also violate Lorentz symmetry, the converse is not true.

The basic idea behind the inclusion of CPT and Lorentz breaking into the SME involves fixed, non-dynamical, tensorial structures that select preferred directions. More specifically, the SME effective Lagrangian $\mathcal{L}_{\mathrm{SME}}$ is

$$\mathcal{L}_{\mathrm{SME}} = \mathcal{L}_{\mathrm{SM}} + \mathcal{L}_{\mathrm{EH}} + \delta\mathcal{L}_{\mathrm{SME}}\,, \tag{1}$$

where $\mathcal{L}_{\mathrm{SM}}$ and $\mathcal{L}_{\mathrm{EH}}$ denote the usual Standard-Model and Einstein–Hilbert terms, and $\delta\mathcal{L}_{\mathrm{SME}}$ contains small Lorentz- and CPT-violating contributions:

$$\delta\mathcal{L}_{\mathrm{SME}} = i\overline{\psi}c^{\mu\nu}\gamma_{\mu}\partial_{\nu}\psi - \overline{\psi}a^{\mu}\gamma_{\mu}\psi + \ldots\,. \tag{2}$$

Here, $c^{\mu\nu}$ and a^{μ} are SME coefficients, where a^{μ} controls certain types of both Lorentz and CPT breakdown, while the Lorentz-violating $c^{\mu\nu}$ is CPT symmetric. Experimental studies seek to measure or constrain these (and other) tensorial coefficients.

The SME has been the framework for numerous experimental and phenomenological investigations of CPT and Lorentz invariance [4], including ones involving cosmic radiation [5], particle colliders [6], resonance cavities [7], neutrinos [8], and precision spectroscopy [9]. A number of theoretical analyses have studied various aspects of the internal consistency of the SME [10], but no problematic issues have been found thus far.

3 Tests with exotic atoms

Exotic atoms provide an excellent test ground for CPT and Lorentz symmetry for two reasons: the possibility for high-precision spectroscopy and the experimental access to types of matter difficult to study otherwise. The prospect for comparing matter to antimatter, and the associated potential for ultra-sensitive CPT tests, is of particular interest in this context. In what follows, we will focus on two specific classes of tests, namely (anti)hydrogen spectroscopy and the interaction of antihydrogen with the gravitational field.

The effects of CPT and Lorentz violation on the spectrum of (anti)hydrogen within the minimal SME have been determined about a decade ago [11] and have previously been discussed at EXA [12]. It is therefore sufficient to give just a brief description of these results in the present work.

A natural candidate transition for CPT tests in (anti)hydrogen is the unmixed 1S–2S transition: the projected sensitivity is at the 10^{-18} level, which is promising in the context of Planck-scale tests. Unfortunately, the highly symmetric composition of these states eliminates first-order effects in the minimal SME. From a theoretical perspective, this transition is therefore less useful for CPT tests. Another option is a measurement of the spin-mixed 1S–2S transition. A study within the minimal SME does show that this transition would exhibit unsuppressed CPT violation. But there may be a practical disadvantage: the inhomogeneities in the trapping B field would lead to a corresponding position dependence of the transition frequency, which represents a drawback for precision studies. A third idea would be the experimental determination of the hyperfine Zeeman splitting within the 1S state itself. An analysis within the minimal SME indeed establishes unsuppressed CPT-breaking effects, and in principle, a magnetic-field independent transition point can be selected. Note in particular that similar transitions of this type, such as the usual Hydrogen-maser line, can be resolved with ultra-high precision.

The EFT description of the interplay between gravity and CPT and Lorentz violation is more involved than in the flat-spacetime limit. For example, the Bianchi identities become nontrivial on a curved manifold and impose conditions on the CPT- and Lorentz-violating background. One important theoretical result in this context is that an explicitly symmetry-breaking background is usually inconsistent [3]. A spontaneous violation of CPT and Lorentz invariance avoids such issues and is assumed in what follows.

For simplicity, we consider a sample scenario in which there is only a single fermion of mass m, and the only non-zero SME coefficients are a^{μ} and $c^{\mu\nu}$. The corresponding terms in the Lagrangian are shown in (2). Our goal is now to extract the effective interaction of two point-like fermions in the non-relativistic limit for weak gravitational fields. This procedure requires various steps [13]. For instance, the aforementioned consistency condition of spontaneous symmetry breakdown requires the initial introduction of a further dynamical field. The dynamics of these additional degrees of freedom must then be interpreted correctly (i.e., they must be integrated out).

After various other considerations, and a further simplification to the isotropic case, only the two SME coefficients a_T and c_{TT} survive. The resulting two-parameter toy model is called the isotropic parachute model (IPM), and its classical non-relativistic point-particle Lagrangian reads [13]

$$L_{\text{IPM}} = \frac{1}{2} m_{\text{inertial}} v^2 + \frac{m_{\text{grav}} M_{\text{grav}}}{r} , \tag{3}$$

where the SME coefficients are contained in the effective inertial and gravitational masses:

$$m_{\text{inertial}} = m(1 + \frac{5}{3}\bar{c}_{TT}) , \qquad m_{\text{grav}} = m + m\bar{c}_{TT} \pm 2\alpha(\bar{a}_{\text{eff}})_T . \tag{4}$$

Here, $\bar{c}_{TT}$ and $(\bar{a}_{\text{eff}})_T$ are effective coefficients for CPT and Lorentz breaking closely related to the original a_T and c_{TT}. The $\pm$ sign distinguishes between fermion and antifermion. The constant α depends on the details of the symmetry-violating field that needed to be introduced for consistency.

Lagrangian (3) is reminiscent of the conventional Lagrangian for a classical non-relativistic point particle in the gravitational field of a second point particle. All effects of CPT and Lorentz violation are absorbed into m_{inertial} and m_{grav}. One can now consider various special regions in our toy model's two-dimensional $(m\bar{c}_{TT}, \alpha(\bar{a}_{\text{eff}})_T)$ parameter space. For example, the special choice $\frac{1}{3}m\bar{c}_{TT} = \pm\alpha(\bar{a}_{\text{eff}})_T$ means the following. For the $+$ sign, the fermion behaves conventionally with $m_{\text{inertial}} = m_{\text{grav}}$, and free-fall modifications exist only for antifermions. For the other sign choice, the situation would be vice versa: novel effects are confined to matter, and antimatter behaves conventionally. The IPM toy model therefore illustrates that certain regions of the SME's parameter space are best tested with antimatter.

Acknowledgements The author wishes to thank the organizers for arranging this stimulating meeting and for the invitation to participate. This work is supported by the Indiana University Center for Spacetime Symmetries and the Portuguese Fundação para a Ciência e a Tecnologia under Grant No. CERN/FP/116373/2010.

References

1. See, e.g., Kostelecký, V.A., Samuel, S.: Phys. Rev. D **39**, 683 (1989); Alfaro, J., Morales-Técotl, H.A., Urrutia, L.F.: Phys. Rev. Lett. **84**, 2318 (2000); Carroll, S.M., et al.: Phys. Rev. Lett. **87**, 141601 (2001); Bjorken, J.D.: Phys. Rev. D **67**, 043508 (2003); Kostelecký, V.A., et al.: Phys. Rev. D **68**, 123511 (2003); Bertolami, O.: et al.: Phys. Rev. D **69**, 083513 (2004)
2. Greenberg, O.W.: Found. Phys. **36**, 1535–1553 (2006)
3. Colladay, D., Kostelecký, V.A.: Phys. Rev. D **58**, 116002 (1998); Kostelecký, V.A., Lehnert, R.: Phys. Rev. D **63**, 065008 (2001); Kostelecky, V.A.: Phys. Rev. D **69**, 105009 (2004); Kostelecký, V.A., Mewes, M.: Phys. Rev. D **80**, 015020 (2009)
4. Kostelecký, V.A., Russell, N.: Rev. Mod. Phys. **83**, 11 (2011)
5. See, e.g., Kostelecký, V.A., Mewes, M.: Phys. Rev. Lett. **87**, 251304 (2001); Phys. Rev. D **66**, 056005 (2002); Lehnert, R., Potting, R.: Phys. Rev. Lett. **93**, 110402 (2004); Phys. Rev. D **70**, 125010 (2004); Altschul, B.D.: Astropart. Phys. **28**, 380–384 (2007); Phys. Rev. D **78**, 085018 (2008)
6. See, e.g., Hohensee, M.A., et al.: Phys. Rev. Lett. **102**, 170402 (2009); Phys. Rev. D **80**, 036010 (2009); Amelino-Camelia, G., et al.: Eur. Phys. J. C **68**, 619 (2010); Bocquet, J.-P., et al. [GRAAL Collaboration]: Phys. Rev. Lett. **104**, 241601 (2010); Altschul, B.D.: Phys. Rev. D **80**, 091901 (2009); Phys. Rev. D **84**, 076006 (2011)
7. Müller, H., et al.: Phys. Rev. D **68**, 116006 (2003); Eisele, Ch., et al.: Phys. Rev. Lett. **103**, 090401 (2009); Tobar, M.E.: et al.: Phys. Rev. D **80**, 125024 (2009)
8. Katori, T., et al.: Phys. Rev. D **74**, 105009 (2006); Díaz, J.S., et al.: Phys. Rev. D **80**, 076007 (2009); Barger, V., et al.: Phys. Rev. D **84**, 056014 (2011); Díaz, J.S., Kostelecký, V.A.: [arXiv:1108.1799 [hep-ph]]
9. Bluhm, R., et al.: Phys. Rev. D **57**, 3932 (1998); Dehmelt, H., et al.: Phys. Rev. Lett. **83**, 4694 (1999); Mittleman, R., et al.: Phys. Rev. Lett. **83**, 2116 (1999); Gabrielse, G., et al.: Phys. Rev. Lett. **82**, 3198 (1999); Bear, D., et al.: Phys. Rev. Lett. **85**, 5038 (2000); Wolf, P., et al.: Phys. Rev. Lett. **96**, 060801 (2006); Altarev, I., et al.: Phys. Rev. Lett. **103**, 081602 (2009); Europhys. Lett. **92**, 51001 (2010)
10. See, e.g., Jackiw, R., Kostelecký, V.A.: Phys. Rev. Lett. **82**, 3572 (1999); Kostelecký, V.A., et al.: Phys. Rev. D **65**, 056006 (2002); Altschul, B., Kostelecký, V.A.: Phys. Lett. B **628**, 106 (2005); Lehnert, R.: Phys. Rev. D **68**, 085003 (2003); J. Math. Phys. **45**, 3399 (2004); Phys. Rev. D **74**, 125001 (2006); Rev. Mex. Fís. **56**(6), 469 (2010); Hariton, A.J., Lehnert, R.: Phys. Lett. A **367**, 11 (2007); Colladay, D., McDonald, P.: Phys. Rev. D **77**, 085006 (2008); Phys. Rev. D **79**, 125019 (2009); Cambiaso, M., et al.: [arXiv:1201.3045v1 [hep-th]]
11. Bluhm, R., et al.: Phys. Rev. Lett. **82**, 2254–2257 (1999)
12. Lehnert, R.: Hyperfine Interact. **193**, 275 (2009)
13. Kostelecký, V.A., Tasson, J.: Phys. Rev. Lett. **102**, 010402 (2009); Phys. Rev. D **83**, 016013 (2011)

Hyperfine Interact (2012) 209:35–41
DOI 10.1007/s10751-012-0560-4

Synthesis of antihydrogen atoms in a CUSP trap

Naofumi Kuroda · Yoshinori Enomoto · Koji Michishio · Chanhyoun Kim · Hiroyuki Higaki · Yugo Nagata · Yasuyuki Kanai · Hiroyuki A. Torii · Maurizzio Corradini · Marco Leali · Evandro Lodi-Rizzini · Valerio Mascagna · Luca Venturelli · Nicola Zurlo · Koki Fujii · Miki Ohtsuka · Kazuo Tanaka · Hiroshi Imao · Yasuyuki Nagashima · Yasuyuki Matsuda · Bertalan Juhász · Eberhard Widmann · Akihiro Mohri · Yasunori Yamazaki

Published online: 20 January 2012

Abstract ASACUSA collaboration has been making a path to realize high precision microwave spectroscopy of ground-state hyperfine transitions of antihydrogen atom in flight for stringent test of the *CPT* symmetry. Recently, we have succeeded in synthesizing our first cold antihydrogen atoms employing a CUSP trap. It is expected that synthesized antihydrogen atoms in the low-field-seeking states are

N. Kuroda (✉) · C. H. Kim · H. A. Torii · K. Fujii · M. Ohtsuka · K. Tanaka · Y. Matsuda
Graduate School of Arts and Sciences, University of Tokyo,
3–8–1 Komaba Meguro-ku, 153–8902 Tokyo, Japan
e-mail: kuroda@phys.c.u-tokyo.ac.jp

Y. Enomoto · Y. Nagata · Y. Kanai · A. Mohri · Y. Yamazaki
RIKEN Advanced Science Institute, 2–1 Hirosawa Wako-shi, 351–0198 Saitama, Japan

K. Michishio · Y. Nagashima
Department of Physics, Tokyo University of Science,
Kagurazaka Shinjuku-ku,
162–8601 Tokyo, Japan

H. Higaki
Graduate School of Advanced Science of Matter, Hiroshima University,
Kagamiyama, Higashi-Hiroshima, 739–8530 Hiroshima, Japan

M. Corradini · M. Leali · E. Lodi-Rizzini · V. Mascagna · L. Venturelli · N. Zurlo
Dipartimento di Chimica e Fisica per l'Ingegneria e per i Materiali,
Università di Brescia & Instituto Nazionale di Fisica Nucleare,
Gruppo Collegato di Brescia, 25133 Brescia, Italy

H. Imao
RIKEN Nishina Center for Accelerator-Based Science,
2–1 Hirosawa Wako-shi, 351–0198 Saitama, Japan

B. Juhász · E. Widmann
Stefan Meyer Institut für Subatomare Physik, Boltzmangasse 3, 1090 Wien, Austria

Table 1 Mass, electric charge, and g-factor difference between e^- and e^+, p and $\bar{\text{p}}$

	$\lvert m-\bar{m}\rvert/m$	$\lvert q+\bar{q}\rvert/\lvert q\rvert$	$\lvert g+\bar{g}\rvert/\lvert g\rvert$
e^- vs. e^+	$< 8\times10^{-9}$	$< 4\times10^{-8}$	$(-0.5\pm2.1)\times10^{-12}$
p vs. $\bar{\text{p}}$	$< 2\times10^{-9}$	$< 2\times10^{-9}$	$(-0.1\pm2.1)\times10^{-3}$

preferentially focused along the cusp magnetic field axis whereas those in the high-field-seeking states are not focused, resulting in the formation of a spin-polarized antihydrogen beam. We report the recent results of antihydrogen atom synthesis and beam production developed with the CUSP trap.

Keywords Antihydrogen • CPT invariance • Atomic beam • Hyperfine splitting

1 Introduction

Antihydrogen ($\bar{\text{H}}$) atom bound state of antiproton ($\bar{\text{p}}$) and positron (e^+) is a favorable target to make stringent tests of the CPT symmetry since physical quantities of its counter part, hydrogen atom bound state of proton (p) and electron (e^-), have been studied via high precision spectroscopic techniques [1, 2]. Several physical quantities can be considered as a candidate to test CPT symmetry as shown in Table 1. The g-factor or the magnetic moment of $\bar{\text{p}}$ is only known in the order of 10^{-3} [3, 4]. On the other hand, the ground state hyperfine splitting between $F=0$ (singlet) state and $F=1$ (triplet) state of hydrogen atom has been measured with atomic beam experiments and with maser experiments which ultimately achieved a high precision of 10^{-12} [2]. Regarding to the test of the CPT symmetry, it is a first step to determine the $\bar{\text{p}}$ magnetic moment experimentally. Therefore the ASACUSA Collaboration targets on the $\bar{\text{p}}$ magnetic moment among them by measuring the ground state hyperfine splitting. The ground state hyperfine splitting is approximately proportional to the proton magnetic moment to the leading order. Below the order of 10^{-6} accuracy, the hyperfine splitting depends on the form factor of p/$\bar{\text{p}}$. The measurement of the ground state hyperfine splitting will also give us an information about $\bar{\text{p}}$ form factor.

The ASACUSA Collaboration proposed an atomic beam spectroscopy to be suitable to determine the magnetic moment of $\bar{\text{p}}$ [5]. With this atomic beam type experiment, the trapping of $\bar{\text{H}}$ atoms is not essential to determine the magnetic moment of $\bar{\text{p}}$, but an efficient extraction of spin-polarized beams is the key. This experiment also can work with a relatively high temperature (50–100 K) atoms, while trapping would require < 1 K $\bar{\text{H}}$ atoms [6]. Under an existence of an external magnetic field, the triplet states of hydrogen atom split into three states $(F, M_F) =$ (1,1), (1,0), (1,−1) according to the total magnetic quantum number, M_F. The states (1,1) and (1,0) are antiparallel to the magnetic field are known as low-field-seeking states, while high-field-seeking states (1,−1) and (0,0) are parallel. Since these two kinds of states can be separated with a magnetic gradient, a microwave spectroscopy of atomic $\bar{\text{H}}$ beams will be applicable by an analogy with classical Stern-Gerlach experiment.

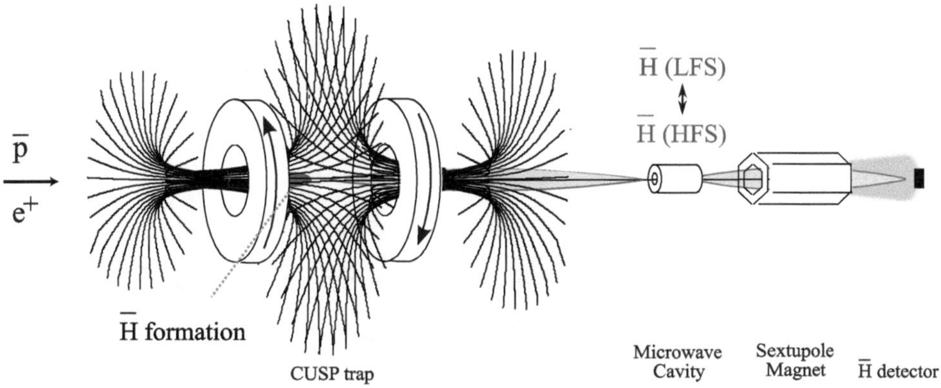

Fig. 1 A conceptual drawing of ground-state hyperfine transition measurement of H̄ atoms with the CUSP trap. *Black lines* corresponds to magnetic field lines. Trajectories of low-field-seeking (LFS) states of H̄ atoms are shown as *red*. The high-field-seeking (HFS) H̄ atoms flipped their spin state at the microwave cavity are shown as *blue*

2 CUSP trap scheme for atomic beam spectroscopy

Figure 1 shows a conceptual drawing of a CUSP trap method to make a microwave spectroscopy of ground-state H̄. The CUSP trap consists of a superconducting anti-Helmholtz coil and multiple ring electrodes (MRE). The currents in coils have opposite direction each other. Since the cusp magnetic field becomes zero at the center point of the symmetry plane, the field strength $|B|$ increases in all directions from the point. The H̄ atoms are not synthesized at the minimum B region but at a higher magnetic field region shown as a red ellipse on the left hand side in Fig. 1. A synthesized H̄ atom in high-field-seeking state deviates from the field axis. On the other hand, a low-field-seeking state atom is once decelerated by the higher magnetic field inside the right hand side coil in Fig. 1, then accelerated along the axis. Thus the cusp field configuration will help to preferentially extract the low-field-seeking state of H̄, resulting in a spin polarized H̄ beam [7].

The microwave cavity installed at the exit of the CUSP trap induces hyperfine transitions from low-field-seeking states to high-field-seeking states when the microwave frequency is on resonance. A sextupole magnet sorts out H̄ atoms in high-field-seeking states from those in low-field-seeking states. The extracted H̄ atoms are then detected by an MCP. When the microwave frequency is tuned to the hyperfine splitting, H̄ atoms in low-field-seeking states flip into high-field-seeking states, which results in decreasing counts at the MCP. Thus the resonance frequency will be determined.

3 Experimental Results of synthesis of H̄

Figure 2a shows a cross sectional view of the CUSP trap with magnetic field lines. The magnetic field strength became its maximum at ± 15 cm from the center. A

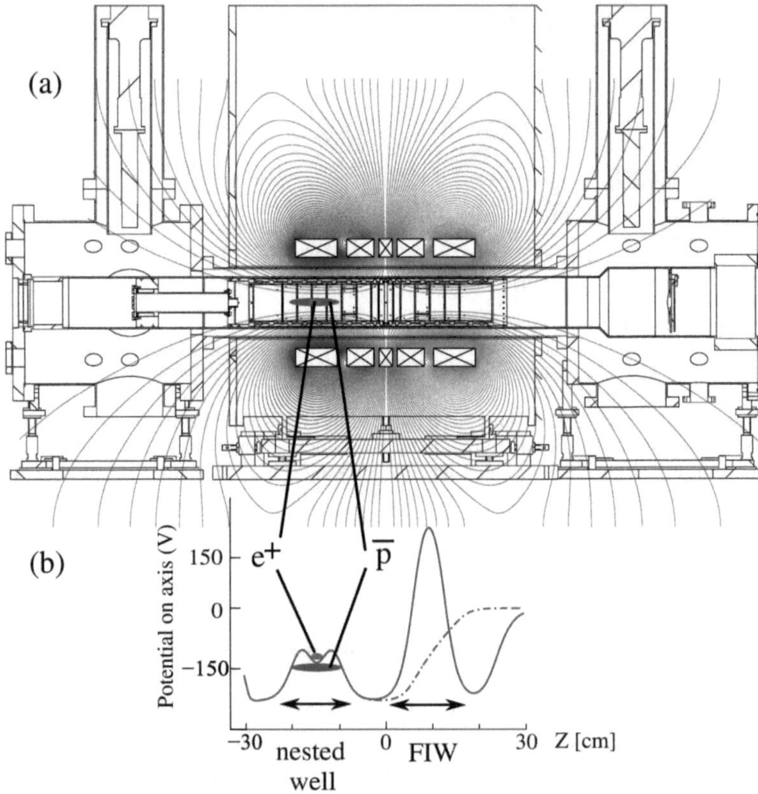

Fig. 2 **a** A cross sectional view of the CUSP trap and a distribution of magnetic field lines. **b** *Solid line* the electric potential along the axis for mixing of p̄ and e^+. *Dashed line* for extraction of p̄ accumulated in the field ionization well (FIW)

multiple-ring electrodes (MRE) installed inside the CUSP bore provided so-called nested well configuration [10] as shown in Fig. 2b.

The procedure to synthesize and detect H̄ atoms consisted of the following steps [9]. Positrons were extracted from a e^+ accumulator [8], and injected into and trapped at the upstream spindle CUSP region. Positrons with its number of 3×10^6 spontaneously cooled via cyclotron radiation resulted in forming a non-neutral plasma. The potential was smoothly varied to the nested well configuration. Antiprotons with its number of 3×10^5 were injected from a p̄ accumulator (Kuroda et al, private communication). Positrons were heated up by injected p̄, then cooled via cyclotron radiation, and eventually captured by p̄s to form H̄ atoms. Although p̄ in the nested well had no chance to move beyond the nested potential, H̄ atoms moved freely because they are neutral, and a part of them reached the field-ionization well (FIW) as shown in Fig. 2b. If H̄ atoms in high Rydberg states were formed via three body recombination process, they were field ionized and their p̄ were accumulated in the FIW [11].

When the FIW was opened, p̄ accumulated in the FIW were released and annihilated. By observing produced pions, the ionized p̄ were counted. Figure 3a shows the

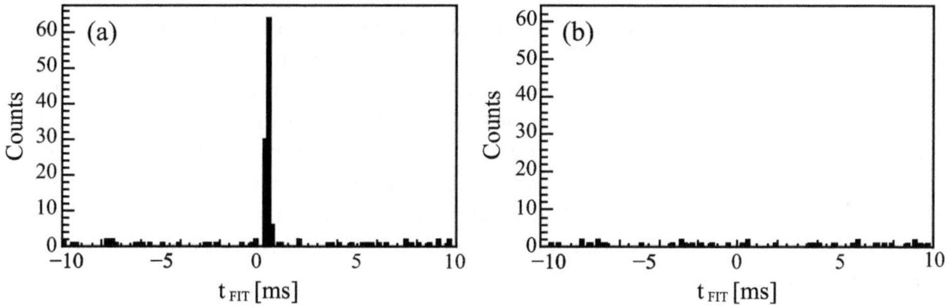

Fig. 3 The time spectrum around the timing of the FIW opening (-10 ms $< t_{FI} <$ 10 ms) **a** with e^+ and **b** without e^+

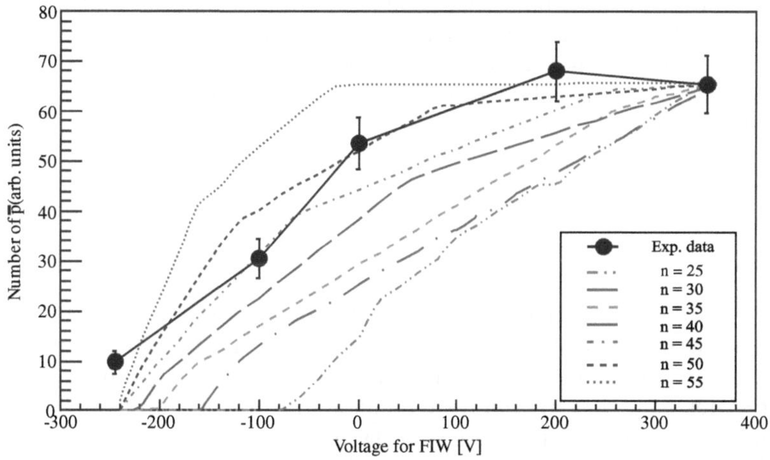

Fig. 4 The number of field-ionized H̄ atoms as a function of the voltage applied to form the FIW. The *solid circles* show the experimental results. *Seven curves* corresponds to field-ionization probabilities are scaled to the experimental value at 350 V

time spectrum of p̄ annihilation at $t_{FI} = 0$ when the FIW was opened after the 50 s mixing of p̄ and e^+. On the other hand, after the same operation but without any e^+s there was no count at the timing as shown in Fig. 3b. The number yielded from one mixing cycle became typically 70. Assuming an isotropic angular distribution of H̄ atoms and taking into account the solid angle covered by the FIW, the total number of H̄ atoms in high Rydberg states was estimated to be about 7×10^3.

In order to estimate the principal quantum number n of the synthesized H̄ atoms, the field strength of the FIW was tuned by changing the voltage applied on electrodes forming the FIW. We calculated the probability for H̄ atoms to be field-ionized and accumulated in the FIW with electric field inside the MRE, assuming they were drifting out isotropically from the nested well with a specific principal quantum number n. We adopted a semi-quantitative formula of field ionization neglecting influence of the magnetic field for the sake of simplicity, $|E| \sim 3.2 \times 10^8 n^{-4}$(V/m), where $|E|$ is the minimum field strength at which the atom in the n state is field ionized [12]. The solid circles in Fig. 4 show the number of field-ionized H̄ atoms as a function of the voltage. It is seen that the number of field ionized $\bar{\mathrm{H}}$ atoms increased as the applying voltage increased, and then saturated. Seven curves in Fig. 4 show such probabilities for $25 \leq \mathrm{n} \leq 55$. The present observation is consistent with the curve for $n \sim 45$ and/or 50. It is noted that the $\bar{\mathrm{H}}$ atoms with $\mathrm{n} \geq 55$ are field ionized by the electric field of the nested well before they reach the FIW. In the case of three body recombination processes, a typical n can be crudely estimated as $n \sim 400/\sqrt{T(\mathrm{K})}$ by equating the binding energy and the e^+ temperature. The temperature of e^+ plasma was estimated to be around 1×10^2 K consistent with our offline measurement.

In order to obtain ground-state $\bar{\mathrm{H}}$ atoms at the microwave cavity, it is not necessary to synthesize ground-state atoms, but just colder $\bar{\mathrm{H}}$ atoms ($\sim 20 - 30$K) are still useful. Because it takes such colder $\bar{\mathrm{H}}$ atom a longer travelling time to reach the cavity than that of 1×10^2 K $\bar{\mathrm{H}}$, which allows to cascade down to the ground-state. In the current mixing scheme discussed in this conference proceeding a $\bar{\mathrm{p}}$ beam was directly injected to a e^+ plasma. If an e^--cooled $\bar{\mathrm{p}}$ is prepared in the nested well, a $\bar{\mathrm{p}}$ cloud with a narrower energy width will be used to form such colder $\bar{\mathrm{H}}$ atoms. One possible scheme is the so-called autoresonance injection developed by the ALPHA collaboration [13].

4 Summary

The CUSP trap was developed to perform a microwave spectroscopy of ground-state hyperfine structure of $\bar{\mathrm{H}}$. We demonstrated the synthesis of cold $\bar{\mathrm{H}}$ atoms employing the CUSP trap. In the present configuration, the estimated number of $\bar{\mathrm{H}}$ in high Rydberg states were 7×10^3 when we assume an isotropic emission. The principal quantum numbers of high Rydberg state $\bar{\mathrm{H}}$ atoms were estimated to be around 45 and/or 50. This success opens a realistic path to make a stringent test of the CPT symmetry with an $\bar{\mathrm{H}}$ beam.

Acknowledgements This work was supported by the Grant-in-Aid for Specially Promoted Research (19002004) of Japanese Ministry of Education, Culture, Sports, Science and Technology (MEXT), Special Research Projects for Basic Science of RIKEN, Università di Brescia, and Instituto Nazionale di Fisica Nucleare.

References

1. Niering, M., et al.: Phys. Rev. Lett. **84**, 2000 (5496)
2. Ramsey, N.: In: T. Kinoshita (ed.) Quantum Electrodynamics, pp. 673–695. World Scientific, Singapore (1990)
3. Kreissl, A., et al.: Z. Phys. Chem. **37**, 557 (1988)
4. Pask, T., et al.: Phys. Lett. B **678**(1), 55 (2009)
5. ASACUSA Collaboration: Tech. Rep. SPSC-P-307 Add.1. CERN-SPSC 2005-006, CERN, Geneva (2005)
6. Andresen, G., et al.: Nature **468**(7324), 673 (2010)
7. Mohri, A., Yamazaki, Y.: Europhys. Lett. **63**(2), 207 (2003)
8. Imao, H., et al.: J. Phys. Conf. Ser. **225**(1), 012018 (2010)

9. Enomoto, Y., et al.: Phys. Rev. Lett. **105**(24), 243401 (2010)
10. Gabrielse, G., et al.: Phys. Lett. A **129**(1), 38 (1988)
11. Gabrielse, G., et al.: Phys. Rev. Lett. **89**(21), 213401 (2002)
12. Gallagher, T.: Rydberg Atoms. Cambridge University Press, New York (1994)
13. Andresen, G.B., et al.: Phys. Rev. Lett. **106**(2), 025002 (2010)

Hyperfine Interact (2012) 209:43–49
DOI 10.1007/s10751-012-0583-x

The AEGIS experiment at CERN

Measuring the free fall of antihydrogen

The AEGIS Collaboration

Published online: 14 March 2012

Abstract After the first production of cold antihydrogen by the ATHENA and ATRAP experiments ten years ago, new second-generation experiments are aimed at measuring the fundamental properties of this anti-atom. The goal of AEGIS (Antimatter Experiment: Gravity, Interferometry, Spectroscopy) is to test the weak equivalence principle by studying the gravitational interaction between matter and antimatter with a pulsed, cold antihydrogen beam. The experiment is currently being assembled at CERN's Antiproton Decelerator. In AEGIS, antihydrogen will be produced by charge exchange of cold antiprotons with positronium excited to a high Rydberg state ($n > 20$). An antihydrogen beam will be produced by controlled acceleration in an electric-field gradient (Stark acceleration). The deflection of the horizontal beam due to its free fall in the gravitational field of the earth will be measured with a moiré deflectometer. Initially, the gravitational acceleration will be determined to a precision of 1%, requiring the detection of about 10^5 antihydrogen atoms. In this paper, after a general description, the present status of the experiment will be reviewed.

Keywords Antimatter · Antihydrogen · Gravity · Matter interferometry · Deflectometry · Weak equivalence principle

1 Introduction

Gravity takes a special place amongst the four fundamental interactions. While the electromagnetic force and the strong and weak nuclear forces are described by quantum field theories, general relativity expresses gravity as a geometric phenomenon: Test bodies travel along geodesics in four-dimensional spacetime, which is

The AEGIS Collaboration
URL: http://cern.ch/aegis

distorted by the presence of massive objects. It follows directly from this geometric approach to gravity that all bodies must behave in exactly the same way in an external gravitational field, regardless of their composition. This postulate is called the weak equivalence principle (WEP) of general relativity. It has been extremely well tested with ordinary matter, at a wide range of length scales from microscopic to astronomical distances [1].

Attempts have been made to formulate a quantum theory of gravity in order to construct a more consistent theoretical framework for all forces, ultimately providing a "Theory of Everything" combining all four interactions. In quantum gravity, the attraction (or repulsion) between two bodies is due to the exchange of virtual bosons, called gravitons, coupling to a gravitational charge. The properties of the force depend on the spins and masses of the exchange bosons as well as the signs of the charges. As a general feature of quantum field theories, the exchange of even-spin particles creates an attractive force between all types of charges, whereas odd-spin particles mediate a repulsive force between like charges.

Ordinary "Newtonian" gravity corresponds to the exchange of massless tensor (spin-2) gravitons. Additionally, vector (spin-1) and scalar (spin-0) gravitons may exist. A hypothetical vector graviton would result in a repulsive force between like charges, producing an opposite effect on matter and antimatter particles—a violation of the WEP. Assuming similar masses and strengths of the vector and scalar components, their effects could cancel out in ordinary matter. They would, however, produce a deviation from Newtonian gravity in the gravitational interaction between matter and antimatter. Such an anomalous "anti-gravity" could result in the segregation of matter and antimatter in different regions of the cosmos and hence explain the apparent absence of antimatter in the observable universe [2].

The effect of gravity on antimatter has never been tested experimentally. Two previous attempts, with positrons [3] and antiprotons [4], were unsuccessful due to the overwhelming effect of stray electromagnetic fields on the charged test particles. Hence, an anomalous gravitational acceleration of antimatter cannot currently be ruled out. Ever since the first production of cold neutral antihydrogen atoms at CERN [5, 6], a high-precision test of antimatter gravity has come within reach. It is the main physics goal of the AEGIS experiment (Antimatter Experiment: Gravity, Interferometry, Spectroscopy), currently being set up at CERN, to measure the effect of gravity on antimatter for the first time [7, 8].

2 Experimental technique

The AEGIS experiment is located at CERN's Antiproton Decelerator (AD) [9]. Its main principle is a measurement of the vertical deflection of a pulsed, cold, horizontal antihydrogen ($\overline{\mathrm{H}}$) beam in the gravitational field of the earth. $\overline{\mathrm{H}}$ will be produced by a charge exchange reaction between highly excited positronium (Ps) and an ensemble of ultracold antiprotons ($\bar{p}$). The design of the apparatus draws heavily on the experience gained with other $\overline{\mathrm{H}}$ experiments, in particular, techniques developed for the capture and cooling of $\bar{p}$ and positrons (e^+) in Penning traps. The experimental procedure consists of three main steps, which will be reviewed in the following sections. An overview sketch of the low-magnetic-field part of the AEGIS apparatus, including the gravimeter, is shown in Fig. 1.

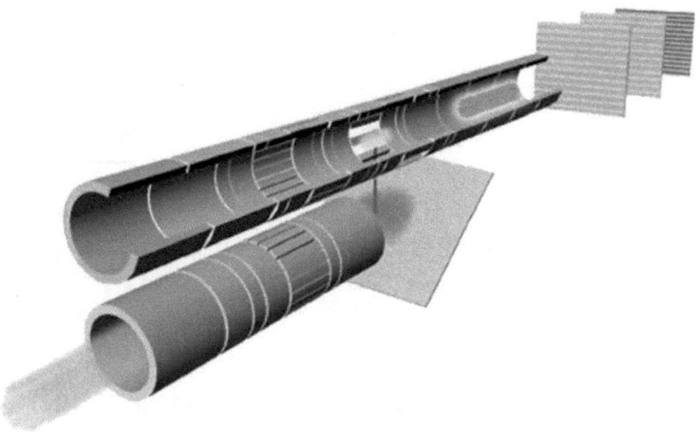

Fig. 1 Three-dimensional cut-open sketch of part of the AEGIS apparatus, showing the low-field Penning traps, the positronium converter, and the deflectometer. Illustration by Internosei, used with permission © Asimmetrie/INFN

2.1 Positronium production and excitation

In the AEGIS experiment, Ps will be produced by implanting e^+ at kinetic energies of several keV into a wafer of nanoporous insulator, which acts as a highly efficient Ps converter [10]. The e^+ scatter off atoms and electrons (e^-) in the bulk and are slowed to eV energies within a few ps. The slow e^+ capture either bound e^- or those released in prior collisions and form Ps. These tend to accumulate in pores of the material due to the reduced dielectric strength. Ps repeatedly bounces off the cavity walls in the pores and is gradually thermalized with the target material. Despite some *ortho*-Ps losses due to so-called pick-off annihilations of e^+ on the cavity walls, a large fraction diffuses out of the film at thermal energies.

Overall, the *ortho*-Ps fraction released from the sample can exceed 20% in silica-based materials cooled to 50 K, as demonstrated by 2–3 gamma ratio of positronium (3γ PAS) measurements [11]. The exact yield and the final velocity distribution depend on the characteristics of the target material, the implantation depth, and the target temperature. Furthermore, it was demonstrated that the energy profile of Ps emitted from the surface of a silica film at room temperature followed a Maxwell–Boltzmann distribution, indicating that a large fraction of the Ps was fully thermalized [12]. Members of the AEGIS Collaboration have been investigating the optimal parameters of a new converter material and the most favorable e^+ energy in terms of *ortho*-Ps yield and temperature, both theoretically and experimentally [13, 14].

The Ps cloud exiting the sample must be excited to Rydberg levels. Laser systems for direct photo-excitation ($\approx$ 180 nm) are not commercially available, hence we will perform a two-step excitation, from the ground to the $n = 3$ state ($\lambda = 205$ nm), and then to a suitable high-n level band ($\lambda \approx 1670$ nm) [15]. The development of the two pulsed laser systems, both based on optical parametric generator crystals with amplification stages and pumped by a Q-switched Nd:YAG laser, has recently been completed [16]. The lasers provide sufficient power to excite the Ps within a few ns. Their bandwidths are tailored to the widths of the transitions, which are broadened by the Doppler effect as well as by level splitting due to the motional Stark and Zeeman effects.

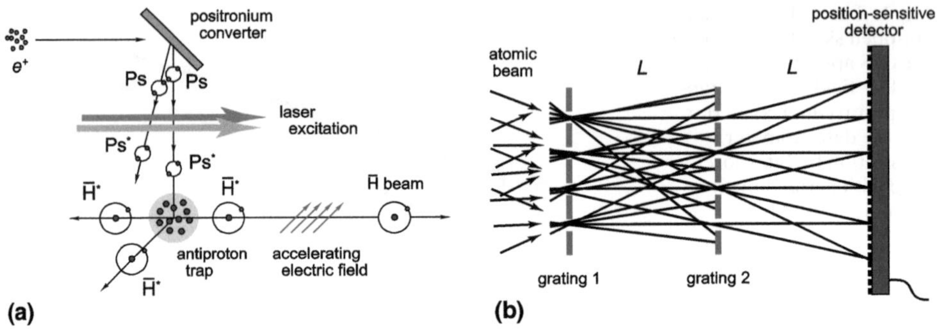

Fig. 2 **a** Method used for $\overline{\mathrm{H}}$ recombination and subsequent acceleration. **b** Principle sketch of the moiré deflectometry technique with two identical gratings and a position-sensitive detector

2.2 Antihydrogen recombination and beam formation

AEGIS will employ an $\overline{\mathrm{H}}$ recombination scheme based on resonant charge exchange with Ps according to the reaction [17]

$$\mathrm{Ps}^* + \bar{p} \longrightarrow \overline{\mathrm{H}}^* + e^-, \tag{1}$$

where stars denote highly excited Rydberg states. The principle is illustrated in Fig. 2a. This reaction has a particularly large cross-section, which scales approximately with the fourth power of the principal quantum number. Most importantly, $\overline{\mathrm{H}}$ formed with $\bar{p}$ at rest is created with a velocity distribution dominated by the $\bar{p}$ temperature, hence the surrounding (cryogenic) environment. A conceptually similar charge exchange technique based on Rydberg cesium [18] has been successfully demonstrated several years ago [19].

The created anti-atoms, produced at velocities of $25\ldots80$ m s^{-1}, will be axially accelerated by exposing them to an electric-field gradient (Stark acceleration). Since the dipole moment scales approximately with the square of the principal quantum number, Rydberg atoms are especially amenable to being manipulated in this way. One of the member groups of AEGIS has recently demonstrated the Stark acceleration of Rydberg hydrogen [20, 21], achieving accelerations of 2×10^8 m s^{-2}. In this way, a horizontal beam of $\overline{\mathrm{H}}$ atoms at a velocity of ≈ 400 m s^{-1} will be projected toward the gravimeter.

2.3 Gravity measurement and data analysis

In the AEGIS experiment, the deflection of the horizontal $\overline{\mathrm{H}}$ beam will be measured with a moiré deflectometer, the classical counterpart of a matter wave interferometer [22, 23]. Three identical material gratings are placed at equal distances L from each other. A particle beam passing through the first two gratings produces a shadow pattern on the third. The change in vertical position of the shadow pattern due to gravity is determined by recording the overall transmission as a function of the position of the third grating. Alternatively, a position-sensitive detector may be used to replace the third grating and detector, as shown in Fig. 2b. A three-grating moiré deflectometer was first used by one of the AEGIS members to measure the local gravitational acceleration to a relative precision of 2×10^{-4} [24].

Under the influence of gravity, the shadow pattern is vertically displaced by a distance

$$\delta x = -gT^2, \tag{2}$$

where g is the local gravitational acceleration and $T = L/v$ is the time of flight between each pair of gratings of a particle beam traveling at velocity v. To extract the value of g from the primary observables (time of flight T and vertical displacement of the fringe pattern δx) the $\overline{\mathrm{H}}$ annihilation events are first binned in symmetric classes of T^2. Secondly, the vertical displacement δx of the fringe pattern is extracted for each of the count classes. Finally, the vertical displacement for all count classes is plotted against the mean time of flight in the class. A quadratic fit to that graph will then yield g. The zero position of the vertical displacement (i.e., without gravity) can be obtained by performing a calibration measurement with the gratings and detector rotated by 90° about the beam axis.

3 Current status and outlook

As of the end of 2011, installation of the AEGIS apparatus at the AD is well advanced. The e^+ source, rare-gas moderator and e^+ trap have been installed, as well as the high-field superconducting magnet and $\bar{p}$ trap. The low-field superconducting magnet has been completed and is awaiting installation. The e^+ accumulator, the dilution refrigerator, and the detector monitoring $\overline{\mathrm{H}}$ production will be installed in the first half of 2012. Finally, the recombination traps, the moiré deflectometer and the position-sensitive detector are currently under development. The system is expected to be ready for $\bar{p}$ capture in 2012 and for a first gravity measurement in 2014, after the CERN accelerator shutdown scheduled for 2013.

Initially, it is planned to measure g of $\overline{\mathrm{H}}$ to a relative precision of 1%, based on the cooling of $\bar{p}$ to $T = 100$ mK with a dilution cryostat. Monte Carlo simulations performed by us indicate that this will require about 10^5 $\overline{\mathrm{H}}$ atoms, corresponding to about 2–3 weeks of data taking. At a later stage of the experiment, we intend to significantly improve the relative precision by cooling the $\bar{p}$ to ultracold temperatures by evaporative cooling or indirect laser cooling. In summary, the AEGIS experiment, currently being set up at the CERN AD, is en route to performing the first measurement of the gravitational interaction of antimatter.

References

1. Heckel, B., et al.: Adv. Space Res. **25**, 1225 (2000). doi:10.1016/S0273-1177(99)00995-3
2. Steigman, G.: Annu. Rev. Astron. Astrophys. **14**, 339 (1976). doi:10.1146/annurev.aa.14.090176.002011
3. Fairbank, W.M., et al.: In: Bertotti, B. (ed.) International School of Physics "Enrico Fermi", p. 310. Academic Press, New York (1974)
4. Holzscheiter, M.H., et al.: Nucl. Phys. A **558**, 709c (1993). doi:10.1016/0375-9474(93)90432-W
5. Amoretti, M., et al.: Nature **419**, 456 (2002). doi:10.1038/nature01096
6. Gabrielse, G., et al.: Phys. Rev. Lett. **89**, 213401 (2002). doi:10.1103/PhysRevLett.89.213401

7. Drobychev, G., et al.: Proposal to the SPS Experiments Committee CERN-SPSC-2007-017, CERN. http://cdsweb.cern.ch/record/1037532 (2007)
8. Kellerbauer, A., et al.: Nucl. Instrum. Methods B **266**, 351 (2008). doi:10.1016/j.nimb.2007.12.010
9. Hémery, J.Y., Maury, S.: Nucl. Phys. A **655**, c345 (1999). doi:10.1016/S0375-9474(99)00223-7
10. Gidley, D.W., Peng, H.G., Vallery, R.S.: Annu. Rev. Mater. Res. **36**, 49 (2006). doi:10.1146/annurev.matsci.36.111904.135144
11. Mariazzi, S., et al.: Appl. Surf. Sci. **255**, 191 (2008). doi:10.1016/j.apsusc.2008.05.207
12. Vallery, R.S., Zitzewitz, P.W., Gidley, D.W.: Phys. Rev. Lett. **90**, 203402 (2003). doi:10.1103/PhysRevLett.90.203402
13. Mariazzi, S., Salemi, A., Brusa, R.S.: Phys. Rev. B **78**, 085428 (2008). doi:10.1103/PhysRevB.78.085428
14. Mariazzi, S., Bettotti, P., Brusa, R.S.: Phys. Rev. Lett. **104**, 243401 (2010). doi:10.1103/PhysRevLett.104.243401
15. Castelli, F., et al.: Phys. Rev. A **78**, 052512 (2008). doi:10.1103/PhysRevA.78.052512
16. Cialdi, S., et al.: Nucl. Instrum. Methods B **269**, 1527 (2011). doi:10.1016/j.nimb.2011.04.108
17. Charlton, M.: Phys. Lett. A **143**, 143 (1990). doi:10.1016/0375-9601(90)90665-B
18. Hessels, E.A., Homan, D.M., Cavagnero, M.J.: Phys. Rev. A **57**, 1668 (1998). doi:10.1103/PhysRevA.57.1668
19. Storry, C.H., et al.: Phys. Rev. Lett. **93**, 263401 (2004). doi:10.1103/PhysRevLett.93.263401
20. Vliegen, E., Merkt, F.: J. Phys. B **39**, L241 (2006). doi:10.1088/0953-4075/39/11/L03
21. Vliegen, E., et al.: Phys. Rev. A **76**, 023405 (2007). doi:10.1103/PhysRevA.76.023405
22. Zehnder, L.: Z. Instrumentenkd. **11**, 275 (1891)
23. Mach, L.: Z. Instrumentenkd. **12**, 89 (1892)
24. Oberthaler, M.K., et al.: Phys. Rev. A **54**, 3165 (1996). doi:10.1103/PhysRevA.54.3165

The AEGIS Collaboration

A. Kellerbauer[1], Y. Allkofer[2], C. Amsler[2], A. S. Belov[3], G. Bonomi[4], P. Bräunig[5], J. Bremer[6], R. S. Brusa[7], G. Burghart[6], L. Cabaret[8], C. Canali[2], F. Castelli[9], K. Chlouba[10], S. Cialdi[9], D. Comparat[8], G. Consolati[11], L. Dassa[4], L. Di Noto[7], A. Donzella[4], M. Doser[6], A. Dudarev[6], T. Eisel[6], R. Ferragut[11], G. Ferrari[12], A. Fontana[13], P. Genova[14], M. Giammarchi[15], A. Gligorova[16], S. N. Gninenko[3], S. Haider[6], J. P. Hansen[16], F. Haug[6], S. D. Hogan[17], L. V. Jørgensen[6], T. Kaltenbacher[6], D. Krasnický[18], V. Lagomarsino[18], S. Mariazzi[19], V. A. Matveev[3], F. Merkt[17], F. Moia[11], G. Nebbia[20], P. Nédélec[21], T. Niinikoski[6], M. K. Oberthaler[5], D. Perini[6], V. Petráček[10], F. Prelz[15], M. Prevedelli[22], C. Regenfus[2], C. Riccardi[14], J. Rochet[2], O. Røhne[23], A. Rotondi[14], M. Sacerdoti[15], H. Sandaker[16], M. Špaček[10], J. Storey[2], G. Testera[24], A. Tokareva[3], D. Trezzi[15], R. Vaccarone[24], F. Villa[9], U. Warring[1], S. Zavatarelli[24], A. Zenoni[4]

[1]Max Planck Institute for Nuclear Physics, Saupfercheckweg 1, 69117 Heidelberg, Germany
[2]Physics Institute, University of Zurich, Winterthurerstrasse 190, 8057 Zürich, Switzerland
[3]Institute for Nuclear Research of the Russian Academy of Sciences, 7a 60th October Anniversary prospect, Moscow 117312, Russia
[4]Department of Mechanical and Industrial Engineering, University of Brescia, Via Branze 38, 25133 Brescia, Italy
[5]Kirchhoff Institute for Physics, University of Heidelberg, Im Neuenheimer Feld 227, 69120 Heidelberg, Germany
[6]Physics Department, European Organisation for Nuclear Research, 1211 Genève 23, Switzerland

[7]Department of Physics, University of Trento, Via Sommarive 14, 38050 Povo (Trento), Italy
[8]Centre national de la recherche scientifique, Laboratoire Aimé Cotton, Campus d'Orsay, 91405 Orsay Cedex, France
[9]Department of Physics, University of Milano, Via Celoria 16, 20133 Milano, Italy
[10]Department of Physics, Czech Technical University in Prague, Břehová 7, 115 19 Praha 1, Czech Republic
[11]Department of Physics, Politecnico di Milano, Piazza Leonardo da Vinci 32, 20133 Milano, Italy
[12]Istituto Nazionale di Ottica, Consiglio Nazionale delle Ricerche, Largo Fermi 6, 50125 Firenze, Italy
[13]Istituto Nazionale di Fisica Nucleare, Sezione di Pavia, Via Agostino Bassi 6, 27100 Pavia, Italy
[14]Department of Nuclear and Theoretical Physics, University of Pavia, Via Agostino Bassi 6, 27100 Pavia, Italy
[15]Istituto Nazionale di Fisica Nucleare, Sezione di Milano, Via Celoria 16, 20133 Milano, Italy
[16]Institute of Physics and Technology, University of Bergen, Alleegaten 55, 5007 Bergen, Norway
[17]Laboratory for Physical Chemistry, ETH Zurich, 8093 Zürich, Switzerland
[18]Department of Physics, University of Genova, Via Dodecaneso 33, 16146 Genova, Italy
[19]Istituto Nazionale di Fisica Nucleare, Gruppo collegato di Trento, Via Sommarive 14, 38050 Povo (Trento), Italy
[20]Istituto Nazionale di Fisica Nucleare, Sezione di Padova, Via Marzolo 8, 35131 Padova, Italy
[21]Institut de Physique Nucléaire de Lyon, Claude Bernard University Lyon 1, 4 Rue Enrico Fermi, 69622 Villeurbanne Cedex, France
[22]Department of Physics, University of Bologna, Via Irnerio 46, 40126 Bologna, Italy
[23]Department of Physics, University of Oslo, Sem Sælands vei 24, 0371 Oslo, Norway
[24]Istituto Nazionale di Fisica Nucleare, Sezione di Genova, Via Dodecaneso 33, 16146 Genova, Italy

Hyperfine Interact (2012) 209:51–56
DOI 10.1007/s10751-011-0538-7

Measurement of Pionic ^{121}Sn atoms at the RI beam factory

Kenta Itahashi · Georg P. A. Berg · Hiroyuki Fujioka · Hans Geissel · Ryugo S. Hayano · Satoru Hirenzaki · Natsumi Ikeno · Naohito Inabe · Satoshi Itoh · Daisuke Kameda · Toshiyuki Kubo · Hiroaki Matsubara · Shinichiro Michimasa · Kenjiro Miki · Hiroyuki Miya · Masaki Nakamura · Takahiro Nishi · Shumpei Noji · Shinsuke Ota · Ken Suzuki · Hiroyuki Takeda · Koichi Todoroki · Kyo Tsukada · Tomohiro Uesaka · Helmut Weick · Koichi Yoshida

Published online: 13 January 2012

Abstract We have measured the energy spectrum of pionic ^{121}Sn atoms by missing-mass spectroscopy of the ^{122}Sn(d, ^{3}He) reaction near the π^- emission threshold. The measurement serves as a pilot experiment for high precision systematic spectroscopy of deeply bound pionic atoms in a new pionic Atom Factory project (piAF) at the RI beam factory (RIBF) of RIKEN. The status of the analysis is reported.

This work is partly supported by the Grants-in-Aid for Scientific Research (22105517, 20540273, 22105510), JSPS fellows (No. 23·2274), and Specially Promoted Research (20002003).

K. Itahashi (✉)
RIKEN Nishina Center, RIKEN, 351-0198 Saitama, Japan
e-mail: itahashi@riken.jp

G. P. A. Berg
JINA and Department of Physics, University of Notre Dame, Indiana 46556, USA

H. Fujioka
Division of Physics and Astronomy, Kyoto University, Kitashirakawa-Oiwakecho, Sakyo-ku, 606-8502 Kyoto, Japan

H. Geissel · H. Weick
GSI—Helmholtzzentrum für Schwerionenforschung GmbH, 64291 Darmstadt, Germany

R. S. Hayano · S. Itoh · T. Nishi · K. Todoroki
Department of Physics, University of Tokyo, 113-0033 Tokyo, Japan

S. Hirenzaki · N. Ikeno
Department of Physics, Nara Women's University, 630-8506 Nara, Japan

N. Inabe · D. Kameda · T. Kubo · M. Nakamura · H. Takeda · T. Uesaka · K. Yoshida
RIKEN Nishina Center, RIKEN, 351-0198 Saitama, Japan

H. Matsubara · S. Michimasa · H. Miya · S. Noji · S. Ota
Center for Nuclear Study, University of Tokyo, 351-0198 Saitama, Japan

K. Miki
RCNP, Osaka University, 567-0047 Osaka, Japan

Keywords Pionic atom · Precision spectroscopy · Chiral condensate

1 Introduction

It is known that the QCD presents different aspects for different energies of the object systems [1]. In the lowest energy regions, the QCD shows its non-perturbative features and hadrons appear as degree of freedom governed by chiral SU(3) dynamics. It is also known that the chiral symmetry is spontaneously broken in the vacuum of low energy density and is parametrized by chiral condensate $< \overline{q}q >$, which varies as a function of the temperature and the density of the medium [2, 3].

Recent experimental spectroscopy of deeply bound pionic atoms in missing-mass measurement of (*d*, ^{3}He) reactions at GSI [4–7] and the theoretical interpretation of the data lead to the first quantitative evaluation of the $< \overline{q}q >$ condensate at the normal nuclear density. It was found to be $\sim$ 30 % smaller than in vacuum [7–9]. However, the evaluation is associated with large statistical and systematic errors, and new high precision data have been long awaited for.

We conducted a first experiment to measure pionic ^{121}Sn atom at RIBF. From the viewpoint of pionic atom spectroscopy using the (*d*, ^{3}He) reaction, RIBF has a huge advantage of a high intensity deuteron beam and BigRIPS as a high-resolution spectrometer with a large acceptance. The objectives of the experiment were: (i) investigation and optimization of the spectral resolution, (ii) test of the detector performance under extremely high rate conditions of $\sim$ 50 MHz, (iii) overall test of the conditions of systematic studies of pionic atoms that will follow. We set our goal to the world highest spectral resolution of $\sim$ 400 keV (FWHM) with high intensity beam of $\sim 10^{12}$ deuterons/sec. Here, we give short summary of the experiment.

2 Pilot experiment

2.1 Experimental conditions

The experiment was carried out in October 2010. We had about three days of data taking during the experimental period of about four days. Figure 1 shows schematic view of the experimental site RIBF [10]. We have used a $T_d = 500$ MeV deuteron beam with a maximum intensity of about 100 pnA. The deuteron beam impinges on the ^{122}Sn target (TA) of a 10 mg/cm^2 thick and 1 mm wide strip. The momentum of the ejected ^{3}He is analyzed by the section TA-F5 of BigRIPS which has a dispersion of about 60 mm/%. The following F5–F7 section is mainly used for particle identification (PiD) purposes. We installed two sets of multi-wire drift chambers (MWDCs) and a segmented scintillation counter near the F5 focal plane and a scintillation counter near the F7 focal plane. The instantaneous count rate

K. Suzuki
Stefan-Meyer-Institute für subatomare Physik, 1090 Vienna, Australia

K. Tsukada
Department of Physics, Tohoku University, 980-8578 Miyagi, Japan

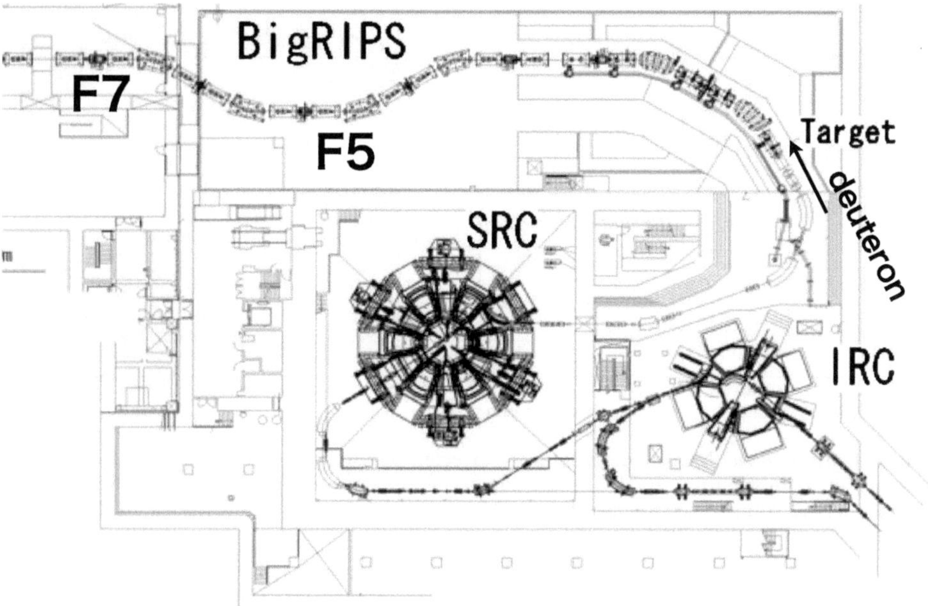

Fig. 1 A schematic view of the experimental site, RIBF [10]. $T_d = 500$ MeV deuteron beam is provided by the cyclotron SRC and impinges on the target. The ^{3}He particles are momentum analyzed by BigRIPS and measured at the dispersive focal plane F5. The F5–F7 section is used for PiD purposes

Fig. 2 Particle identification capability of the detector system to measure the TOF between F5 and F7 and the energy loss in the F5 scintillator. ^{3}He is identified with perfect background rejection

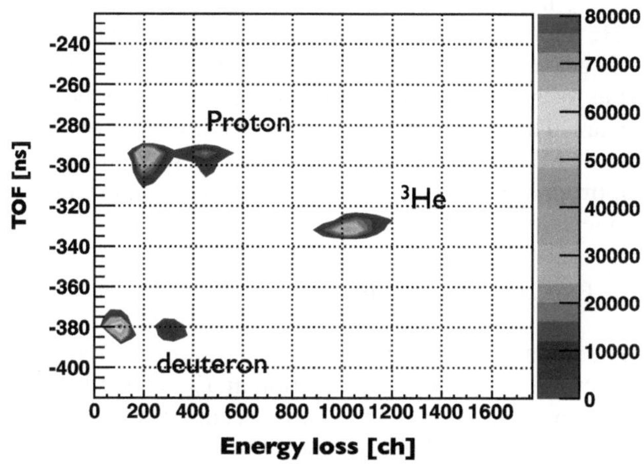

of the F5 segmented scintillation counter amounted to 35 MHz. Most of the events consist of protons produced in the breakup reaction of the incident deuteron beam at the target.

Figure 2 shows the particle identification capability of the detector system. The ordinate is the time of flight measured between the F5 and F7 scintillation counters and the abscissa is the measured energy loss of the particles in the scintillator at F5.

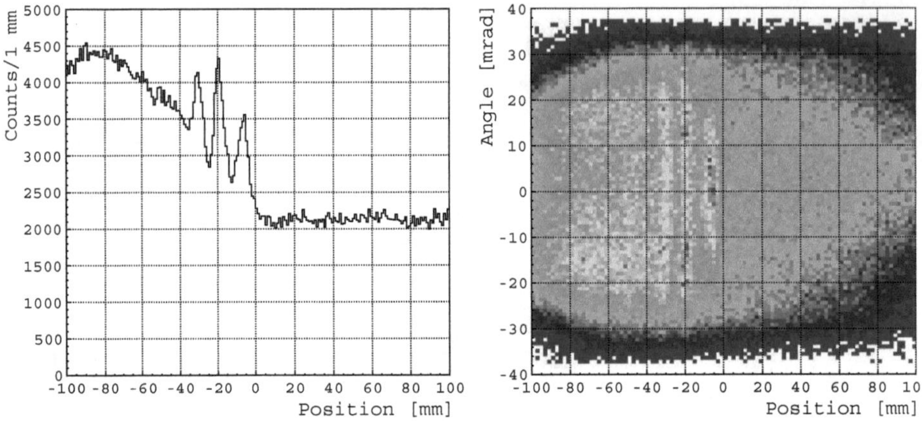

Fig. 3 *Left-side panel* Measured acceptance corrected position spectrum of the $(d, {}^3\mathrm{He})$ reaction near the pion-emission threshold. The left side events show the unbound component and the right side the bound component. *Right-side panel* Measured reaction angle dependence of the position spectrum for $(d, {}^3\mathrm{He})$ reaction

We can clearly identify ^{3}He without contamination. The scintillation counter also provided the trigger for the data acquisition system. We have constructed a very fine-tuned timing trigger for the F5–F7 coincidence measurements, and the trigger rate was limited to less than 500 Hz even at the highest rate of 35 MHz background protons.

We have developed a new analysis method to achieve focal plane spectrum for ^{3}He under very high multiplicity conditions. We observed by a single trigger 7 tracks (6 background protons and one ^{3}He) in the MWDCs. This high multiplicity is mainly due to the break-up protons at the target produced by the high intensity deuteron beam. The new method employs the information on the energy loss of the particles in the MWDCs obtained by measuring the leading and the trailing signals. Setting an appropriate threshold, this method worked perfectly to eliminate proton tracks and to select the ^{3}He track.

2.2 Experimental results

The left-side panel of Fig. 3 shows the measured acceptance corrected F5 position spectrum of the $(d, {}^3\mathrm{He})$ reaction at $T_d = 500$ MeV near the π^- emission threshold. The spectrum shows the events accumulated in a 16 hour long measurement. We are presently working on the calibrations of the abscissa to the ^{3}He energy and the ordinate to the cross section of the reaction, thus the precise position of the threshold is not known yet. However, the peak structures observed in the position [−40 mm, 0 mm] is due to the first experimental observation of pionic ^{121}Sn atoms. The right-side panel of Fig. 3 shows the emission angle at the target vs. the position measured at F5. This is the first observation of the angular dependence of the formation cross section of the pionic atom in the $(d, {}^3\mathrm{He})$ reaction.

The left-side panel of Fig. 4 depicts the measured acceptance corrected F5 position spectrum with the condition of < 15 mrad for the reaction angle. The peaks to the left are smaller compared to the right-most peak. The right-side panel of Fig. 4

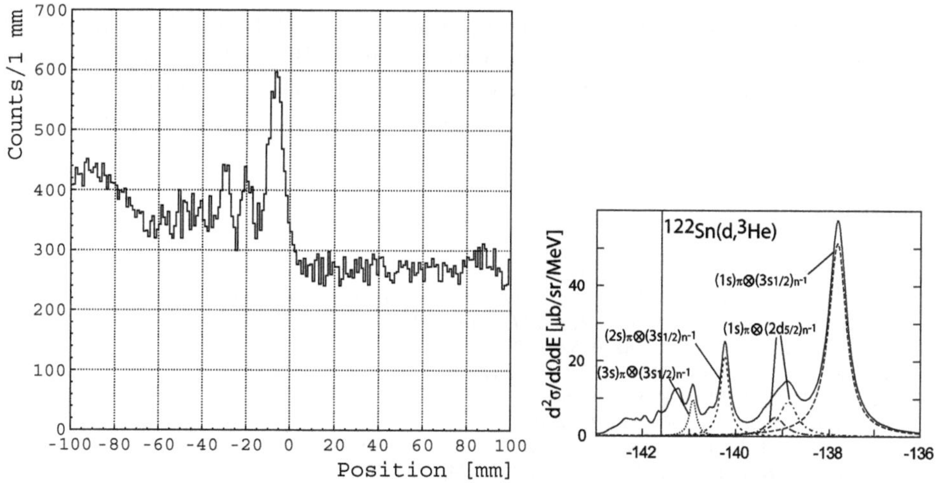

Fig. 4 *Left-side panel* Measured acceptance corrected position spectrum with a condition for the reaction angle < 15 mrad. *Right-side panel* Theoretical calculation by Ikeno with an assumed resolution of 150 keV (FWHM) [11]. The quasi-free component is not included in the calculation. Comparison of the panels lead to the assignment of the largest peak in the left panel to a configuration of $(1s)_\pi \otimes (3s_{1/2})_{n^{-1}}$

shows theoretically calculated Q-value spectrum of the $(d, {}^3\text{He})$ reaction assuming an experimental resolution of 150 keV (FWHM) [11] for comparison. Note that the quasi-free component is not included in the calculation. Overall structures in the panels agree well, and the peak at ~ -8 mm is assigned to a configuration of $(1s)_\pi \otimes (3s_{1/2})_{n^{-1}}$.

3 Conclusion and future perspectives

We have successfully measured for the first time a spectrum of the $^{122}\text{Sn}(d, {}^3\text{He})$ reaction near the π^- emission threshold. A comparison between experimental data and theoretical predictions shown in Fig. 4 indicates good agreement over a wide spectral range. The 1s-pionic state was clearly observed as a prominent peak.

The experimental resolution is estimated to be comparable to the previous value of 400 keV(FWHM) achieved at GSI. This good resolution also indicates that the dispersion matching beam optics was successfully implemented between the primary beam transfer line and the BigRIPS spectrometer to eliminate the effect of the intrinsic momentum spread of about 1 MeV of the incident beam on the spectral resolution.

We are presently working on higher-order aberration corrections of the beam optics, acceptance correction, and calibration of the spectra. The quality of the data is very good and we may have the chance to determine the angular dependence of the formation cross section of the pionic atom of the $(d, {}^3\text{He})$ reaction.

With the presented pilot experiment we have overcome a first hurdle toward our series of planned experiments to perform a systematic study of high-precision spectroscopy of pionic atoms of many isotopes. Note that the spectrum presented in this report was measured in relatively short time of 16 hours. The experiment revealed an aspect of the potential capability of the RIBF facility as a pionic-atom factory. We are going to optimize the experimental procedure in order to achieve higher spectral resolution with smaller experimental errors in the near future.

References

1. Yao, W.M., et al.: J. Phys. **G33**, 1 (2006)
2. Vogl, U., Weise, W.: Prog. Part. Nucl. Phys. **27**, 195 (1991); and references therein
3. Hatsuda, T., Kunihiro, T.: Phys. Rep. **247**, 221 (1994); and references therein
4. Gilg, H., et al.: Phys. Rev. **C62**, 025201 (2000)
5. Itahashi, K., et al.: Phys. Rev. **C62**, 025202 (2000)
6. Geissel, H., et al.: Phys. Rev. Lett. **88**, 122301 (2002)
7. Suzuki, K., et al.: Phys. Rev. Lett. **92**, 072302 (2004)
8. Kolomeitsev, E.E., Kaiser, N., Weise, W.: Phys. Rev. Lett. **90**, 092501 (2003)
9. Hatsuda, T., Kunihiro, T.: Prog. Theor. Phys. **74**, 765 (1985)
10. Kubo, T., et al.: Nucl. Instrum. Methods Phys. Res. B **204**, 97 (2003)
11. Ikeno, N., et al.: Prog. Theor. Phys. **126**, 483 (2011)

Hyperfine Interact (2012) 209:57–62
DOI 10.1007/s10751-011-0524-0

Pionic hydrogen and deuterium

Detlev Gotta · F. D. Amaro · D. F. Anagnostopoulos · P. Bühler · H. Gorke · D. S. Covita · H. Fuhrmann · A. Gruber · M. Hennebach · A. Hirtl · T. Ishiwatari · P. Indelicato · E.-O. Le Bigot · J. Marton · M. Nekipelov · J. M. F. dos Santos · S. Schlesser · Ph. Schmid · L. M. Simons · Th. Strauch · M. Trassinelli · J. F. C. A. Veloso · J. Zmeskal

Published online: 18 December 2011

Abstract The ground-state level shifts and broadenings of the hydrogen isotopes caused by the strong interaction have been redetermined by using a high-resolution crystal spectrometer. An additional measurement of muonic hydrogen reveals properties of the de-excitation cascade of such electrically neutral exotic atoms, in particular Coulomb de-excitation, the understanding of which is essential for the analysis of the hadronic-atom data.

D. Gotta (✉) · H. Gorke · M. Hennebach · M. Nekipelov · Th. Strauch
Forschungszentrum Jülich GmbH and JHCP, 52425 Jülich, Germany
e-mail: d.gotta@fz-juelich.de

F. D. Amaro · D. S. Covita · J. M. F. dos Santos
Department of Physics, Coimbra University, 3000 Coimbra, Portugal

D. F. Anagnostopoulos
Department of Materials Science and Engineering,
University of Ioannina, Ioannina 45110, Greece

P. Bühler · H. Fuhrmann · A. Gruber · A. Hirtl · T. Ishiwatari · J. Marton · Ph. Schmid · J. Zmeskal
SMI, Austrian Academy of Sciences, 1090 Vienna, Austria

D. S. Covita · L. M. Simons
Paul Scherrer Institut (PSI), CH 5232 Villigen, Switzerland

D. S. Covita · J. F. C. A. Veloso
Department of Physics, Aveiro University, 3810 Aveiro, Portugal

P. Indelicato · E.-O. Le Bigot · S. Schlesser · M. Trassinelli
LKB, UPMC-Paris 6, ENS, CNRS; Case 74, 4 place Jussieu, 75005 Paris, France

Present Address:
M. Trassinelli
INS, CNRS, UPMC-Paris 6, 75015 Paris, France

Keywords Exotic atoms · Low-energy pion-nucleon interaction · X-ray spectroscopy

1 Introduction

The measurement of the strong-interaction effects in hadronic atoms by means of X-ray spectroscopy is equivalent to a scattering experiment at relative energy zero (threshold) [1, 2]. In the case of pionic hydrogen (πH), the atomic ground-state level shift $\epsilon_{1s}^{\pi H}$ and broadening $\Gamma_{1s}^{\pi H}$ provide the two independent pion-nucleon (πN) scattering lengths, usually expressed in terms of an isoscalar and an isovector part a^+ and a^- and related to $\pi N \to \pi N$ reaction channels by [3, 4]

$$\epsilon_{1s}^{\pi H} \propto a_{\pi^- p \to \pi^- p} \quad = a^+ + a^- \ + \ ... \tag{1}$$

$$\Gamma_{1s}^{\pi H} \propto (a_{\pi^- p \to \pi^0 n})^2 \propto \quad (a^-)^2 \quad + \ .... \tag{2}$$

Ellipses stand for electromagnetic and strong isospin and non-isospin breaking corrections, which are essential to extract a^+ and a^- in a well defined way from the data, e.g., within the framework of chiral perturbation theory (χPT). For a recent review see [5].

The level shift in pionic deuterium (πD), originating in leading order from the scattering on a proton and a neutron, provides a constraint on the πH data:

$$\epsilon_{1s}^{\pi D} \propto \mathrm{Re}\, a_{\pi D} \ = \ a_{\pi^- p \to \pi^- p} + a_{\pi^- n \to \pi^- n} \ + \ ... \ = \ 2a^+ \ + \ .... \tag{3}$$

Here, $a_{\pi D}$ denotes the complex pion-deuteron scattering length. Ellipses include, in addition, terms arising from multiple scattering and absorptive contributions [6–8].

The scattering lengths a^+ and a^- are related to important quantities in the low-energy description of QCD as are the $\pi N\,\sigma$ term [9], the πN coupling constant [10, 11] and with that to the Goldberger-Treiman discrepancy [12] or the weak induced pseudoscalar coupling [13] as well as charged pion photo production [14].

An independent information is obtained from the πD level broadening $\Gamma_{1s}^{\pi D}$, which is mainly due to the true absorption reaction $\pi^- d \to nn$. Hence, when exploiting detailed balance and after correction for radiative capture $\pi^- d \to nn\gamma$ ($\sim$25%), the absorption strength also measures the s-wave pion-production strength α on isospin zero nucleon pairs [15]:

$$\Gamma_{1s}^{\pi D} \ \propto \ \mathrm{Im}\, a_{\pi D} \ \propto \ \alpha\,. \tag{4}$$

Systematic uncertainties involved in the extraction of α from atom data are much smaller than in the extrapolation of cross-sections to threshold. Furthermore, normalisation problems cease to exist, and Coulomb bound state corrections are well under control [8]. First attempts have been made to calculate α rigorously by the methods of χPT [16].

2 Experiment

The new experiment was performed at the high-intensity low-energy pion beam line πE5 of the Paul Scherrer Institut (PSI) [17]. Pions were stopped in the center of the cyclotron trap [18, 19] in a cylindrical gas cell, where the density of H_2 and D_2 gas

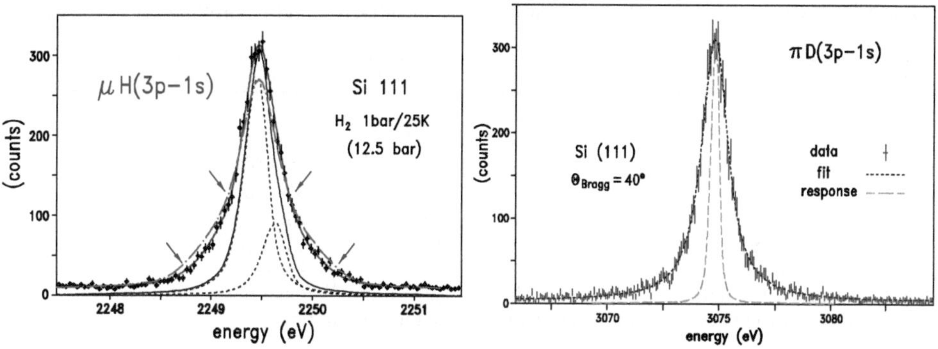

Fig. 1 $(3p-1s)$ transition in muonic hydrogen (from [25]) (*left*) and pionic deuterium (from [28, 29]) (*right*). The *narrow lines* indicate the response of the crystal spectrometer. The additional increase in line width is due to Doppler broadening from Coulomb de-excitation transitions in the case of μH and the strong interaction in the case of πD

was adjusted by means of a cold finger. X-rays emitted after pionic-atom formation were measured by using a Bragg spectrometer equipped with spherically bent silicon and quartz crystals characterized precisely by various methods [20, 21]. As X-ray detector, an array of charge-coupled devices was used [22, 23].

Collisional effects during the life time of the pionic atom play a decisive role when extracting the hadronic broadening. Most important is the case of Coulomb de-excitation [24], where the energy gained during the collision is converted into kinetic energy of the πH and an H atom from a H_2 molecule resulting in a significantly increased line width by Doppler broadening. Hence, the measured line shape is a convolution of spectrometer response, a Lorentz contribution owing to the natural line width (dominated by Γ_{1s}) and, in general, several Doppler induced components from preceding Coulomb de-excitation transitions $(n \rightarrow n')$.

The correction for such a Doppler broadening is essential when extracting Γ_{1s} with the envisaged precision of a few per cent. For that reason, Coulomb de-excitation was studied in a dedicated measurement in muonic hydrogen (μH), where hadronic effects are absent.

3 Results

3.1 Line width

In order to identify Doppler-induced contributions to the line width, the kinetic energy distribution of πH or μH was modeled by narrow intervals at energies corresponding to Coulomb de-excitation transitions expected to dominate the Doppler broadening of the X-ray lines. Non accelerated or thermalised systems are modeled by a low-energy component of [0–2] eV. Intensity and position of the components was searched for by means of a χ^2 analysis of the measured spectra (Fig. 1).

In μH (Fig. 1—left), the low-energy part was found to contribute to about 60%. The rest is shared by the $(5-4)$ and $(4-3)$ components with comparable

Table 1 Transition energies and results for the hadronic effects ϵ and Γ (preliminary in the case of πH)

Transition	E_{QED} / eV	ϵ / meV	Γ / meV	Energy calibration	α / μb
μH($3p-1s$)	2249.461	–	–	–	–
πH($2p-1s$)	2429.5050	–	900 ± 35	–	–
πH($3p-1s$)	2878.8303	$+7.091\pm14$	880 ± 70	πO($6h-5g$)	–
		$+7.110\pm29$	740 ± 60	Zn Kα_1	–
πH($4p-1s$)	3036.0921	$+7.114\pm61$	790 ± 40	Ga Kα_2	–
πD($3p-1s$)	3077.939	-2.356 ± 31	1171 ± 40	Ga Kα_2	251 ± 11

The hadronic shift is defined here by $\epsilon \equiv E_{exp} - E_{QED}$, the difference between the measured and calculated transition energy assuming electromagnetic interaction only

magnitude [25]. Alternatively, using the result of a recent calculation of collision cross sections [25, 26] to predict the kinetic energy distribution [27], reproduces very well the measured line shape. As a side result, the assumed statistical population of the μH hyperfine ground states was confirmed experimentally for the first time.

In the case of the πH($4p-1s$) transition, one high-energy component, $(5-4)$ having about 50% intensity, could be identified. For the πH($3p-1s$) line, also only the preceding Coulomb transition $(4-3)$ was found, again contributing to about 50%. Analyzing the πH($2p-1s$) transition, singly the broadening stemming from the $(4-3)$ Coulomb de-excitation transition is distinguishable. Here, the $(3-2)$ Coulomb de-excitation transition may be already too slow in comparison with other decay channels. Earlier predictions for the kinetic energy distributions of πH, just as for μH, overestimate contributions from $\Delta n = 2$ Coulomb de-excitation [30–32] and, consequently, lead to an unsatisfactory description of the line shape.

The determination of the Doppler induced contributions to the line width limits the accuracy achievable for the hadronic part Γ_{1s} (Table 1). The large value extracted from the πH($2p-1s$) measurements indicates an insufficient treatment of high-energy components of the kinetic energy distribution by such a model independent analysis. Hence, the result from the πH($4p-1s$) transition might be regarded as an upper limit for Γ_{1s}. Even taking into account finally all lines measured in this experiment the precision will hardly exceed 3–4%.

Improvement, however, may become feasible by applying constraints from cascade theory, in particular, by predicting the intensity of the low-energy component with sufficient accuracy. Cross-section calculations are in progress [33], which used together with enhanced cascade codes [30–32], may enable to replace then also for πH the rather insensitive model independent approach described above. Eventually, the analysis of all ground-state transitions should lead to the same value for Γ_{1s} within the errors.

Noteworthy, that for the πD($3p-1s$) line (Fig. 1—right), no high-energy components was identified at the 10% level [28, 29]. Such a behaviour is beyond the change of kinetic energies caused by Coulomb de-excitation only due to the different nuclear mass and not understood at present.

The value achieved for the pion production strength α, as derived from $\Gamma_{1s}^{\pi D}$, is about one order of magnitude more precise than the typical variation of results from the extrapolation of pion production cross sections to threshold [28, 29].

3.2 Line energy

In contrast to earlier experiments [34–40]—using the argon Kα fluorescence lines—as energy calibration for the πH shift measurement served the $(6h - 5g)$ transition in pionic oxygen. In this way, the contribution to the error from the uncertainty of the pion mass cancels in leading order. Besides the πO$(6h - 5g)$ transition, various fluorescence X-rays have been used to check the energy calibration. Both methods result in consistent values for $\epsilon_{1s}^{\pi H}$ (Table 1). In the case of πD, no pionic-atom line close in energy is available, and the Ga Kα_2 fluorescence line had to be used.

The pure electromagnetic transition energies have been recalculated [41] yielding differences of up to 22 meV to the values used in the analysis of the previous experiments. Taken into account, results of previous precision measurements [34–40] for the hadronic shifts in πH and πD agree within the errors.

4 Summary

The accuracy of the measured strong-interaction broadening in pionic hydrogen and deuterium is limited by Doppler broadening to a few per cent. Progress may come from forthcoming cascade calculations, which provide improved kinetic energy distributions used as constraints in the line shape analysis. In contrast, the accuracy achieved for the strong-interaction shift is in the per mille range and exceeds the precision of presently available theoretical calculations.

References

1. Deser, S., Goldberger, L., Kaufmann, K., Thirring, W.: Phys. Rev. **96**, 774 (1954)
2. Gotta, D.: Prog. Part. Nucl. Phys. **52**, 133 (2004) (and references therein)
3. Lyubovitskij, V.E., Rusetsky, A.: Phys. Lett. B **494**, 9 (2000)
4. Rasche, G., Woolcock, W.S.: Nucl. Phys. A **405**, 381 (1982)
5. Gasser, J., Lyubovitskij, V.E., Rusetsky, A.: Phys. Rep. **456**, 167 (2008)
6. Meißner, U.-G., Raha, U., Rusetsky, A.: Phys. Lett. B **639**, 478 (2006)
7. Lensky, V., et al.: Phys. Lett. B **648**, 46 (2007)
8. Baru, V., et al.: Phys. Lett. B **694**, 473 (2011). arXiv:nucl-th/1003.4444v2
9. Gasser, J., Leutwyler, H.: Phys. Rep. **87**, 77 (1982)
10. Goldberger, M.L., Miyazawa, H., Oehme, R.: Phys. Rev. **99**, 986 (1955)
11. Ericson, T.E.O., Loiseau, B., Thomas, A.W.: Phys. Rev. C **66**, 014005 (2002)
12. Scherer, S.: Introduction to chiral perturbation theory. Adv. Nucl. Phys. **27**, 277 (2003). arXiv:hep-ph/0210398v1
13. Gorringe, T., Fearing, H.W.: Rev. Mod. Phys. **76**, 31 (2004)
14. Bernard, V., Kaiser, N., Meißner, U.-G.: Phys. Lett. B **383**, 116 (1996)
15. Brückner, K.A.: Phys. Rev. **98**, 769 (1955)
16. Lensky, V., et al.: Eur. Phys. J. A **27**, 37 (2006)
17. PSI proposal R-98-01: www2.fz-juelich.de/ikp/exotic-atoms
18. Simons, L.M.: Physica Scr. **T22**, 90 (1988)
19. Simons, L.M.: Hyperfine Interact. **81**, 253 (1993)
20. Anagnostopulos, D.F., et al.: Nucl. Instrum. Methods A **545**, 217 (2005)
21. Covita, D.S., et al.: Rev. Sci. Instrum. **79**, 033102 (2008)
22. Nelms, N., et al.: Nucl. Instrum. Methods A **484**, 419 (2002)
23. Indelicato, P., et al.: Rev. Sci. Instrum. **77**, 043107 (2006)
24. Bracci, L., Fiorentini, G.: Nuovo Cim. A **43**, 9 (1978)
25. Covita, D.S., et al.: Phys. Rev. Lett. **102**, 023401 (2009)
26. Popov, V.P., Pomerantsev, V.N.: arXiv:0712.3111v1 [nucl-th] (2007)

27. Jensen, T.S., Pomerantsev, V.N., Popov, V.P.: arXiv:0712.3010v1 [nucl-th] (2007)
28. Strauch, Th., et al.: Phys. Rev. Lett. **104**, 0142503 (2010)
29. Strauch, Th., et al.: Eur. Rev. J. A **47**, 88 (2011)
30. Jensen, T.S., Markushin, V.E.: Eur. Phys. J. D **19**, 165 (2002)
31. Jensen, T.S., Markushin, V.E.: Eur. Phys. J. D **21**, 261 (2002)
32. Jensen, T.S., Markushin, V.E.: Eur. Phys. J. D **21**, 271 (2002)
33. Popov, V.P., Pomerantsev, V.N.: Kinetic Energy Distributions of Muonic and Pionic Hydrogen Atoms (2011)
34. Sigg, D., et al.: Phys. Rev. Lett. **75**, 3245 (1995)
35. Sigg, D., et al.: Nucl. Phys. A **609**, 269 (1996)
36. Sigg, D., et al.: Erratum A **617**, 526 (1997)
37. Chatellard, D., et al.: Phys. Rev. Lett. **74**, 4157 (1995)
38. Chatellard, D., et al.: Nucl. Phys. A **625**, 855 (1997)
39. Hauser, P., et al.: Phys. Rev. C **58**, R1869 (1998)
40. Schröder, H.-Ch., et al.: Eur. Phys. J C **21**, 473 (2001)
41. Schlesser, S., et al.: Phys. Rev. C **84**, 015211 (2011)

Hyperfine Interact (2012) 209:63–68
DOI 10.1007/s10751-012-0559-x

Influence of epithermal muonic molecule formation on kinetics of the μCF processes in deuterium

Andrzej Adamczak · Mark P. Faifman

Published online: 19 January 2012

Abstract The non-resonant formation of $dd\mu$ molecules in the loosely bound state in collisions of non-thermalized $d\mu$ atoms with deuterium molecules D_2 has been considered. The process of such a type is possible only for collision energies exceeded the ionization potential of D_2. The calculated rates of $dd\mu$ formation in the above-threshold energy region are about one order of magnitude higher than obtained earlier. The role of epithermal non-resonant μ-molecule formation for the kinetics of μCF processes in D_2 gas was studied. It was shown that the non-resonant $dd\mu$ formation by $d\mu$ atoms accelerated during the cascade can be directly observed in the neutron time spectra at very short initial times.

Keywords Muonic molecule · Non-resonant formation · Kinetics of μCF in deuterium

1 Introduction

The studies of various reactions with negative μ^- muons in a deuterium target and the muon catalyzed fusion (μCF) phenomenon in particular give rise to special interest in the $dd\mu$-molecule-formation processes (see works [1–3] and references therein). In collisions of the $d\mu$ atoms with the deuterium D_2 molecules, the $dd\mu$ molecules are formed in one of the five bound states, which are defined by the

A. Adamczak (✉)
Institute of Nuclear Physics, Polish Academy of Sciences,
Radzikowskiego 152, 31342 Kraków, Poland
e-mail: andrzej.adamczak@ifj.edu.pl

M. P. Faifman
Research Coordinative Center "MUCATEX", Moscow, 123098, Russia
e-mail: mark@rogova.ru

different rotational (J) and vibrational (υ) quantum numbers [4]. The loosely bound state with binding energy $|\varepsilon_{J=1,\upsilon=1}| = 1.975$ eV refers to $dd\mu$ formation by the resonant reaction:

$$d\mu + D_2 \rightarrow [(dd\mu)_{11}\, dee]^*_{K\nu}\,, \tag{1}$$

where the released energy $\varepsilon \approx 2$ eV is transferred to the excitation of rovibronic ($K\nu$) states of the molecular complex $[(dd\mu)dee]$, according to the resonance mechanism [5]. At the temperature conditions of the majority of previous experiments, the processes (1) mainly occur for thermalized $d\mu$ atoms in the ground state. The rate $\lambda_{dd\mu}$ of resonance reaction (1) depends on the target temperature T and is on the order of 10^6 s^{-1} [3, 6] for the room temperature $T = 300$ K.

In any other ($J\upsilon$) state, the $dd\mu$ molecules are formed via the non-resonant process (2):

$$d\mu + D_2 \rightarrow [(dd\mu)_{J\upsilon}\, de]^+ + e^-\,, \tag{2}$$

with conversion of the released energy into electron ionization of the D_2 molecule (see references in reviews [1, 2]). The rates of transitions to all existing $dd\mu$ rovibronic states have been calculated in [7]. It has been shown that collisions of the thermalized $d\mu$ atoms with the D_2 molecules lead to non-resonant formation of $dd\mu$ molecules with the rates $\lambda_{dd\mu} \sim 10^4$ s^{-1}, whereas these rates are on the order of 10^6 s^{-1} for non-thermalized $d\mu$'s.

It should be noted that for thermalized $d\mu$ atoms the non-resonant $dd\mu$ formation (1) in the state $J = \upsilon = 1$ is impossible. However, as presented in this work, such a process is realized for the non-thermalized $d\mu$ atoms. Also, it is shown that the calculated non-resonant rates are much higher than the data obtained in [7].

2 Non-resonant formation of $dd\mu$ molecule

For thermalized $d\mu$ atoms, collision energies ε are usually much less than the ionization potential $I_e = 15.46$ eV of the D_2 molecule ($\varepsilon \ll I_e$). Then the non-resonant $dd\mu$ formation (2) in the loosely bound state with electron conversion is impossible ($\varepsilon_{11} < I_e$), and only resonant formation (1) is realized. However, the reactions (2) take place for $dd\mu$ formation in deeper bound states with binding energies $\varepsilon_{J\upsilon} \geq I_e$. Besides, when the non-thermalized $d\mu$ atoms have quite high energies $\varepsilon \geq I_e$, the non-resonant formation of $dd\mu$ molecules in the loosely bound state ($J = 1, \upsilon = 1$) also becomes possible.

A method for calculating the non-resonant formation rates in collisions of the epithermal $d\mu$ atoms with the D_2 molecules is analogous to the method developed in [7]. There it was shown that the dominating transitions from the scattering states of the $d\mu + d$ system to the bound states of muonic molecules, with the total orbital angular momenta $J = 1$, are the electric $E1$ transitions only. Such transitions have been considered in this work. The corresponding rates are presented in Fig. 1a as functions of collision energy ε in the center of mass of the system $d\mu$+D_2 (additionally, the analogous dependencies of non-resonant formation rates of the $dt\mu$ molecules are shown in the Fig. 1b). At $\varepsilon = I_e - |\varepsilon_{11}|$, the plotted rates have a typical

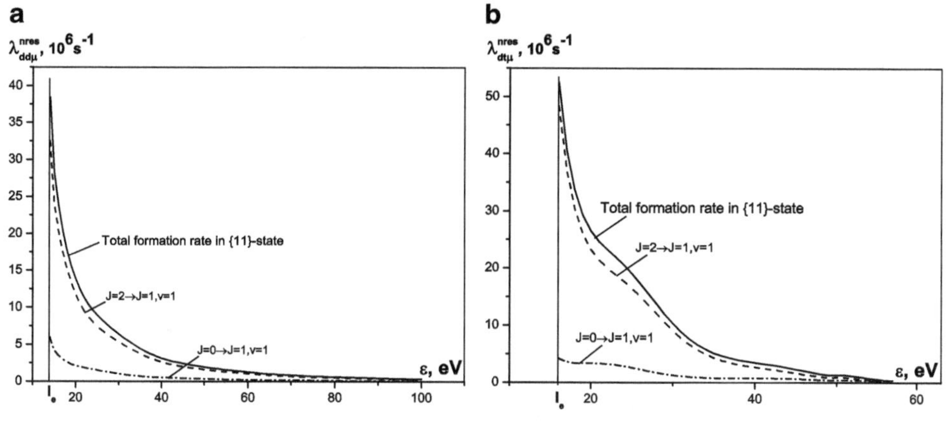

Fig. 1 Non-resonant formation rates of $dd\mu$ (**a**) and $dt\mu$ (**b**) in the state ($J = \upsilon = 1$). The *dash-dotted* and *dashed lines* denote the $E1$-transitions from the initial states $J = 0$ and $J = 2$, respectively

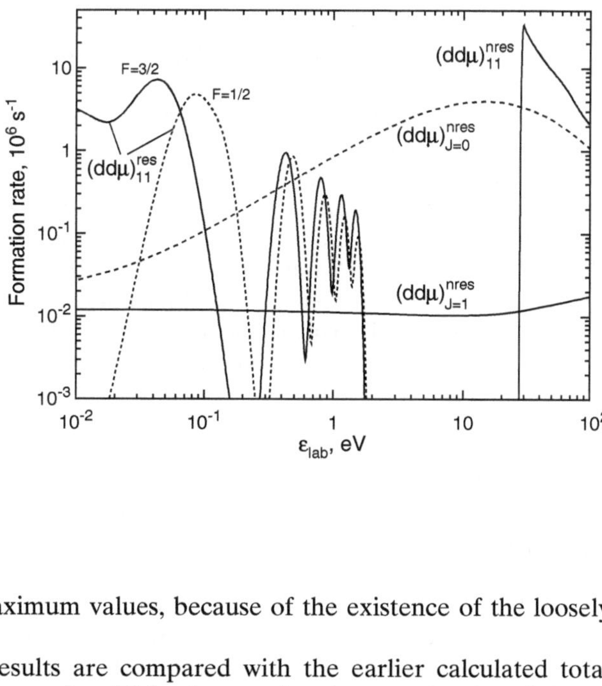

Fig. 2 Resonant and non-resonant $dd\mu$ formation rates versus $d\mu$ kinetic energy in laboratory system

threshold peculiarity and maximum values, because of the existence of the loosely bound state (11).

In Fig. 2, the obtained results are compared with the earlier calculated total rates of non-resonant $dd\mu$ formation [7] in the rotational states $J = 1$ and $J = 0$, as well as resonant formation in the ($J = 1, \upsilon = 1$) state [8]. It is apparent that in the above-threshold energy region the calculated rates of $dd\mu$ formation in the loosely bound (11) state are about one order of magnitude higher than the rates [7] of $dd\mu$ formation in the lower state $J = 0$. Also, they are more than three orders of magnitude higher for formation in the $J = 1$ ($\upsilon \neq 1$) state. This follows from the fact that the overlap of the wave functions of the initial and final states of the $d\mu$+D_2 system is much stronger in the case of non-resonant $dd\mu$ formation in the loosely bound ($J = 1, \upsilon = 1$) state than in the state ($J = 1, \upsilon \neq 1$).

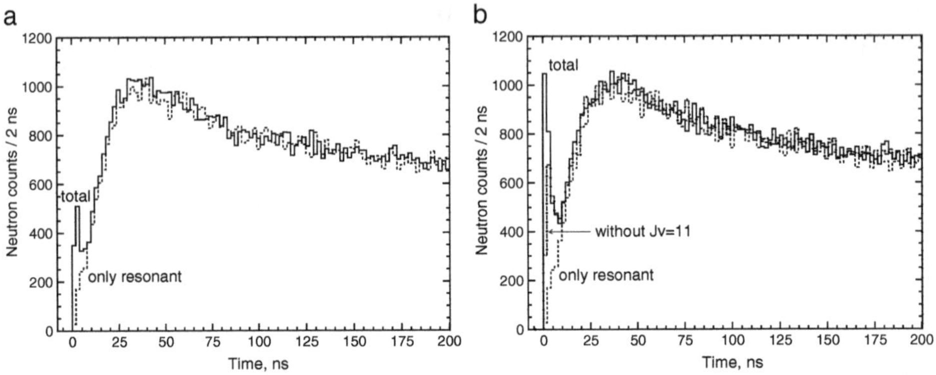

Fig. 3 Neutron spectra from *dd* fusion in D_2 at $T = 40$ K and $\phi = 0.05$ (*solid line*) and the analogous spectrum obtained assuming absence of nonresonant $dd\mu$ formation (*dashed line*): **a** the initial mean energy of non-thermalized $d\mu$ atoms $\varepsilon_{avg} = 10$ eV, **b** $\varepsilon_{avg} = 50$ eV. The *dot-dashed line* represents the spectrum without a contribution from the non-resonant $dd\mu$ formation in the state $J = \upsilon = 1$

3 Demonstration of the epithermal $dd\mu$ formation effect

The kinetics of μCF processes in a pure D_2 gas has been studied in order to take into account effects of the presence of non-thermalized $d\mu$ atoms. For this purpose, the kinetic-energy distributions of $d\mu$ atoms in different atomic states, which are established just after cascade de-excitations of the formed $d\mu$'s, have been calculated using method [9]. These calculations confirmed that most of the $d\mu$ atoms in the final $1S$ state are not thermalized, due to collisions in the cascade process [10–12]. For simplicity of further numerous calculations, an assumption of a simple two-Maxwell shape of the initial energy distribution of $d\mu$ atoms [8, 10] has been employed. One of the Maxwell components of this distribution corresponds to the non-thermalized atoms, while the second component describes the thermalized atoms. The time spectra of neutrons from the *dd* fusion in $dd\mu$ have been calculated by means of Monte-Carlo simulations of the kinetics of μCF processes [8]. They are shown in Fig. 3a and b for the D_2-gas target at temperature $T = 40$ K and density $\phi = 0.05$ (in the liquid-hydrogen-density units). Since the accuracy of calculating the kinetic-energy distribution of $1S$ $d\mu$ atoms is still insufficient, the two average energies $\varepsilon_{avg} = 10$ eV (Fig. 3a) and $\varepsilon_{avg} = 50$ eV (Fig. 3b) of the non-thermalized Maxwell component [8] have been chosen. The neutron spectra with both the resonant and non-resonant $dd\mu$ formation taken into account are represented by the solid lines, while the dashed lines have been calculated without the presence of the non-resonant formation processes.

The significant effect of the epithermal non-resonant $dd\mu$ formation is displayed in Fig. 3 as a peak at the time $t \approx 5$ ns. According to Fig. 2, it is clear that such an effect can be revealed only at short times ($t \lesssim 20$ ns), when most of the $d\mu$ atoms are not yet slowed down to the energy region corresponding to the high peaks of resonant formation. In Fig. 3a, the contribution to the prompt peak due to non-resonant formation in loosely bound (11)-state is negligible, since the initial kinetic energies of $d\mu$'s are rather low. In Fig. 3b, the prompt peak is much more pronounced, since the non-thermalized fraction of initial $d\mu$'s corresponds to a higher mean energy

$\varepsilon_{avg} = 50$ eV. In particular, it is apparent that the contribution from the (11) state is strong (compare the solid and the dot-dashed lines). The neutron yield in the first 4 ns is about two times greater, when the non-resonant formation in this loosely bound state is taken into account.

The enhancement of the neutron yield from *dd* fusion at short times, which was already observed in μCF experiments in D_2 gas (Fig. 11 in [3]) and was even more pronounced in HD gas (Fig. 18 in [3]), confirms the nature of phenomenon considered above. A further consistent comparison between the measured and calculated neutron time spectra for the μCF in D_2 target would enable drawing a final conclusion about the mean energy of $d\mu$ atoms in the $1S$ state, which are accelerated in the cascade processes.

4 Conclusions

A study of the kinetics of μCF processes in D_2 gas, in particular the neutron spectra at short times, revealed the significant role of the non-resonant formation of $dd\mu$ molecules at kinetic energies characteristic to the non-thermalized ground-state $d\mu$ atoms, which are accelerated during the atomic cascade.

Besides the well-known processes (1) and (2) of μ-molecule formation, a new possibility of non-resonant formation of the $dd\mu$ molecule in the loosely bound state in the presence of non-thermalized $d\mu$-atoms has been considered. The calculated rates of such formation reach the magnitude $\lambda_{dd\mu} \sim 10^6\ \mathrm{s}^{-1}$, near the energy threshold of reaction (2). Therefore, these reactions should also be taken into account in analyses of μCF kinetics in deuterium, in particular, at low target densities. Moreover, detailed information on the experimental short-time neutron spectra would allow to extract the mean energy of $d\mu$ atoms in the ground state and to estimate the reliability of various cascade-characteristics calculations.

Acknowledgements The authors are grateful to Profs. L.I. Ponomarev and A.A. Vorobyov for helpful discussions. It is a pleasure to thank Drs. M. Jeitler and N.I. Voropaev for their keen interest.

References

1. Zel'dovich, Ya.B., Gershtein, S.S.: The nuclear reactions in cold hydrogen. Usp. Fiz. Nauk **71**, 581 (1960) [Sov. Phys. Uspekhi **3**, 593 (1961)]
2. Ponomarev, L.I.: Muon catalyzed fusion. Contemp. Phys. **31**, 219 (1991)
3. Balin, D.V., et al.: High precision study of muon catalyzed fusion in D_2 and HD gas. Phys. El. Part. At. Nucl. **42**, 185 (2011)
4. Korobov, V.I., Puzynin, I.V., Vinitsky, S.I.: Bound states of hydrogen mesic molecular ions: variational approach. Muon Catal. Fusion **7**, 63 (1992)
5. Vesman, E.A.: Concerning one possible mechanism of production of the mesic-molecular ion $(dd\mu)^+$. Pis'ma Zh. Eksp. Teor. Fiz. **5**, 113 (1967) [Sov. Phys. JETP Lett. **5**, 91 (1967)]
6. Men'shikov, L.I., et al.: Resonant formation of $dd\mu$ muonic molecules. Zh. Eksp. Teor. Fiz. **92**, 1173 (1987) [Sov. Phys. JETP **65**, 656 (1987)]
7. Faifman, M.P.: Nonresonant formation of hydrogen isotope mesic molecules. Muon Catal. Fusion **4**, 341 (1989)
8. Adamczak, A., Faifman, M.P.: Monte Carlo simulations of the μCF processes kinetics in deuterium gas. Eur. Phys. J. D **51**, 341 (2009)

9. Faifman, M.P., Men'shikov, L.I.: Study of cascade processes in $p\mu^-$ and pK^- atoms based on a new approach. In: Proc. of Int. Conf. MCF-07 on Muon Catalyzed Fusion and Related Topics, Dubna, p. 233, 18–21 June 2007
10. Abbott, D.J., et al.: Diffusion of muonic deuterium and hydrogen atoms. Phys. Rev. A **55**, 214 (1997)
11. Faifman, M.P., Menshikov, L.I.: Cascade processes in muonic hydrogen atoms. Hyperfine Interact. **138**, 61 (2001)
12. Jensen, T.S., Markushin, V.E.: Collisional deexcitation of exotic hydrogen atoms in highly excited states. Eur. Phys. J. D **21**, 271 (2002)

Hyperfine Interact (2012) 209:69–74
DOI 10.1007/s10751-011-0516-0

Induced absorption and annihilation in hadronic hydrogen atoms

Vladimir N. Pomerantsev · Vladimir P. Popov

Published online: 10 December 2011

Abstract The induced absorption or annihilation in the collisions of the hydrogen hadronic atoms in the excited states with ordinary hydrogen have been described in a unified manner with the elastic scattering, Stark transitions, and Coulomb de-excitation in the framework of a close-coupling approach including both the open and closed channels corresponding to both the stationary and non-stationary states of hadronic atom. The general features of the induced absorption cross sections have been studied in a wide range of the complex energy-shift values. The total and differential cross sections of all processes have been calculated for $\pi^- p$, $K^- p$, and $\bar{p}p$ atoms with the principal quantum numbers $n = 2 - 8$ and kinetic energy from 0.001 eV up to 100 eV.

Keywords Hadronic atoms · Induced absorption · Non-stationary states in scattering

1 Introduction

The close-coupling approach (CCA) is widely used to describe the collisional processes of the composite particles in the atom, molecular, and nuclear physics. In the approach the total wave function of the system is expanded in terms of the basis functions corresponding to the stationary states of the subsystems. However, the excited states of all real quantum-mechanical systems and even some ground states (e.g., in exotic atoms) have a finite lifetimes, i.e., are non-stationary ones. It is usually assumed that the lifetimes τ of the basis states are much larger as compared with

This work was supported by Russian Foundation for Basic Researches, grants No. 10-02-01096, 11-02-08381.

V. N. Pomerantsev (✉) · V. P. Popov
Institute of Nuclear Physics, Moscow State University, 119991 Moscow, Russia
e-mail: vpomeran@mail.ru

the collision time, $\tau \gg \tau_{\rm coll} \sim R_0/v$ (R_0 is an effective radius of interaction and v is the relative velocity of the colliding subsystems). In the most problems of atomic and nuclear collisions this qualitative condition is fulfilled, but in case, when one of the colliding subsystems is an hadronic hydrogen atom (h^-a) ($h^- = \pi^-, K^-$, *etc.*; a is a hydrogen isotope), the stationarity condition may be strongly violated in the channels corresponding to the low angular-momentum states (*ns*, *np*). The strong hadron-nucleon interaction results in the lifetimes of these states $\tau_{nl} = \hbar/\Gamma_{nl}$ ($\Gamma_{nl} = -2{\rm Im}\Delta E_{nl}$, where ΔE_{nl} is a complex shift of the level energy) comparable with the time of collision. Besides, the above mentioned condition has no sense in case the closed channels correspond to unstable states.

The problem of inclusion of the short-lived states in the stationary scattering theory arises in any theoretical approach to the description of the similar systems. According to our knowledge, the nuclear absorption during hadronic hydrogen—hydrogen atom collisions was considered in the framework of semiclassical approximation [1, 2] which is valid if the number of essential partial waves is large. First attempt to treat the problem within the quantum close-coupling method was done in paper [3] in which the authors assumed that the widths Γ_{ns} are turned off at some distances ($R > R_0$) between two atoms. It is well known that in the potential scattering theory the procedure to turn on/off the interaction is ambiguous and leads in general case to scale-dependent results. Recently, the problem was considered in the framework of CCA more consistently [4] with the correct boundary conditions. However, the effects of the closed channels corresponding to the unstable states were not taken into account in [3, 4]. The inclusion of these closed channels in case of kaonic and antiprotonic hydrogen are very important for the realistic description of both the elastic scattering and induced absorption at the wide energy range below $np - ns$ thresholds.

The main purpose of the present paper was to give *ab initio* quantum-mechanical treatment of the hadronic hydrogen – hydrogen atom collisional processes including the induced absorption in $(\pi^- p)$ and $(K^- p)$ atoms or annihilation in $(\bar{p}p)$ atom which is the least studied theoretically until now.

2 Close-coupling approach

In the present work the CCA developed earlier by the authors in a number of the papers [5–7] was generalized and applied to treat the collision-induced absorption in the hadronic hydrogen—hydrogen atom collisions $(h^-a)_{nl} + H \rightarrow (h^-a)_{n'l'} + H$ in a unified manner together with the elastic scattering ($n' = n,\ l' = l$), Stark transitions ($n' = n,\ l' \neq l$), and Coulomb deexcitation ($n' \leq n - 1$).

The total wave function $\Psi_E^{JM\pi}(\boldsymbol{\rho}, \mathbf{r}, \mathbf{R})$ of the four-body system $(ah^- + be^-)$ at fixed total energy E and the definite quantum numbers of the total angular momentum (J, M) and parity $\pi = (-1)^{l+L}$ is expanded in terms of the basis states $|1s, nl, L : JM\rangle$ constructed from the hydrogen-like wave functions of the ordinary hydrogen in its ground state, hadronic hydrogen in the $|n, l, m\rangle$ state, and the eigenfunctions $|L, \lambda\rangle$ of the orbital angular momentum of the relative motion. The expansion leads to the coupled second-order differential equations for the radial functions $G_{nlL}^{EJ\pi}(R)$:

$$\left(\frac{d^2}{dR^2} + k_{nl}^2 - \frac{L(L+1)}{R^2}\right) G_{nlL}^{EJ\pi}(R) = 2M_r \sum_{n'l'L'} W_{nlL,n'l'L}^{J\pi}(R)\, G_{n'l'L'}^{EJ\pi}(R), \qquad (1)$$

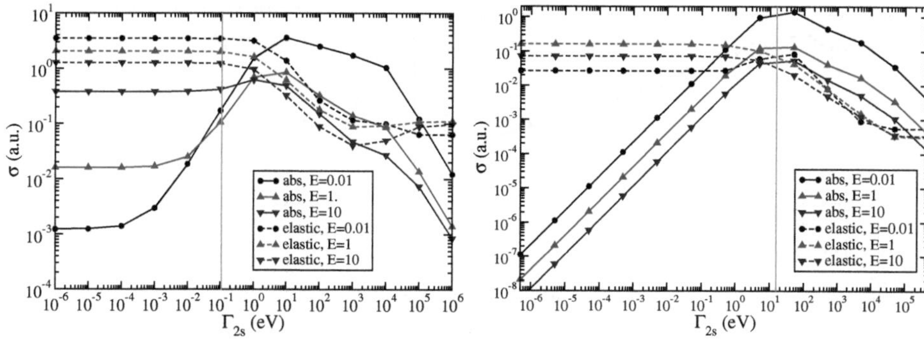

Fig. 1 *Left* The Γ-dependence of the elastic (*dashed lines*) and absorption (*solid lines*) cross-sections for $(\pi^- p)_{2p} + H$ collisions at three values of kinetic energy. *Right* The same for $(K^- p)_{2p} + H$ collisions. The *vertical lines* show the physical values of the widths

where $k_{nl}^2 = 2M_r(E_{\rm cm} - \Delta\varepsilon_{nl,n_1l_1})$ specifies the channel wave number, $E_{\rm cm}$ is the relative motion energy in the entrance (n_1l_1) channel and $\Delta\varepsilon_{nl,n_1l_1}$ is the difference of the current (nl) and entrance (n_1l_1) channel thresholds: $\Delta\varepsilon_{nl,n_1l_1} = \varepsilon_{nl} - \varepsilon_{n_1l_1}$. Here ε_{nl} is the complex energy of the (n, l)-state with both the strong interaction and vacuum polarization shifts taken into account. The interaction potential matrix $W^J_{nlL,n'l'L'}$ couples the asymptotic $(nlL; J)$ and $(n'l'L'; J)$ channels. The wave numbers of the closed channels and the channels corresponding to the non-stationary states have the imaginary parts (Im$k > 0$) and the incoming waves can not be presented in these channels (they are exponentially increased at $R \to \infty$). Thus the functions of unstable-state channels behave like functions of the closed channels and have to satisfy the similar boundary conditions [7].

It is noted, that, in contrast to the division into the open and closed channels, the boundary between the channels with the stationary and non-stationary states is washed out and non-stationary states with the negligible decay widths can be considered as the stationary ones depending on the physical conditions and the kinetic energy of the collision. The partial-wave equations (1) are solved by numerical integration using the propagator matrix method (for the details see [7] Appendix A). The scattering matrix is defined in the subspace of the open channels (with the real and positive wave numbers) from the boundary conditions at $R \to \infty$. Due to non-stationary of the low angular-momentum states (*ns* and *np*) the S matrix of the transitions is not unitary. The unitary defect allows us to define the cross sections of the induced absorption (or annihilation) from the *nl*-state ($l \geq 1$):

$$\sigma_{nl}^{\rm ind}(E) = \frac{\pi}{k_{nl}^2}\frac{1}{2l+1}\sum_J (2J+1)\sum_{LL'n'l'}\left(\delta_{nlL,n'l'L'} - \left|S^J_{nlL\to n'l'L'}(E)\right|^2\right). \quad (2)$$

3 Some results

In the calculations we used the following values of the strong interaction shifts and widths of hadronic atoms: $\epsilon_{1s}^{\rm had} = -7.11$ eV [8], $\Gamma_{1s} = 0.76$ eV [9] (for pionic hydrogen); $\epsilon_{1s}^{\rm had} = 193$ eV, $\Gamma_{1s} = 249$ eV [10] (for kaonic hydrogen); and $\epsilon_{1s}^{\rm had} = 714$ eV, $\Gamma_{1s} = 1097$ eV, $\Gamma_{2p} = 38$ meV [11] (for antiprotonic hydrogen). Examples of our results are shown in Figs. 1, 2 and 3.

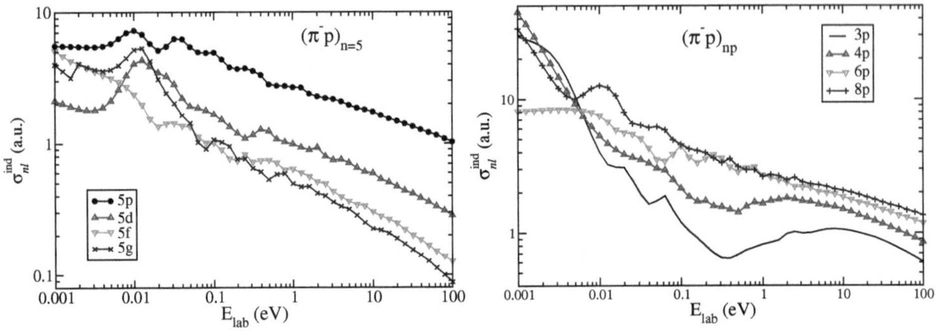

Fig. 2 The E-dependence of the cross sections of induced absorption in $(\pi^{-p})_{nl} + H$ collisions for the different states: $5p$, $5d$, $5f$, $5g$ (*left*) and $3p$, $4p$, $6p$, $8p$ (*right*)

In Fig. 1 we illustrate the Γ_{2s}—dependence of the elastic scattering and induced absorption cross sections in case of pionic and kaonic hydrogen atoms. In case of $(\pi^- p)$ the channel with the unstable 2s state is opened, while in $(K^- p)$ the corresponding channel is closed. The behavior of the cross-sections as a function of Γ_{2s} reveals a few **general features**.

At small values of Γ_{2s} in comparison with both the diagonal and non-diagonal interaction potentials the collision-induced cross section increases linearly with Γ_{2s} and can be described as follows:

$$\sigma_{nl}^{\text{ind}}(E,\, \Gamma_{2s}) = \sigma_{nl}^{\text{ind}}(E,\, \Gamma_{2s} = 0) + \beta\Gamma_{2s}, \tag{3}$$

where $\sigma_{nl}^{\text{ind}}(E, \Gamma_{2s} = 0)$ is a cross-section of the Stark transition $2p \to 2s$ (in the case of the closed channel $\sigma_{nl}^{\text{ind}}(E, \Gamma_{2s} = 0) \equiv 0$); β is a function of collision energy, Re $\Delta\varepsilon_{2p,2s}$, and some other parameters of the scattering problem. At the same time the elastic cross section practically does not depend in this range on the value of Γ_{2s}. Such a behavior can be qualitatively explained in the simple one-channel model with a complex potential. Scattering by this potential results in complex phase shifts $\delta = \delta_1 + i\delta_2$ (with $\delta_2 > 0$). Since Γ_{2s} can be considered as perturbation, than $\delta_2 \propto \Gamma_{2s}$. Using this condition we obtain:

$$\sigma_{2p}^{el} \propto 4 \sin^2 \delta_1,\ \sigma_{2p}^{\text{ind}} \propto 4\Gamma_{2s}.$$

In contrast, at Γ_{2s} much greater than all interaction potentials and $(\text{Re}(E_{\text{cm}} + \Delta\varepsilon_{2p,2s}))$ the cross section of the induced absorption is sharply decreased and behaves like $\sigma_{2p}^{\text{ind}} \propto 1/\Gamma_{2s}$, i.e., proportionally to a lifetime of the unstable state. In this case the lifetime of the unstable state is much less than the collision time and the channels corresponding to both the stable $2p$-state and unstable $2s$-state become uncoupled.

In the intermediate region the values of Γ are comparable with the all interaction potentials and the Γ-dependence of the induced absorption cross section can not be described by a simple form and the detailed calculations are needed.

Figure 2 shows an examples of the energy dependence of absorption cross sections for $(\pi^- p)_{nl}$ with different values n and l. It is seen from Fig. 2 (left) that the absorption from the states with $l = 1$ is more probable than for $l \geq 2$. Besides, in spite of that $\Gamma_{ns} = \Gamma_{1s}/n^3$ we don't observe the corresponding suppression in the absorption cross sections (see Fig. 2 right).

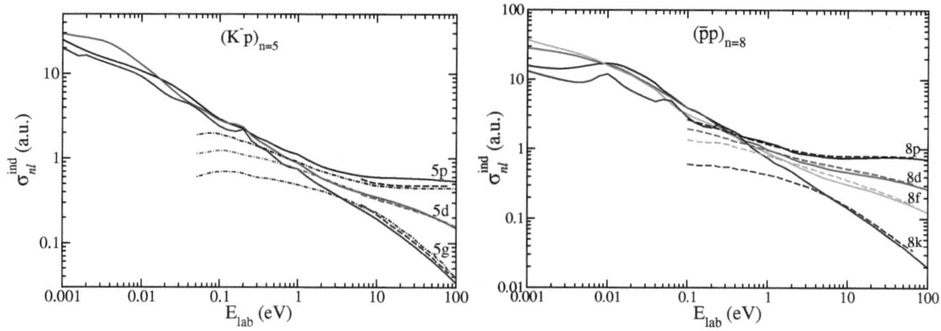

Fig. 3 *Left* Comparison of the present results (*solid lines*) for the absorption cross sections for $(K^-p)_{5l}$ states ($l = 1, 2, 4$) scattering from hydrogen with the results of quantum-mechanical (*dashed lines*) and semiclassical (*dashed-dotted lines*) calculations [3]. *Right* The cross sections of the induced annihilation for $(\bar{p}p)_{8l}$ states ($l = 1, 2, 3, 7$) scattering from hydrogen: the present results are shown with *solid lines*, the results of semiclassical calculations [3] are shown with *dashed lines*)

Contrary to the (π^-p)-atom, the energy shifts of the *ns* states in (K^-p) and $(\bar{p}p)$ atoms are repulsive, hence, the $nl \to ns$ transitions are closed below the corresponding thresholds. According to our knowledge, in the case of the kaonic and antiprotonic hydrogen there are no quantum-mechanical calculations of the absorption cross sections at the energy range below $(ns - nl)$ thresholds until now. Figure 3 shows an example of the energy dependence of absorption cross sections for $(K^-p)_{n=5}$ and $(\bar{p}p)_{n=8}$ at the different values of l. For kinetic energies above $(ns - nl)$ thresholds the present results are as a whole in good agreement with the results of quantum-mechanical and semiclassical description in [3]. Below the thresholds the semiclassical description breaks down and significantly underestimates the absorption cross sections.

4 Summary

The induced absorption or annihilation in the collisions of the hadronic hydrogen atoms in the excited states with the ordinary hydrogen have been described, for the first time, *in a unified manner* with the elastic scattering, Stark transitions, and Coulomb de-excitation processes in the framework of the close-coupling approach. The approach was generalized to include *both the open and closed channels* corresponding to *the stationary and non-stationary states* of hadronic atoms. The general features of the induced absorption cross-sections have been studied in a wide range of the complex energy-shift values. The total and differential cross sections of all processes have been calculated for π^-p, K^-p, and $\bar{p}p$ atoms in the states with $n = 2 - 8$ and kinetic energy from 0.001 eV up to 100 eV. The present results are very important for the kinetics of atomic cascade in hadronic atoms. In particular, the effect of the induced absorption on the values of the X ray yields and the kinetic energy distributions of the hadronic atoms must be taken into account to analyze the high precision experiments on the spectroscopy of the lowest levels of (π^-p), (K^-p), and $(\bar{p}p)$ atoms.

References

1. Leon, M., Bethe, H.A.: Phys. Rev. **127**, 636 (1962)
2. Terada, T.P., Hayano, R.S.: Phys. Rev. C **55**, 73 (1997)
3. Jensen, T.S., Markushin, V.E.: Eur. Phys. J. D **19**, 165 (2002)
4. Korenman, G., Yudin, S.: J. Phys.: Conf. Ser. **88**, 012060 (2007), arXiv:0711.4048
5. Korenman, G.Ya., Pomerantsev, V.N., Popov, V.P.: JETP Lett. **81**, 543 (2005)
6. Pomerantsev, V.N., Popov, V.P.: JETP Lett. **83**, 331 (2006); Phys. Rev. A **73**, 040501(R) (2006)
7. Popov, V.P., Pomerantsev, V.N.: Phys. Rev. A **83**, 032516 (2011)
8. Schroeder, H.Ch., et al.: Eur. Phys. J. C **21**, 473 (2001)
9. Hirtl, A., et al.: Hyperfine Interact. **193**, 153 (2008)
10. Beer, G.: Phys. Rev. Lett. **94**, 212302 (2005)
11. Gotta, D.: Prog. Part. Nucl. Phys. **52**, 133 (2004)

Hyperfine Interact (2012) 209:75–81
DOI 10.1007/s10751-011-0514-2

Kinetic energy distributions of muonic and pionic hydrogen atoms

Vladimir P. Popov · Vladimir N. Pomerantsev

Published online: 12 December 2011

Abstract The kinetic energy distributions of $\mu^- p$ and $\pi^- p$ atoms at a time of the radiative $np \to 1s$ transitions and charge-exchange reaction (in case of $\pi^- p$) have been studied in the improved version of extended standard cascade model (ESCM). *Ab initio* quantum-mechanical calculations of the differential and integral cross sections of the elastic scattering, Stark transitions, Coulomb deexcitation (CD), and induced absorption (in case of pionic hydrogen) have been performed in a framework of the close-coupling approach for the states of exotic atoms with $n \leqslant 8$ and relative motion energies $E \geqslant 0.0001$ eV. The calculated X-ray yields and kinetic energy distributions are in good agreement with the known experimental data. The initial (n, l, E)-distributions of the exotic atoms and target motion are explicitly taken into account.

Keywords Muonic and pionic hydrogen · Scattering processes · Atomic cascade

1 Introduction

The formation of light exotic atoms ($\mu^- p$, $\pi^- p$, etc.) is followed by the atomic cascade in which the initial distributions in the quantum numbers (n, l) and kinetic energy E of exotic atom change due to various processes: radiative transitions, external Auger effect, elastic scattering, Stark transitions, Coulomb deexcitation (CD), and nuclear absorption (in case of hadronic atoms). The development of the cascade models beginning from the standard cascade model [1] has a long history and the significant progress was achieved in a framework of the extended

This work was supported by Russian Foundation for Basic Researches, grants No. 10-02-01096, 11-02-08381.

V. P. Popov (✉) · V. N. Pomerantsev
Institute of Nuclear Physics, Moscow State University, 119991 Moscow, Russia
e-mail: popov@nucl-th.sinp.msu.ru

standard cascade model (ESCM) (see [2, 3] and references therein). Nevertheless, a number of experimental observations [4–9] for $\mu^- p$ and $\pi^- p$ atoms did not have a realistic theoretical explanation in the previous ESCM until recently, in particular, population, lifetime, mechanism of the collisional quenching of $(\mu^- p)_{2s}$ below $2p$ threshold, the yield of high-energy $(\mu^- p)_{1s}$ with kinetic energy 0.9 keV, high-energy components 105 eV and 209 eV in the neutron-time-of-flight spectra, and, finally, Doppler broadening of the X-ray lines in muonic and pionic hydrogen. These experiments as well as the precision shift/width experiments [10] in pionic hydrogen require a more sophisticated approach to a reliable theoretical description of the cascade processes.

2 The improved extended standard cascade model

The main improvement of the present version of ESCM is *ab initio* fully quantum-mechanical calculations of the differential and integral cross-sections of the elastic scattering, Stark transitions, CD, and induced absorption (in case of hadronic atom) in the exotic atom collisions with hydrogen one. The processes have been described by an unified manner in the framework of the close-coupling approach (CCA) applied by the authors earlier [11–14]. The cross-sections were calculated in a wide range of the principle quantum number values ($n = 2 \div 8$) and relative energies $E_{\rm lab} \geqslant 0.0001$ eV for muonic and pionic hydrogen atoms and were used as input data in the present version of ESCM. The most of the calculations were done with the extended basis set including all states of exotic atom with $n \leq 20$. The vacuum polarization and strong interaction shifts (for hadronic atoms) were taken into account in the present CCA. Thus, a significant improvement of the present ESCM in comparison with the previous ones was achieved. The model does not employ any fitting parameters and the additional assumptions concerning the collisional cross sections and the kinetics of atomic cascade is treated more reliably.

The knowledge of the initial conditions is also needed for the realistic study of atomic cascade. The initial conditions are determined by the distributions of exotic atom in the quantum numbers n, l, and kinetic energy at the instant of their formation. The simplest picture of the exotic atom formation is usually used: the initial principal quantum number is fixed at $n_i \sim \sqrt{M}$ (M is a reduced mass of exotic atom in atomic units), and the statistical l-distribution $P_l = (2l+1)/n^2$ is assumed. More elaborate approaches [15–17] taking the molecular effects of the target into account predict the distribution in the initial n with the peak shifted towards lower values than n_i and non-statistical l-distribution. In the limit of the lowest target densities the atomic cascade is mainly determined by the lifetime of exotic particle and the rates of radiative transitions. Therefore, the information about the initial (n, l)- and E-distributions is conserved up to the end of the cascade and can be extracted from the experimental data obtained at very low target densities. In the present study the initial distributions are described by: Gaussian n-distribution, modified statistical l-distribution $P_l \propto (2l+1)e^{-\alpha_l(2l+1)}$, and two-exponential E-distribution. The parameters of the initial (n, l, E) - distributions are obtained at very low target densities from the comparison of the calculated relative K_α-line yield, i.e. the ratio of the absolute K_α-line yield to the total yield of K-line, (see Fig. 1 left) and integrated kinetic energy distribution I($E_{\rm lab}$) (in percent) of $(\mu^- p)_{1s}$ (see Fig. 1 right) with

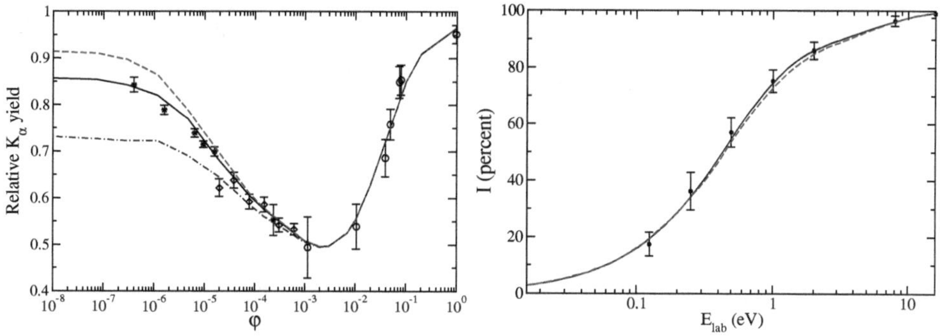

Fig. 1 *Left* The density dependence of relative K_α yield in muonic hydrogen for different initial l-distributions: statistical (*dashed line*) and modified statistical with $\alpha_l = 0.08$ (*solid line*) and $\alpha_l = 0.2$ (*dashed-dotted line*). The experimental data are from [18] (*asterisks*), [19] (*diamonds*), and [20] (*open circles*). *Right* The integrated E-distribution (*in percent*) of $(\mu^- p)_{1s}$ at target density $\varphi = 7.53 \times 10^{-8}$: fit of the experimental data [4] (*solid line*) and the present cascade calculation (*dashed line*)

the experimental data [18–20] and [4], respectively. Besides, we use the Maxwell distribution to describe the thermal motion of the target.[1]

According to our study (see Fig. 1 left), the modified statistical l-distribution ($\alpha_l = 0.08$) leads as a whole to an excellent agreement between theoretical predictions and experimental data [18–20] for all densities under consideration contrary to the statistical l-distribution which results in a significant disagreement with the experimental data [18] at densities $\varphi \lesssim 10^{-5}$. The experimental energy distribution [4] at $\varphi = 7.53 \times 10^{-8}$ was analytically approximated by the two-exponential formula (for detail see [11]) and used as an initial one in the cascade model. As it is seen (Fig. 1 right), the calculated E-distribution of $(\mu^- p)_{1s}$ practically coincides with the experimental data [4] and the calculated mean kinetic energy 1.25 eV is in a perfect agreement with the experimental value 1.3 ± 0.8 eV [4]. The initial (n, l, E)-distributions determined above were used in our further cascade calculations.

3 Some results and discussion

The kinetic energy distribution (E-distribution) of the exotic atom changes during the cascade and is a more refined probe of the theoretical approaches to the description of the cascade processes. Using the cascade model described above we calculated the E-distributions of **muonic hydrogen** at a time of radiative $np \to 1s$ transitions ($n = 2 \div 4$) for the different values of the target densities (see Figs. 2 and 3). For the reliable comparison (or analysis) of experimental data with the results of theoretical calculations the real experimental conditions (density and temperature of the target) must be taken into account. In particular, the effect of the target motion (temperature) on the E-distribution of $\mu^- p$ at a time of radiative transition may be very significant (see Fig. 2 left).

[1]The present cascade calculations have been done in the density range $\varphi = 10^{-8} \div 1$ (in units of Liquid Hydrogen Density, $LHD = 4.25 \cdot 10^{22}$ atoms/cm^3).

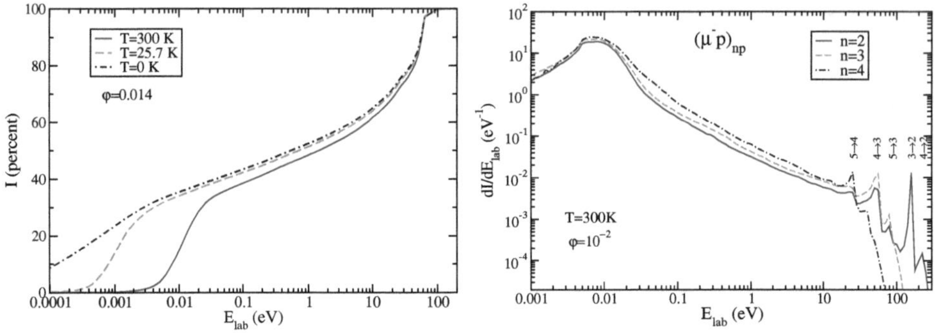

Fig. 2 *Left* The integrated E-distributions of $\mu^- p$ at the instant of radiative transition $3p \to 1s$ (effect of the target motion). *Right* The differential E-distribution of $\mu^- p$ at a time of radiative $np \to 1s$ transitions ($n = 2 \div 4$)

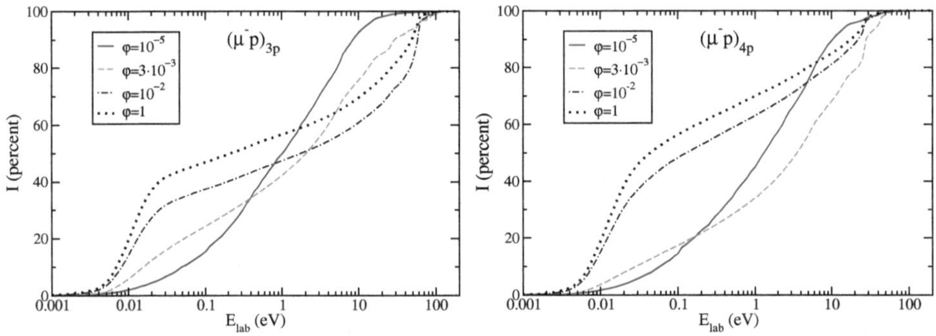

Fig. 3 Density dependence of the integrated E-distributions for $\mu^- p$ at a time of radiative $3p \to 1s$ (*left*) and $4p \to 1s$ (*right*) transitions. Target temperature $T = 300$ K

The differential E-distributions of $\mu^- p$ at a time of the radiative $np \to 1s$ transitions ($n = 2 \div 4$) calculated at fixed values of the target density and temperature are shown in Fig. 2 (right). We see a quite similar behavior of the E-distributions at energy below $\approx 10 \div 20$ eV with regular increasing of the so-called "cold fraction" and distinctive high-energy structures arising from the different CD processes with $\Delta n \geq 1$ preceding the radiative transition. The complicated shapes of these structures can be explained by the interplay of three factors: the kinetic energy distribution of exotic atom before CD, the anisotropy of the angular distribution in the CD process, and, finally, the deceleration due to the elastic scattering and Stark transitions after CD but before the radiative transition.

The typical examples of the muonic-hydrogen E-distributions (integrated) calculated at the instant of radiative $np \to 1s$ transitions calculated at the different values of target density are shown in Fig. 3. The results given in Fig. 3 demonstrate a strong dependence of the E-distributions on both the initial state n and the target density. Since the profiles of the $np \to 1s$ X-ray lines are included a Doppler broadening due to the non-thermal motion of exotic atoms at the instant of radiative transition, the corresponding E-distribution of exotic atoms must be taken into account in the precision analysis of the experimental data. Besides, we can obtain an additional

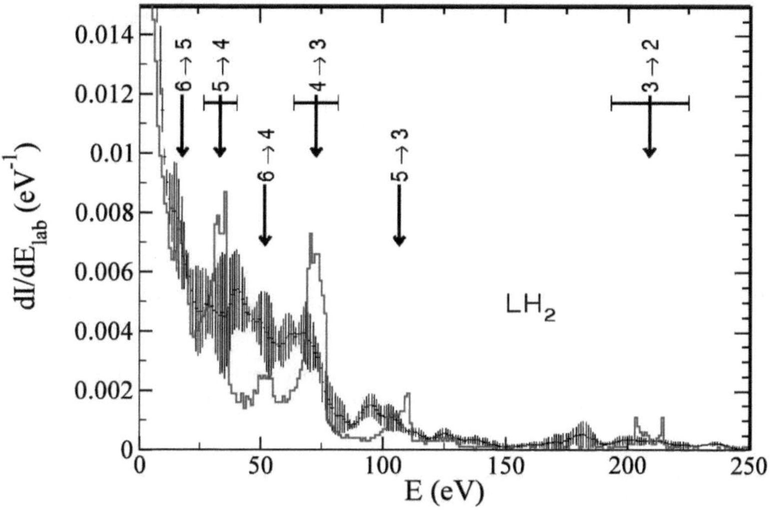

Fig. 4 The kinetic energy distribution of $\pi^- p$ at the instant of charge-exchange reaction (at LHD): the experimental data [7]) in comparison with the results of the present cascade calculations (*solid line*)

test of the reliability of our theoretical predictions. Recently, the profile of K_β line in muonic hydrogen was described using our preliminary E-distribution for the analysis of the line shape of $(\mu^- p)$ $3p \rightarrow 1s$ hyperfine transitions [10]. The good agreement of the obtained values (194 ± 12) meV (for the hyperfine splitting) and $(2.94 \pm 0.24) : 1$ (for the triplet-to-singlet intensity ratio) with the calculated value (182.725 ± 0.062) meV [21] and a relative statistical population of $3 : 1$ (for the triplet and singlet components) serves a confirmation of the significantly improved description of the kinetics of atomic cascade in the present version of ESCM.

The present study of the atomic cascade in **pionic hydrogen** was focused on the description of the experimental data for the neutron time-of-flight spectrum [7] obtained at LHD and E-distributions at a time of radiative $np \rightarrow 1s$ transitions $(n = 2 - 4)$. The E-distribution of pionic hydrogen at a time of charge-exchange reaction in liquid hydrogen is shown in Fig. 4. For the first time the theoretical description explains in details both the kinetic energy distribution of the $\pi^- p$ atoms at the instant of charge-exchange reaction and fine structure of this distribution. In particular, our results explain the high-energy components ~ 105 eV and ~ 209 eV (due to $5 \rightarrow 3$ and $3 \rightarrow 2$ CD, respectively) and are in very good agreement with the experimental weights of these components [7].

As in case of muonic hydrogen, the typical examples of the integrated E-distributions of pionic hydrogen calculated at the instant of the radiative $np \rightarrow 1s$ transitions are presented in Figs. 5 and 6. In spite of the some similarity between the muonic and pionic hydrogen E-distributions calculated at the same target conditions, the details of these distributions can play a crucial role in the proper description of the Doppler broadening corrections. It is especially important for the analysis of the shift/width experiments [9] in pionic hydrogen. If the Doppler corrections will be properly taken into account, then the measurements for the radiative $np \rightarrow 1s$

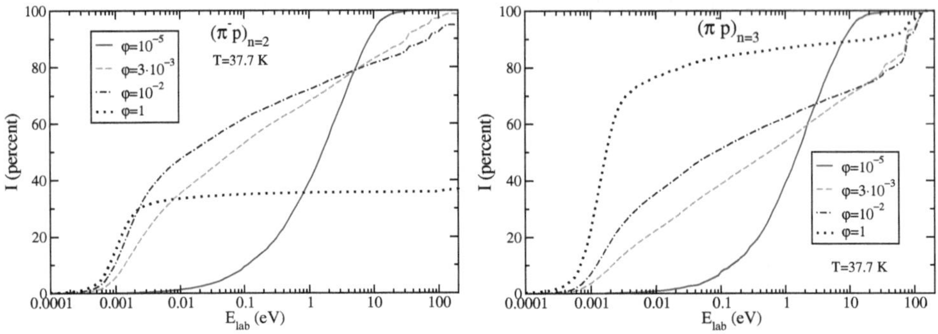

Fig. 5 The integrated energy distributions of pionic hydrogen at the instant of radiative transitions: $2p \to 1s$ (*left*), and $3p \to 1s$ (*right*) at different densities of the target

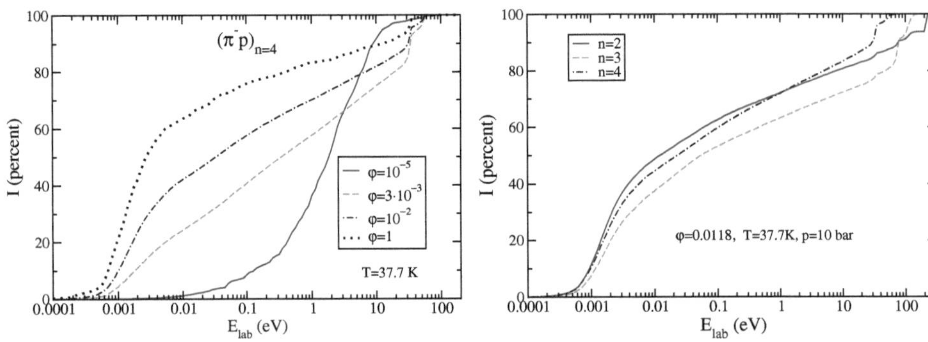

Fig. 6 The integrated energy distributions of pionic hydrogen at the instant of radiative transitions: $4p \to 1s$ at different densities of the target (*left*), and $np \to 1s$ ($n = 2 \div 4$) at the target density 10 bar (*right*)

transitions at different target densities must result in a good agreement for the values of the ground state width of pionic hydrogen.

4 Conclusion

The present version of the cascade model significantly improves the previous cascade calculations due to the self-consistent treatment of the elastic scattering, Stark transitions, CD, and induced absorption (in case of hadronic atoms). The inclusion of the results of the detailed quantum mechanical calculations of the differential and integral cross sections for $nl \to n'l'$ collisional transitions in the cascade code allowed us to describe the kinetics of the atomic cascade without employing any fitting parameters and more accurately than previously. The results of the cascade calculations are in good agreement with the experimental data: quenching rate of the metastable 2s state in muonic hydrogen, observed kinetic energy distribution of $\pi^- p$ atoms at the instant of nuclear absorption, and the yield of hot $(\mu^- p)_{1s}$. The calculated kinetic energy distributions at the time of the radiative $np \to 1s$

transitions in muonic and pionic hydrogen allow to extract Doppler broadening from the measured profiles of the X-ray lines and determine the strong interaction width of the ground state of pionic hydrogen.

References

1. Leon, M., Bethe, H.A.: Phys. Rev. **127**, 636 (1962)
2. Markushin, V.E.: Phys. Rev. A **50**, 1137 (1994)
3. Jensen, T.S., Markushin, V.E.: Eur. Phys. J. D **21**, 271 (2002)
4. Pohl, R.: Ph.D. thesis, ETH Zurich No. 14096 (2001)
5. Pohl, R., et al.: Hyperfine Interact. **138**, 35 (2001)
6. Kottman, F., et al.: Hyperfine Interact. **138** 55 (2001)
7. Badertscher, A., et al.: Europhys. Lett. **54**, 313 (2001)
8. Pohl, R., et al.: Phys. Rev. Lett. **97**, 193402 (2006)
9. Covita, D.S., et al.: Phys. Rev. Lett. **102**, 023401 (2009)
10. Gotta, D.: Prog. Part. Nucl. Phys. **52**, 133 (2004)
11. Popov, V.P., Pomerantsev, V.N.: Phys. Rev. A **83**, 032516 (2011)
12. Korenman, G.Ya., Pomerantsev, V.N., Popov, V.P.: JETP Lett. **81**, 543 (2005)
13. Pomerantsev, V.N., Popov, V.P.: JETP Lett. **83**, 331 (2006)
14. Pomerantsev, V.N., Popov, V.P.: Phys. Rev. A **73**, 040501(R) (2006)
15. Korenman, G.Ya., Popov, V.P.: AIP Conf. Proc. **181**, 145 (1988)
16. Korenman, G.Ya., Popov, V.P., Fesenko, G.A.: Muon Catal. Fusion **7**, 179 (1992)
17. Cohen, J.S.: Rep. Prog. Phys. **67**, 1769 (2004)
18. Anderhub, H., et al.: Phys. Lett. B **143**, 65 (1984)
19. Bregant, N., et al.: Phys. Lett. A **241**, 344 (1998)
20. Lauss, B., et al.: Phys. Rev. Lett. **80**, 3041 (1998)
21. Martynenko, A.P., Faustov, R.N.: JETP **98**, 39 (2004)

Hyperfine Interact (2012) 209:83–92
DOI 10.1007/s10751-012-0576-9

Atoms in flight and the remarkable connections between atomic and hadronic physics

Stanley J. Brodsky

Received: 16 November 2011 / Accepted: 25 January 2012 / Published online: 16 February 2012

Abstract Atomic physics and hadron physics are both based on Yang Mills gauge theory; in fact, quantum electrodynamics can be regarded as the zero-color limit of quantum chromodynamics. I review a number of areas where the techniques of atomic physics provide important insight into the theory of hadrons in QCD. For example, the Dirac-Coulomb equation, which predicts the spectroscopy and structure of hydrogenic atoms, has an analog in hadron physics in the form of light-front relativistic equations of motion which give a remarkable first approximation to the spectroscopy, dynamics, and structure of light hadrons. The renormalization scale for the running coupling, which is unambiguously set in QED, leads to a method for setting the renormalization scale in QCD. The production of atoms in flight provides a method for computing the formation of hadrons at the amplitude level. Conversely, many techniques which have been developed for hadron physics, such as scaling laws, evolution equations, and light-front quantization have equal utility for atomic physics, especially in the relativistic domain. I also present a new perspective for understanding the contributions to the cosmological constant from QED and QCD.

Keywords Quantum electrodynamics · Atomic physics · Hadron physics · Light-front

1 Introduction

Quantum Electrodynamics, the fundamental theory of leptons and photons which underlies all of atomic and molecular physics, and Quantum Chromodynamics, the quark and gluon theory with three colors underlying hadronic and nuclear physics, are both derived from Yang-Mills gauge theory. The Yang-Mills Lagrangian for

S. J. Brodsky (✉)
SLAC National Accelerator Laboratory, Stanford University, Stanford, CA 94309, USA
e-mail: sjbth@slac.stanford.edu

$SU(N_C)$ is invariant under arbitrary color rotations and phases at each point of space and time. In fact, in the limit where the number of colors N_C vanishes, with $\alpha_s C_F = \alpha$ held fixed ($C_F \equiv (N_C^2 - 1)/2N_C$), QCD becomes equivalent to Abelian gauge theory [1]. This analytic connection as a function of N_C between QCD and QED provides a valuable link between the two fields; processes and analyses in QCD must connect at zero color to the analogous reactions and procedures of QED.

In this paper, I will review a number of areas where the techniques of atomic physics give important insight into the theory of hadrons, the color-singlet bound states of quarks and gluons in QCD. For example, the Dirac-Coulomb equation, which predicts the spectroscopy and structure of hydrogenic atoms has an analog in hadron physics in the form of relativistic frame-independent equations of motion derived from light-front holography [2] which give a remarkable first approximation to the spectroscopy, dynamics, and structure of light hadrons. The renormalization scale for the running coupling which is unambiguously set in QED leads to a solution for setting the renormalization scale in QCD. The production of atoms in flight provides a method for computing the formation of hadrons at the amplitude level. Conversely, many techniques and theorems developed for hadron physics, such as scaling laws, evolution equations, and light-front quantization have equal utility for atomic physics, especially in the relativistic domain.

2 Production of exotic atoms in flight and hadronization at the amplitude level

Relativistic antihydrogen was first produced in 1995 at CERN-LEAR [3] and at the Fermilab Antiproton Accumulator [4]. The production mechanism [5] is illustrated in Fig. 1a. The incident antiproton beam produces a Bethe-Heitler electron-positron pair in the Coulomb field of a target nucleus $\bar{p}Z \to \bar{p}e^+e^-Z \to [\bar{p}e^+]Z$. The comoving off-shell $\bar{p}$ and e^+ then coalesece into antihydrogen atoms via the Schrödinger Coulomb wavefunction which connects the off-shell state to the on-shell anti-atom. The atom is dominantly in its 1S ground state. In principle, one can measure its "anti-Lamb-Shift" using the Robiscoe level-crossing method [6].

The production of antihydrogen in flight provides important insight into the dynamics of hadron production in QCD. For example, the $\Lambda(sud)$ baryon can be produced at high longitudinal momentum fraction x_F in $pp \to \Lambda X$ reactions by the coalescence of the ud valence quarks of the beam with a comoving strangeness quark. This method can be generalized to produce heavy hadrons such as $\Lambda_c(cud)$, Λ_b, double charmed baryons, etc., using the high x intrinsic heavy quarks which exist in the higher Fock states of the proton wavefunction [7].

The analog of intrinsic charm in hadrons is the $\mu^+\mu^-$ content of positronium. The $|e^+e^-\mu^+\mu^- >$ Fock state appears through the cut of the muon-loop light-by-light contribution to the self energy of the positronium eigenstate. In this Fock state, the muons carry almost all of the momentum of the moving atom since the off-shell virtuality is minimal at equal velocity. In QED the probability for intrinsic leptons $L\bar{L}$ exist in positronium scales as $1/m_L^4$ whereas in QCD the probability of intrinsic heavy quarks in the wavefunction of a light hadron scales as $1/m_Q^2$ because of its non-Abelian couplings [8, 9].

The production of a $q\bar{q}$ meson in an e^+e^- annihilation event is illustrated in Fig. 1b. One first calculates the T matrix element for the production of off-shell

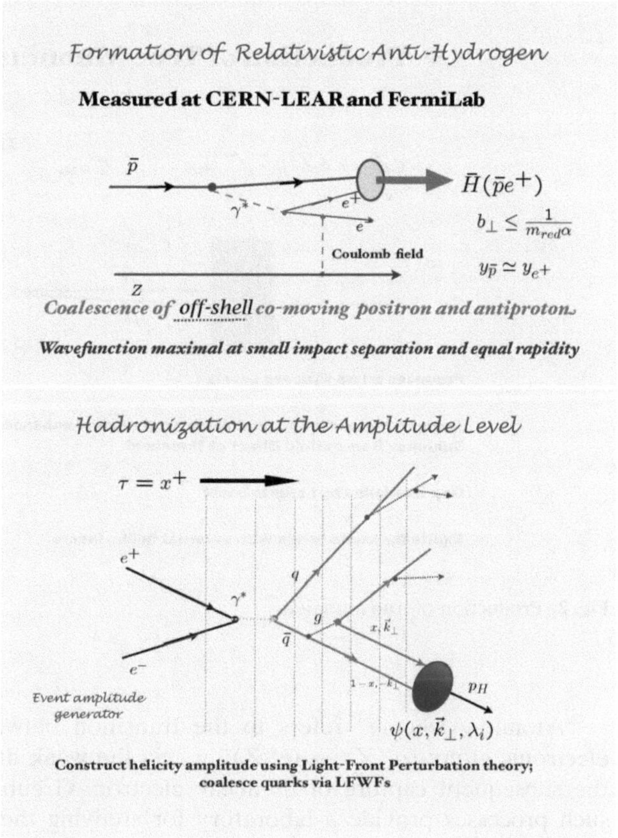

Fig. 1 **a** Production of relativistic antihydrogen. **b** Hadronization at the amplitude level in electron-positron annihilation

quarks and gluons at the amplitude level using light-front time-ordered perturbation theory. The light-front wavefunction of the meson then converts the off-shell comoving $q\bar{q}$ pair into the final-state meson. The confined colored quarks thus never appear on-shell. This first-principle method for forming hadrons in QCD [10] can replace phenomenological jet hadronization models such as PYTHIA. The light-front wavefunction required for calculating "hadronization at the amplitude level" [10, 11] is the frame-independent analog of the Schrödinger wavefunction of atomic physics. It is obtained from the eigensolution of the QCD light-front Hamiltonian quantized at fixed light-front time τ which can be determined by solving the Heisenberg matrix $H_{LF}^{QCD}|\Psi_H>= M_H^2|\Psi_H>$ using a method such as discretized light-cone quantization (DLCQ) [12] or using the AdS/QCD approach together with Light-Front Holography [2].

It is very interesting to produce "true muonium", the $[\mu^+\mu^-]$ bound state Lebed and I [13] have discussed QED production and decay mechanisms, such as electroproduction of relativistic true muonium below the $\mu^+\mu^-$ threshold via $e^-Z \to [\mu^+\mu^-]e^-Z$ or $e^+e^- \to [\mu^+\mu^-]\gamma$. See Fig. 2. The APEX electroproduction experiment [14], which will search for dark matter candidates at Jefferson Laboratory, could be the first to see this exotic atom. Studying the precision spectroscopy of the $[\mu^+\mu^-]$ atom is important in view of the anomalies seen in the muon $g-2$ [15] and the μ^-p Lamb shift [16].

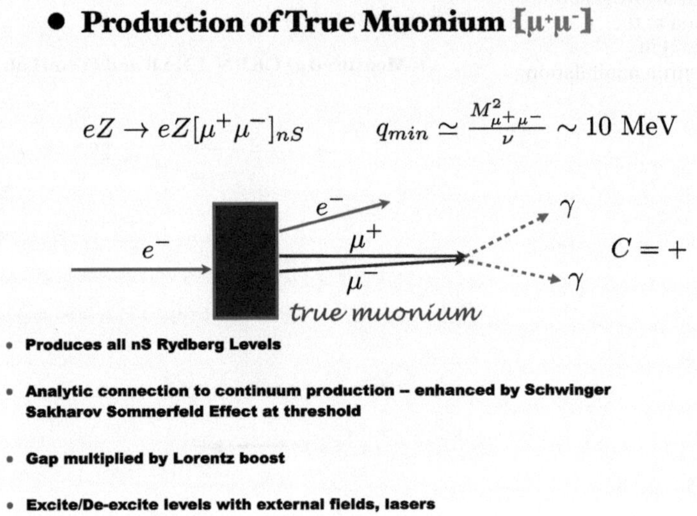

Fig. 2 Production of true muonium

"Atomic Alchemy" refers to the transition between a muonic atom into an electronic atom: $(\mu^- Z) \to (e^- Z)\bar{\nu}_e \nu_\mu$ via the weak decay of the bound muon and the subsequent capture of its decay electron. Greub et al. [17] have shown that such processes provide a laboratory for studying the relativistic high momentum tail of wavefunctions in atomic physics; in addition, they provide a simple toy model for investigating analogous exclusive heavy hadronic decays in quantum chromodynamics such as $B \to \pi e \nu$.

The QCD analog of a molecule in QCD is a bound state of heavy quarkonium with a nucleus such as $[J/\psi A]$ [18, 19]. The binding occurs through two-gluon exchange, the hadronic analog of the Van der Waals interaction. Since the kinetic energy of the J/ψ and the nucleus are both small, one expects to produce these exotic hybrid states at threshold. Examples of nuclear-bound quarkonium are the $|uuduudss\bar{s}>$ and $|uuduudc\bar{c}>$ resonances which apparently appear as intermediate states in $pp \to pp$ elastic exchange. These resonances can account [20] for the large spin-spin A_{NN} correlations [21] observed at the strangeness $E_{cm} \simeq 3$ GeV and $E_{cm} \simeq 5$ GeV and charm thresholds.

At high energies, Compton scattering on an atom $\gamma A \to \gamma A$ is dominated by the Thomson amplitude—the elastic scattering of the photon on the atomic electrons. The analog in hadron physics is the scattering of photons on quarks $\gamma q \to \gamma q$ via a local seagull or instantaneous light-front term which gives an energy-independent contribution to the Compton amplitude proportional to the charge squared of the struck quark—a contribution which has no analog in hadron scattering reactions. In ref. [22] we show that this local contribution has a real phase and is universal, giving the same contribution for real or virtual Compton scattering for any photon virtuality and skewness at fixed momentum transfer squared t. The t-dependence of this $J = 0$ fixed Regge pole is parameterized by a yet unmeasured even charge-conjugation form factor of the target nucleon. The $t = 0$ limit gives an important constraint on the dependence of the nucleon mass on the quark mass through the Weisberger relation. The same $J = 0$ amplitude enters the two-photon exchange contribution to muon-proton scattering, and thus also could contribute an important contribution to the μp Lamb Shift.

3 Renormalization scale setting

A key difficulty in making precise perturbative predictions for QCD is the uncertainty in determining the renormalization scale μ of the running coupling $\alpha_s(\mu^2)$. In the standard Gell-Mann–Low scheme for QED, the renormalization scale is simply the virtuality of the virtual photon [23]. Although the *initial* choice of renormalization scale t_0 is arbitrary, the *final* scale t which sums the vacuum polarization corrections is unique and unambiguous. The resulting perturbative series is identical to the conformal series with zero β-function. In the case of muonic atoms, the modified muon-nucleus Coulomb potential is precisely $-Z\alpha(-\vec{q}^{\,2})/\vec{q}^{\,2}$; i.e., $\mu^2 = -\vec{q}^2$. Again, the renormalization scale is unique. The same principle underlying renormalization scale-setting in QED for $N_C = 0$ must also hold in QCD since the n_F terms in the QCD β function have the same role as the lepton N_ℓ vacuum polarization contributions in QED. Thus the same scale-setting procedure must be applicable to all renormalizable gauge theories.

The purpose of the running coupling in any gauge theory is to sum all terms involving the β function; in fact, when the renormalization scale μ is set properly, all non-conformal $\beta \neq 0$ terms in a perturbative expansion arising from renormalization are summed into the running coupling. The remaining terms in the perturbative series are then identical to that of a conformal theory; i.e., the theory with $\beta = 0$. The divergent "renormalon" series of order $\alpha_s^n \beta^n n!$ does not appear in the conformal series. Thus as in quantum electrodynamics, the renormalization scale μ is determined unambiguously by the "Principle of Maximum Conformality (PMC)" (Brodsky and Di Giustino 2011, to be published; Brodsky and Wu 2011, to be published). This is also the principle underlying BLM scale setting [24]. An important feature of the PMC is that its QCD predictions are independent of the choice of renormalization scheme. The PMC procedure also agrees with QED scale-setting in the $N_C \to 0$ limit.

4 Light-front quantization

The distributions of electrons within an atom are determined in QED using the Schrödinger wavefunction, the eigenfunction of the QED Hamiltonian. In principle, one could calculate hadronic spectroscopy and wavefunctions by solving for the eigenstates of the QCD Hamiltonian: $H|\Psi\rangle = E|\Psi\rangle$ at fixed time t. However, this traditional method—called the "instant form" by Dirac [25], is plagued by complex vacuum and relativistic effects, as well as by the fact that the boost of such fixed-t wavefunctions away from the hadron's rest frame is an intractable dynamical problem. However, there is an extraordinarily powerful non-perturbative alternative—quantization at fixed light-front (LF) time $\tau = t + z/c = x^+ = x^0 + x^3$—the "front-form" of Dirac [25]. In this framework each hadron H is identified as an eigenstate of the QCD Hamiltonian $H_{LF}^{QCD}|\Psi_H\rangle = M_H^2|\Psi_H\rangle$, where $H_{LF}^{QCD} = P_\mu P^\mu = P^- P^+ - P_\perp^2$ is derived directly from the QCD Lagrangian or action. The eigenvalues of this Heisenberg equation give the complete mass spectrum of hadrons. The eigensolution $|\Psi_H\rangle$ projected on the free Fock basis provides the set of valence and non-valence light-front Fock state wavefunctions $\Psi_{n/H}(x_i, k_{\perp i}, \lambda_i)$, which describe the hadron's momentum and spin distributions and the direct measures of its structure at the quark and gluon level. If one quantizes the gluon field in light-cone gauge $A^+ = A^0 + A^3 = 0$, the gluons have physical polarization $S^z = \pm 1$, there are no ghosts, so that one has a physical interpretation of the quark and gluon constituents. The constituents of a bound state in a light-front wavefunction are measured at the same light-front time τ—along the front of a light-wave, as in a flash picture. In contrast, the constituents of a bound state in an instant form wavefunction must be measured at the same instant time t—this requires the exact synchrony in time of many simultaneous probes.

A remarkable feature of LFWFs is the fact that they are frame independent; i.e., the form of the LFWF is independent of the hadron's total momentum $P^+ = P^0 + P^3$ and $P_\perp$. The boost invariance of LFWFs contrasts dramatically with the complexity of boosting the wavefunctions defined at fixed time t [26]. Light-front quantization is thus the ideal framework to describe the structure of hadrons in terms of their quark and gluon degrees of freedom. The constituent spin and orbital angular momentum properties of the hadrons are also encoded in the LFWFs. The total angular momentum projection [27] $J^z = \sum_{i=1}^n S_i^z + \sum_{i=1}^{n-1} L_i^z$ is conserved Fock-state by Fock-state and by every interaction in the LF Hamiltonian. The constituent spin and orbital angular momentum properties of the hadrons are thus encoded in their LFWFs. The empirical observation that quarks carry only a small fraction of the nucleon angular momentum highlights the importance of quark orbital angular momentum. In fact the nucleon anomalous moment and the Pauli form factor are zero unless the quarks carry nonzero L^z.

Hadron observables, e.g., hadronic structure functions, form factors, distribution amplitudes, GPDs, TMDs, and Wigner distributions can be computed as simple convolutions of light-front wavefunctions (LFWFs). For example, one can calculate the electromagnetic and gravitational form factors $< p+q|j^\mu(0)|p >$ and $< p+q|t^{\mu\nu}(0)|p >$ of a hadron from the Drell-Yan-West formula—i.e., the overlap of LFWFs. The anomalous gravitomagnetic moment $B(0)$ defined from the spin-flip matrix element $< p+q|t^{\mu\nu}(0)|p >$ at $q \to 0$ vanishes—consistent with the equivalence theorem of gravity. In contrast, in the instant form, the overlap of instant time wavefunctions is not sufficient. One must also couple the photon probe to currents arising spontaneously from the vacuum which are connected to the hadron's constituents. The Light-Front method is directly applicable for describing atomic bound states in both the relativistic and nonrelativistic domains; it is particularly useful for atoms in flight since the LFWFs are frame-independent. It also satisfies theorems such as cluster decomposition.

One can solve the LF Hamiltonian problem for theories in one-space and one-time by Heisenberg matrix diagonalization. For example, the complete set of discrete and continuum eigensolutions of mesons and baryons in QCD(1+1) can be obtained to any desired precision for general color, multiple flavors, and general quark masses using the discretized light-cone quantized (DLCQ) method [28, 29]. The DLCQ approach can in principle be applied to QED(3+1) and QCD(3+1); however, in practice, the huge matrix diagonalization problem is computational challenging.

5 AdS/QCD light-front holography

Recently a new nonperturbative QCD approach has been developed which leads to an elegant analytical and phenomenologically compelling first approximation to the full LF Hamiltonian method—"Light-Front Holography" [2]. Light front holographic methods allow one to project the functional dependence of the wavefunction $\Phi(z)$ computed in the AdS fifth dimension to the hadronic frame-independent light-front wavefunction $\psi(x_i, b_{\perp i})$ in $3+1$ physical space-time. The variable z maps to a transverse LF variable $\zeta(x_i, b_{\perp i})$. The result is a single-variable light-front Schrödinger equation which determines the eigenspectrum and the LFWFs of hadrons for general spin and orbital angular momentum. The transverse coordinate ζ is closely related to the invariant mass squared of the constituents in the LFWF and its off-shellness in the LF kinetic energy, and it is thus the natural variable to characterize the hadronic wavefunction. In fact ζ is the only variable to appear in the relativistic light-front Schrödinger equations predicted from holographic QCD in the limit of zero quark masses. The coordinate z in AdS space is thus uniquely identified with a Lorentz-invariant coordinate ζ which measures the separation of the constituents within a hadron at equal light-front time.

The hadron eigenstates generally have components with different orbital angular momentum; e.g., the proton eigenstate in LF holographic QCD with massless quarks has $L = 0$ and $L = 1$ light-front Fock components with equal probability. Higher Fock states with extra quark-anti quark pairs also arise. The resulting LFWFs then lead to a new range of hadron phenomenology, including the possibility to compute the hadronization of quark and gluon jets at the amplitude level. The soft-wall model also predicts the form of the non-perturbative effective coupling and its β-function [30].

6 Lensing and the Sivers effect

A well-known phenomenon in QED rescattering via final-state Coulomb interactions. Although the Coulomb phase for a given partial wave is infinite, the interference of Coulomb phases arising from different partial waves leads to observable effects.

The calculation of the Sivers single-spin asymmetry in deep inelastic lepton scattering in QCD is illustrated in Fig. 3. The analysis requires two different orbital angular momentum components: S-wave with the quark-spin parallel to the proton spin and P-wave for the quark with anti-parallel spin; the difference between the final-state "Coulomb" phases leads to a $\vec{S} \cdot \vec{q} \times \vec{p}$ correlation of the proton's spin with the virtual photon-to-quark production plane [31]. Thus, as it is clear from its QED analog, the final-state gluonic interactions of the scattered quark lead to a T-odd non-zero spin correlation of the plane of the lepton-quark scattering plane with the polarization of the target proton [31]. This leading-twist Bjorken-scaling "Sivers effect" is nonuniversal since QCD predicts an opposite-sign correlation [32, 33] in Drell-Yan reactions due to the initial-state interactions of the annihilating antiquark. The $S-$ and P-wave proton wavefunctions also appear in the calculation of the Pauli form factor quark-by-quark. Thus one can correlate the Sivers asymmetry for each struck quark with the anomalous magnetic moment of the proton carried by

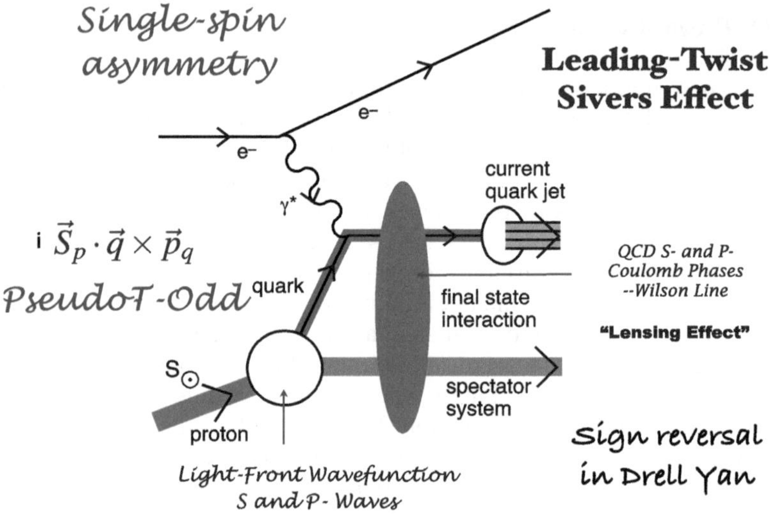

Fig. 3 Origin of the Sivers single-spin asymmetry in deep inelastic lepton scattering

that quark [34], leading to the prediction that the Sivers effect is larger for positive pions. The physics of the lensing dynamics involves nonperturbative quark-quark interactions at small momentum transfer, not the hard scale Q^2 of the virtuality of the photon. It would interesting to see if the strength of the soft initial- or final- state scattering can be predicted using the confining potential of AdS/QCD.

7 Vacuum condensates and the cosmological constant

It is important to distinguish two very different concepts of the vacuum in quantum field theories such as QED and QCD. The vacuum is normally defined as the lowest energy eigenstate of the instant-form Hamiltonian—the vacuum defined by quantizing at fixed time t. In QED, the instant-time vacuum is saturated with quantum loops of leptons and photons. In calculations of physical processes one must normal-order the vacuum and divide the S-matrix elements by the disconnected vacuum loops. In contrast, the front-form (light-front) vacuum is defined as the lowest mass eigenstate of light-front Hamiltonian defined by quantizing at fixed $\tau = t - z/c$. The vacuum is remarkably simple in light-front quantization because of the restriction $k^+ \geq 0$. For example QED vacuum graphs such as $e^+e^-\gamma$ loops do not arise. The LF vacuum thus coincides with the vacuum of the free LF Hamiltonian. The front-form vacuum and its eigenstates are Lorentz invariant; whereas the instant form vacuum depends on the observer's Lorentz frame. The instant-form vacuum is a state defined at the same time t at all spatial points in the universe. In contrast, the front-from vacuum only senses phenomena which are causally connected; i.e., or within the observer's light-cone. Causality in quantum field theory follows the fact that commutators vanish outside the light-cone. In fact in the LF analysis the spatial support of QCD condensates is restricted to the interior of hadrons, physics which arises due to the interactions of confined quarks and gluons. The condensate physics is replaced by the dynamics of higher non-valence Fock states as shown by Casher and Susskind [35]. In particular, chiral symmetry is broken in a limited domain of size $1/m_\pi$, in analogy to the limited physical extent of superconductor phases. This novel description of chiral symmetry breaking in terms of "in-hadron condensates" has also been observed in Bethe-Salpeter studies [36, 37]. The usual argument for a quark vacuum condensate is the Gell-Mann–Oakes–Renner formula: $m_\pi^2 = -2m_q\langle 0|\bar{q}q|0\rangle/f_\pi^2$. However, in the Bethe-Salpeter and light-front formalisms, where the pion is a $q\bar{q}$ bound-state, the GMOR relation is replaced by $m_\pi^2 = -2m_q\langle 0|\bar{q}\gamma_5 q|\pi\rangle/f_\pi$, where $\rho_\pi \equiv -\langle 0|\bar{q}\gamma_5 q|\pi\rangle$ represents a pion decay constant via an an elementary pseudoscalar current.

The cosmological constant measures the matrix element of the energy momentum tensor $T^{\mu\nu}$ in the background universe. It corresponds to the measurement of the gravitational interactions of a probe of finite mass; it only senses the causally connected domain within the light-cone of the observer. If the universe is empty, the appropriate vacuum state is thus the LF vacuum since it is causal. One automatically obtains a vanishing cosmological constant from the LF vacuum. Thus, as argued in [38–41] the 45 orders of magnitude conflict of QCD with the observed value of the cosmological condensate is removed, and a new perspective on the nature of quark and gluon condensates in QCD is thus obtained [38–40].

Acknowledgements I am grateful to the organizers of EXA2011 for their invitation to this outstanding interdisciplinary conference. I thank all of my collaborators whose work has been cited in this report, particularly Guy de Teramond, Rich Lebed, Ivan Schmidt, and Dae Sung Hwang. This research was supported by the Department of Energy contract DE–AC02–76SF00515.

References

1. Brodsky, S.J., Huet, P.: Phys. Lett. B **417**, 145. arXiv:hep-ph/9707543 (1998)
2. de Teramond, G.F., Brodsky, S.J.: Phys. Rev. Lett. **102**, 081601. arXiv:0809.4899 [hep-ph] (2009)
3. Baur, G., et al.: Phys. Lett. B **368**, 251 (1996)
4. Blanford, G., Christian, D.C., Gollwitzer, K., Mandelkern, M., Munger, C.T., Schultz, J., Zioulas, G.: E862 Collaboration. Phys. Rev. Lett. **80**, 3037 (1998)
5. Munger, C.T., Brodsky, S.J., Schmidt, I.: Phys. Rev. D **49**, 3228 (1994)
6. Robiscoe, R.T.: Phys. Rev. **138**, A22 (1965)
7. Brodsky, S.J., Peterson, C., Sakai, N.: Phys. Rev. D **23**, 2745 (1981)
8. Brodsky, S.J., Collins, J.C., Ellis, S.D., Gunion, J.F., Mueller, A.H.: Intrinsic Chevrolets at the SSC. DOE/ER/40048-21 P4, C84/06/23. p. 10. In: Proc. Of Conference: C84-06-23 (1984) (Snowmass Summer Study 1984:0227)
9. Franz, M., Polyakov, M.V., Goeke, K.: Phys. Rev. D **62**, 074024. arXiv:hep-ph/0002240 (2000)
10. Brodsky, S.J., de Teramond, G., Shrock, R.: AIP Conf. Proc. **1056**, 3. arXiv:0807.2484 [hep-ph] (2008)
11. Brodsky, S.J., de Teramond, G.F.: Light-front holography and QCD hadronization at the amplitude level. SLAC-PUB-13504. Jan 2009. p. 5. In: Proceedings of the 18th International Conference On Particles And Nuclei (PANIC 08). Eilat, Israel, 9–14 Nov 2008 arXiv:0901.0770 [hep-ph] (2009)
12. Pauli, H.C., Brodsky, S.J.: Phys. Rev. D **32**, 1993 (1985)
13. Brodsky, S.J., Lebed, R.F.: Phys. Rev. Lett. **102**, 213401. arXiv:0904.2225 [hep-ph] (2009)
14. Abrahamyan, S., et al.: APEX Collaboration. Phys. Rev. Lett. **107**, 191804. arXiv:1108.2750 [hep-ex] (2011)
15. Bennett, G.W., et al.: Muon G-2 Collaboration. Phys. Rev. D **73**, 072003. arXiv:hep-ex/0602035 (2006)
16. Pohl, R., Antognini, A., Nez, F., Amaro, F.D., Biraben, F., Cardoso, J.M.R., Covita, D.S., Dax, A., et al.: Nature **466**, 213 (2010)

17. Greub, C., Wyler, D., Brodsky, S.J., Munger, C.T.: Phys. Rev. D **52**, 4028. arXiv:hep-ph/9405230 (1995)
18. Brodsky, S.J., Schmidt, I.A., de Teramond, G.F.: Phys. Rev. Lett. **64**, 1011 (1990)
19. Luke, M.E., Manohar, A.V., Savage, M.J.: Phys. Lett. B **288**, 355. arXiv:hep-ph/9204219 (1992)
20. Brodsky, S.J., de Teramond, G.F.: Phys. Rev. Lett. **60**, 1924 (1988)
21. Court, G.R., et al.: Phys. Rev. Lett. **57**, 507 (1986)
22. Brodsky, S.J., Llanes-Estrada, F.J., Szczepaniak, A.P.: Phys. Rev. D **79**, 033012. arXiv:0812.0395 [hep-ph] (2009)
23. Gell-Mann, M., Low, F.E.: Phys. Rev. **95**, 1300 (1954)
24. Brodsky, S.J., Lepage, G.P., Mackenzie, P.B.: Phys. Rev. D **28**, 228 (1983)
25. Dirac, P.A.M.: Rev. Mod. Phys. **21**, 392 (1949)
26. Brodsky, S.J., Primack, J.R.: Ann. Phys. **52**, 315 (1969)
27. Brodsky, S.J., Hwang, D.S., Ma, B.Q., Schmidt, I.: Nucl. Phys. B **593**, 311. arXiv:hep-th/0003082 (2001)
28. Pauli, H.C., Brodsky, S.J.: Phys. Rev. D **32**, 2001 (1985)
29. Hornbostel, K., Brodsky, S.J., Pauli, H.C.: Phys. Rev. D **41**, 3814 (1990)
30. Brodsky, S.J., de Teramond, G.F., Deur, A.: Phys. Rev. D **81**, 096010. arXiv:1002.3948 [hep-ph] (2010)
31. Brodsky, S.J., Hwang, D.S., Schmidt, I.: Phys. Lett. B **530**, 99. arXiv:hep-ph/0201296 (2002)
32. Collins, J.C.: Phys. Lett. B **536**, 43. arXiv:hep-ph/0204004 (2002)
33. Brodsky, S.J., Hwang, D.S., Schmidt, I.: Nucl. Phys. B **642**, 344. arXiv:hep-ph/0206259 (2002)
34. Lu, Z., Schmidt, I.: Phys. Rev. D **75**, 073008. arXiv:hep-ph/0611158 (2007)
35. Casher, A., Susskind, L.: Phys. Rev. D **9**, 436 (1974)
36. Maris, P., Roberts, C.D., Tandy, P.C.: Phys. Lett. B **420**, 267. arXiv:nucl-th/9707003 (1998)
37. Maris, P., Roberts, C.D.: Phys. Rev. C **56**, 3369. arXiv:nucl-th/9708029 (1997)
38. Brodsky, S.J., Shrock, R.: Phys. Lett. B **666**, 95. arXiv:0806.1535 [hep-th] (2008)
39. Brodsky, S.J., Shrock, R.: Proc. Natl. Acad. Sci. **108**, 45. arXiv:0803.2554 [hep-th] (2011)
40. Brodsky, S.J., Roberts, C.D., Shrock, R., Tandy, P.C.: Essence of the vacuum quark condensate. SLAC-PUB-14107, CP3-ORIGINS-2010-18, p. 5. Phys. Rev. C **82**, 022201. arXiv:1005.4610 [nucl-th] (2010)
41. Chang, L., Roberts, C.D., Tandy, P.C.: Expanding the concept of in-hadron condensates, Sep 2011, p. 5. Phys. Rev. C **85**, 012201. arXiv:1109.2903 [nucl-th] (2012)

Hyperfine Interact (2012) 209:93–98
DOI 10.1007/s10751-011-0548-5

The PANDA physics programme

Simonetta Marcello on behalf of the PANDA Collaboration

Published online: 10 February 2012

Abstract The PANDA experiment at the future FAIR facility in Darmstadt, Germany, will use interactions of antiprotons with nucleons or nuclei to investigate fundamental questions of hadron and nuclear physics. Here, a brief overview of the physics programme is given, focusing on a few selected topics.

Keywords Charmed mesons · Exotic mesons · Multi-strange baryons · Proton form factors

1 Introduction

PANDA [1] is one of the major experiments at the future FAIR facility in Europe [2]. Gluonic excitations and the physics of strange and charm quarks will be accessible with unprecedented accuracy, allowing high-precision tests of strong interaction. The FAIR facility is under construction on the area of the GSI Laboratory in Germany. A primary beam of protons up to 30 GeV will be provided. A secondary beam of $\bar{p}$ in the momentum range from 1.5 to 15 GeV/c will be available at the High Energy Storage Ring (HESR), where the PANDA detector will be installed. Commissioning of PANDA is foreseen to start in 2017, but pre-assembling will start at COSY in Jülich in 2014.

The HESR will work into two different operation modes: high momentum resolution mode with $\delta p/p \sim 10^{-5}$ at a moderate luminosity $\mathcal{L} = 10^{31}\,\mathrm{cm}^{-2}\mathrm{s}^{-1}$, by electron cooling, and high luminosity mode $\mathcal{L} = 2 \times 10^{32}\,\mathrm{cm}^{-2}\mathrm{s}^{-1}$, but with greater momentum spread ($\delta p/p \sim 10^{-4}$), by stochastic cooling.

S. Marcello (✉)
Dipartimento di Fisica Sperimentale, Università di Torino,
via P. Giuria 1, 10125 Torino, Italy
e-mail: marcello@ph.unito.it

S. Marcello
INFN Sezione di Torino, 10125 Torino, Italy

In the last decade most of the results on meson spectroscopy in the region of charm has been achieved at e^+e^- colliders. Nevertheless, in order to shed light on the recently discovered narrow states, a very high mass resolution is required. The use of $\bar{p}p$ annihilations allows to achieve an excellent resonance mass resolution, as was successfully demonstrated by experiments at FNAL [3]. Indeed, the precision of the resonance mass and width depends on the energy spread of the beam and not on the resolution of the detector, which is definitely used to identify and reconstruct the final state and for an efficient background rejection. In PANDA a resolution below 100 keV can be achieved, which is unbeatable with respect to the one of a few MeV of experiments at e^+e^- colliders. Moreover, all measurements will profit from the high yield of $\bar{p}$ induced reactions (2×10^7/s) and from the fact that states with all quantum numbers allowed for a $q\bar{q}$ system can be directly populate, in contrast to e^+e^- reactions, which select states with $J^{PC} = 1^{--}$. Of course, states with exotic quantum number will be observed in production measurements.

2 The PANDA physics programme

PANDA has a wide range physics programme, both in collisions of $\bar{p}p$ and $\bar{p}-$ *nucleus*, which will allow to extend our knowledge on hadron structure, quark-gluon dynamics and nuclear physics. A comprehensive discussion of the programme, with feasibility studies and benchmark channels, can be found in the PANDA physics performance report [4]. Since, the available c.m. energy for the $\bar{p}p$ system will be in the range 2.2–5.5 GeV, one of the main goals is the study of charmonium and open charm spectroscopy, where the high resolution can contribute in searching and understanding of gluonic excitations, such as glueballs and hybrids. Many other different topics, addressed to answer to fundamental questions of hadron and nuclear physics, will be pursued as well. For instance, properties of mesons with hidden and open charm in the nuclear medium will be studied to understand the origin of hadron masses [5]. In $\bar{p}p$ annihilations the production of baryon–antibaryon pairs allows to study the excitation spectrum of strange and charmed hyperons, where no data are available, and at the same time the dynamics of the $s\bar{s}$ and $c\bar{c}$ creation can be studied in the non perturbative regime of QCD. Since in PANDA the Ξ and $\bar{\Xi}$ hyperons will be copiously produced, both single-Λ and double-Λ hypernuclei can be generated allowing to study nuclear structure and hyperon-nucleon and hyperon-hyperon interactions [6] with high statistics, comparable to the one achievable at J-PARC [7].

Last, but not least, the study of nucleon structure will be investigated in electromagnetic final states (e^+e^-, $\gamma\gamma$), for instance measuring the proton form factors in the time-like region up to high q^2 values, and in Drell-Yan processes. Such an ambitious programme requires a high performance spectrometer. For this reason a multipurpose set-up, which includes innovative detectors, has been designed and it is described in details in [8].

The following sections are focused on a few selected topics.

3 Spectroscopy of charmonium and charmed mesons

In the last years charmonium spectroscopy has enjoyed a renaissance due to the discovery of several missing states and several unexpected ones at B-factories, where very large data sample have been collected.

After the discovery of J/Ψ in 1974, QCD-motivated quark potential models were proposed. Such models were successful in describing the gross features of the spectrum and many predicted states were confirmed by experiments. Thus, below $\bar{D}D$ threshold all states are established and the spectrum is more clear. New precise measurement of η_c mass has been achieved and the η_c' has been unambiguously observed. Nevertheless, the width of h_{1c} has not yet been measured and the high angular momentum states, which are not populated in e^+e^-, need to be measured using $\bar{p}p$ interactions. On the contrary, above the $\bar{D}D$ threshold the picture is still incomplete, only a few states among the ones predicted by potential models have been identified and many unexpected states, called "X, Y, Z", have been observed by different experiments, mainly in the hadronic decays of B mesons [9]. Interpretation of these states from the decay patterns and from their properties is still unclear and the possibility they are not simple $q\bar{q}$ states, such as molecules, tetraquarks, glueballs or hybrid mesons is under debate. Moreover, the open charm spectrum contains states for which quark model expectations do not hold. For instance, the charmed-strange mesons $D^*_{s0}(2317)$ and $D_{s1}(2460)$, discovered in 2003, are narrow states with masses below the DK and D^*K thresholds, respectively, instead of lying at higher masses as predicted. Performing fine energy scans, PANDA will be able to observe the cross section close to the threshold and to measure the width as explained in [10].

The $\bar{p}p$ annihilations is intrinsically a gluon rich process, indeed, most of the promising results for gluonic hadrons comes from $\bar{p}$ annihilation experiments. In the light meson sector more than 10 states have been classified as exotic candidates, among them the $f_0(1500)$ is the best candidate for the ground state glueball with $J^{PC} = 0^{++}$. In the charm sector glueballs and hybrid ($c\bar{c}g$) mesons are expected to be narrower than in the light meson sector, therefore identification should be easier. Lattice QCD models [11] predict heavy glueballs with $J^{PC} = 2^{+-}$ and 0^{+-} at masses $\sim$4.1 GeV/c^2 and $\sim$4.7 GeV/c^2, respectively, and hybrid charmonia with $J^{PC} = 1^{-+}$ and 0^{+-} in the mass range $\sim$4.2–4.5 GeV/c^2. The identification of these states requires high statistics to have reliable spin-parity analysis. The $\bar{p}p$ production cross sections of these exotic states in the charm sector are expected to be similar to the ones of conventional states ($\sim$100 pb).

In 2003 Belle first observed the $X(3872)$ narrow state [12], which is now well established. Soon after a series of new hadronic structures, like Ys, have been observed by different experiments at mass values higher than the open charm threshold, with unusual decay patterns. For instance, the $Y(4260)$ [13] does not decay into open charm, but in final states containing J/Ψ, even if it lies above the $D\bar{D}$ threshold. Therefore, they are candidate for charmonium hybrid or multiquark states. Among the "XYZ" exotics the $Z^+(4430)$ is unique [14]. In fact, since it has a non-zero electric charge, it cannot be a simple $c\bar{c}$ state nor a $c\bar{c}g$ hybrid meson and it is a prime candidate for a multiquark meson.

PANDA will profit of the $\bar{p}$ beam at high intensity and high momentum resolution for searching new states, with high statistics and high accuracy measurements, and to determine the nature of these states, which do not fit in the expected spectrum of conventional charmonium states [4].

4 Baryon-antibaryon production

4.1 Baryon spectroscopy

In $\bar{p}p$ interactions, a large fraction of the inelastic cross section is associated to channels with a baryon-antibaryon pair in the final state, therefore a comprehensive study of baryon spectroscopy with |S|= 2, 3 and even |C|= 1 is possible in PANDA. In fact, since the HESR kinematic limit is $\sqrt{s} = 5.5$ GeV, the excitation energy for multistrange baryons reaches the continuum region and it is ~ 1 GeV, for the charmed hyperon Λ_c^+. Most of existing data is in the nucleon sector and comes from high energy pp scattering experiments. Scarce information is available for Ξ and Ω excited states for which spin-parity quantum numbers are still uncertain or unknown.

A look at the available data for excitation spectra of baryons shows that some low-lying states are not at the predicted energies and many predicted states are missing. A possible explanation is a quark-diquark structure of hadrons, which, reducing the number of degrees of freedom, would reduce the number of states. Another possibility from the chiral coupled channel dynamics approach is that the excited baryon states are dynamically generated resonances [15].

Since production cross sections for Ξ resonances in $\bar{p}p \rightarrow \bar{Y}Y^*$ interactions are expected to be of the same order as for ground state Ξ production ($\sim 2\mu b$), the study of Ξ resonances with good statistics is feasible. Final states with a baryon-antibaryon pair offer the opportunity to independently investigate the baryon excitation pattern into two charge conjugate systems reducing the systematic uncertainties.

4.2 Non-perturbative QCD dynamics

In hadron physics the identification of relevant degrees of freedom is important for an effective description of the reactions. In the quark picture hyperon pair production involves either the creation of a $q\bar{q}$ pair or the knock-out of such pairs out of the nucleon sea. Hence, the creation mechanism of $q\bar{q}$ pairs and their rearrangement into hadrons can be studied by measuring the reactions of the type $\bar{p}p \rightarrow \bar{Y}Y$ with strange or charmed hyperons and antihyperons in the final states. By comparing several reactions involving different quark flavours, OZI rule violation effects can be tested.

The parity violating weak decay of hyperons introduces an asymmetry in the distribution of the decay particles, which gives access to spin degrees of freedom for these reactions (polarisation and spin correlations), with the exception of Σ^- and the Ω^- for which the decay asymmetry is close to zero.

The $\bar{p}p \rightarrow \bar{\Lambda}\Lambda$ reaction has been investigated at LEAR by PS185 experiment [16] from threshold up to 2 GeV/c and the spin analysis has shown that the $s\bar{s}$ pair is predominantly produced in a triplet state. Data for this reaction and for other channels with strangeness are scarce above 2 GeV/c and they are missing above

7 GeV/c. No data exist for channels with charmed hyperons. The LEAR data were taken near threshold, it is interesting to verify whether the same features are valid increasing the momentum transfer into more perturbative region. In PANDA the same spin analysis can be extended to other $\bar{Y}Y$ channels and also for $\bar{\Lambda}_c^- \Lambda_c^+$ channel, to study whether the creation of $c\bar{c}$ pair shows the same features as the $s\bar{s}$ pair for $\bar{\Lambda}\Lambda$ channel.

A systematic measurement of total and differential cross sections for these reactions will give new information on single and multiple strangeness or charm production mechanism and its dependence on spin observables. This may help to disentangle the perturbative contributions from the non-perturbative ones, since the charm production probes mainly the hard processes, while the strangeness one is influenced by non-perturbative effects.

5 Proton electromagnetic form factors in the time-like region

Nucleon electric (G_E) and magnetic (G_M) form factors play a fundamental role in our understanding of hadron dynamics. Therefore, for over 40 years many attempts have been made to get a theoretical description in the whole complex q^2 plane and to measure them with different experiments. In the time-like region ($q^2 > 0$) the proton form factors can be studied using the crossed reactions $\bar{p}p \leftrightarrow e^+e^-$. Since analiticity requires a continuous transition of the form factors from space-like to time-like region and equality for $|q^2| \to \infty$, the measurements of the form factors at large q^2 allow to verify the asymptotic behaviour. The possibility to cover a large kinematical domain is interesting to investigate the transition region from soft to hard scattering mechanisms dominated by perturbative QCD, which describes the nucleon in terms of quark and gluon degrees of freedom. Therefore, different theoretical approaches can be tested. The high intensity and high momentum of the $\bar{p}$ beam at HESR, together with the performance of the PANDA detector will allow to investigate the proton form factors in the time-like region up to large values of q^2, where the data are scarce and affected by poor statistics [17].

Recent experiments of the form factors in the space-like region ($q^2 < 0$), using polarised beam or targets in elastic ep scattering, are in disagreement with unpolarised measurements. Indeed, at large values of $-q^2$ the ratio $\mu G_E/G_M$ shows a linear trend as a function of $-q^2$. Such a deviation has been attributed to radiative corrections, as two photons exchange or higher order ones, but needs further investigations in both regions [18, 19].

In the time-like region an independent determination of G_E and G_M form factors requires the measurement of the angular distribution of the outgoing leptons at fixed energy. Previous experiment have measured cross sections up to $q^2 = 18$ (GeV/c)^2 and extracted G_M by using the hypothesis $G_E = G_M$ or $G_E = 0$ (affecting G_M up to 30%). Attempts to determine the ratio $\mathcal{R} = |G_E|/|G_M|$ have been done by the old PS170 experiment at LEAR and, recently, in the BABAR experiment through initial-state radiation reaction. However, the results are affected by large errors and show a different trend. The sensitivity to $\mathcal{R}$ decreases at increasing values of q^2, due to the falling of the cross section and to the relative weight of the magnetic term, which is growing as q^2. With the PANDA experiment such a measurement can be performed up to $q^2 \sim 14$(GeV/c)^2, with a precision at low-q^2 one order of magnitude better than for existing data. At increasing q^2, where $\mathcal{R}$ cannot be determined, it will be possible to extract an effective form factor, under the usual hypothesis of $\mathcal{R} = 1$, up to $q^2 \sim 28$ (GeV/c)^2.

6 Conclusions

The PANDA experiment will study many physics topics in the field of QCD, aiming to achieve a better understanding of the structure and dynamics of hadrons, using the $\bar{p}$ beam with high intensity and high momentum resolution of the HESR at the future FAIR facility. This will allow to have an unprecedented accuracy in various key experiments such as measurements of meson spectroscopy or of proton form factors in the time-like region.

Acknowledgements This work is partially supported by Helmholtz Association through funds provided to the Virtual Institute "Spin and Strong QCD" (VH-VI-231).

References

1. The PANDA Collaboration: Technical Progress Report for PANDA, pp. 1–383. http://www-panda.gsi.de/auto/_home.htm (2005)
2. Rosner, G.: The facility fo antiproton and ion research. http://www.fair-center.org (2011, this volume)
3. Andreotti, M., et al.: Nucl. Phys. B 717, 34 (2005)
4. The PANDA Collaboration: Physics Performance Report for PANDA, pp. 1–204. arXiv:0903.3905v1 [hep-ex] (2009, preprint)
5. Bühler, P.: Studying hadrons in matter with PANDA (2011, this volume)
6. Achenbach, P., et al.: Hypernuclear physics at PANDA. Hyperfine Interactions J (2011, this volume)
7. Nagae, T.: Hadron physics at J-PARC. Hyperfine Interactions J. (2011, this volume)
8. Würschig, T.: The PANDA detector at FAIR. Hyperfine Interactions J. (2011, this volume)
9. Brodzicka, J.: Charm physics at Belle. Hyperfine Interactions J. (2011, this volume)
10. Mertens, M.C.: Determination of the Ds0*(2317) width with the PANDA detector. Hyperfine Interactions J. (2011, this volume)
11. Bali, G.: Lattice calculations of hadron properties. Eur. Phys. J. **A19**, 1–8 (2004) and references quoted therein
12. Choi, S.-K., et al.: Observation of a narrow Charmoniumlike state in exclusive $B^{\pm} \to K^{\pm}\pi^{+}\pi^{-}J/\psi$ decays. Phys. Rev. Lett. **91**, 262001 (2003)
13. Aubert, B., et al.: Observation of a broad structure in the $\pi^{+}\pi^{-}J/\psi$ mass spectrum around 4.26 GeV/c^2. Phys. Rev. Lett. **95**, 142001 (2005)
14. Choi, S.-K., et al.: Observation of a resonancelike structure in the $\pi^{+-}\psi'$ mass distribution in exclusive $B \to K\pi^{+-}\psi'$ decays. Phys. Rev. Lett. **100**, 142001 (2008)
15. Kolomeitsev, E.E., Lutz, M.F.M.: On baryon resonances and chiral symmetry. Phys. Lett. **B585**, 243 (2004)
16. Barnes, P.D., et al.: Observables in high-statistics measurements of the reaction $p^{-}p \to \wedge^{-}\wedge$. Phys. Rev. **C54**, 1877 (1996)
17. Sudol, M., et al.: Feasibility studies of the time-like proton electromagnetic form factor measurements with PANDA at FAIR. Eur. Phys. J. **A44**, 373–384 (2010) and references quoted therein
18. Bernauer, J.C.. et al.: High-precision determination of the electric and magnetic form factors of the proton. Phys. Rev. Lett. **105**, 242001 (2010)
19. Puckett, A.J.R., et al.: Recoil polarization measurements of the proton electromagnetic form factor ratio to $Q^2 = 8.5\text{GeV}^2$. Phys. Rev. Lett. **104**, 242301 (2010)

Hyperfine Interact (2012) 209:99–104
DOI 10.1007/s10751-012-0571-1

Hypernuclear physics at $\overline{\text{P}}$ANDA

Experimental challenges

Patrick Achenbach · Sebastian Bleser · Josef Pochodzalla · Alicia Sanchez Lorente · Marcel Steinen

Published online: 2 February 2012

Abstract Hypernuclear research will be one of the main topics addressed by the $\overline{\text{P}}$ANDA experiment at the planned Facility for Anti-proton and Ion Research FAIR at Darmstadt, Germany. A copious production of Ξ-hyperons at a dedicated internal target in the stored anti-proton beam is expected, which will enable the high-precision γ-spectroscopy of double strange systems for the first time. In addition to the general purpose $\overline{\text{P}}$ANDA setup, the hypernuclear experiments require an active secondary target of silicon layers and absorber material as well as high purity germanium (HPGe) crystals as γ-detectors. The design of the setup and the development of these detectors is progressing: a first HPGe crystal with a new electromechanical cooling system was prepared and the properties of a silicon strip detector as a prototype to be used in the secondary target were studied. Simultaneously to the hardware projects, detailed Monte Carlo simulations were performed to predict the yield of particle stable hypernuclei. With the help of the Monte Carlo a procedure for $\Lambda\Lambda$-hypernuclei identification by the detection and correlation of the weak decay pions was developed.

Keywords Strangeness · Double hypernuclei · HPGe detectors

Supported by the Research Infrastructure Integrating Activity "Study of Strongly Interacting Matter" HadronPhysics2 (SPHERE) under the 7th Framework Programme of EU.
We acknowledge financial support from the Bundesministerium für Bildung und Forschung (bmb+f) under contract number 06MZ9182.

P. Achenbach (✉) · S. Bleser · J. Pochodzalla · M. Steinen
Institut für Kernphysik, Johannes Gutenberg-Universität,
J.J.-Becherweg 45, 55099 Mainz, Germany
e-mail: patrick@kph.uni-mainz.de

A. Sanchez Lorente
Helmholtz-Institut Mainz, 55099 Mainz, Germany

1 Introduction

The "standard model" is the fundamental theory which unites weak, electromagnetic, and strong interactions. Strong processes are formulated in Quantum Chromodynamics (QCD), the field theory for the dynamics of quarks and gluons. QCD has been thoroughly probed in strong interactions at very high energies. However, at the energy scale of the nucleon mass hadrons are complex many-body systems. Even though they interact strongly, the description by the fundamental QCD equations is complicated by the non-perturbative nature of the theory. The investigation of strange hadrons carrying an additional flavour degree-of-freedom is essential for understanding the low-energy regime of QCD.

A very interesting phenomenon in nuclear physics is the existence of nuclei containing strange baryons. The lightest hyperons are stable against strong and electromagnetic decays, and as they do not suffer from Pauli blocking by other nucleons they can live long enough in the nuclear cores to become bound. When a hyperon, specifically a Λ-hyperon, replaces one of the nucleons in the nucleus, the original nuclear structure changes to a system composed by the hyperon and the core of the remaining nucleons. The existence of double strange nuclear systems like Ξ^- or $\Lambda\Lambda$ hypernuclei is directly linked to the strength of the attractive hyperon–nucleon (YN) and unknown hyperon–hyperon (YY) interactions. Models for the YY interactions have been constructed from the expansion of the NN interaction and limited YN scattering data. This approach needs to be validated against double hypernuclei binding energies and excitation spectra.

The study of strange nuclear systems provides invaluable information on both, on the structure of nuclei as many-body hadronic systems and on strange baryons in the nuclear medium. Although single and double Λ-hypernuclei were discovered many decades ago in cosmic ray interactions studied by the emulsion technique, only few double Λ-hypernuclear isotopes are presently known. In particular, $\Lambda\Lambda$-hypernuclei formed in anti-proton beams are the only practical systems among all strange baryons for investigating the strong nuclear interaction.

2 Production and detection of hypernuclei at $\overline{\text{P}}$ANDA

The planned Facility for Anti-proton and Ion Research (FAIR) near Darmstadt will include the High Energy Storage Ring (HESR) to store anti-protons of several GeV/c momentum in an intense and high quality beam. With a dedicated internal target in the storage ring a copious production of Ξ-hyperons is expected which can be stopped in dedicated absorbers to form bound states of Ξ hypernuclei. The latter can be used as a gateway to form $\Lambda\Lambda$ hypernuclei, which will enable the high-precision γ-spectroscopy of double strange systems for the first time [1, 2].

The $\overline{\text{P}}$ANDA experiment (AntiProton ANnihilations at DArmstadt) planned at the HESR storage ring is a next-generation hadron physics experiment. In addition to the general purpose $\overline{\text{P}}$ANDA setup for charged particle detection, the hypernuclear experiments require an active secondary target of silicon layers and absorber material in addition to high purity germanium (HPGe) crystals as γ-detectors. Some technical and practical aspects currently being studied by the $\overline{\text{P}}$ANDA hypernuclear groups are

1. design and fabrication of the primary target,
2. design and development of the secondary target, and
3. design and operation of the HPGe γ-array.

In the following we highlight some of the experimental challenges in realizing the setup. The expected performance of the proposed experiment with this setup was simulated with the help of a micro-canonical decay model predicting the yield of particle stable double hypernuclei.

3 The challenge for a nuclear internal storage ring target

On a nuclear internal target low momentum Ξ pairs can be produced in $\bar{p}p \rightarrow \Xi^- \overline{\Xi}^+$ or $\bar{p}n \rightarrow \Xi^- \overline{\Xi}^0$ reactions. The advantage at HESR in the Ξ production rate as compared to kaon beam induced reactions is the fact that the anti-proton is stable and can be retained in the storage ring. The largest $\bar{p}$ production rate achievable at HESR is of the order of $10^7\,\bar{p}$/s with a maximum of approximately 10^{11} stored anti-protons in the HESR ring [2]. This allows a rather high luminosity even with very thin primary targets, either with the standard hydrogen target of the $\overline{\text{P}}$ANDA experiment, or with a dedicated target for hypernuclear spectroscopy.

The beam–target interactions will reduce the life-time of the stored anti-proton beam. Consequently, a reasonable compromise between $\bar{p}$ beam preparation time and beam loss rate during one HESR cycle needs to be found. Obviously, this compromise is dependent on the target material and thickness. The possibility of steering a low density region of the transverse beam profile over the target along with the gradual consumption of anti-protons will be an important feature. These are conditions posing a real challenge for the design of the nuclear internal target inside the storage ring.

At present, techniques for the manufacturing of μm-thin synthetic diamond filaments cut from a membrane as internal target are being explored [3]. Synthetic diamonds can be grown from a hydrocarbon gas mixture by chemical vapour deposition and can be supported by a silicon ring structure matching the beam pipe. The thermal conductivity of synthetic diamond is very high that prevents the materials from overheating. The lateral size of the synthetic diamond filament could be of the order of 100 μm or less. The target could be designed with an empty region inside the support ring providing space during beam preparation and for the high-density beam center during the anti-proton storage. Such geometries can be made very precisely by cutting the diamond membrane with fs-pulsed high-powered lasers [3].

4 The challenge for an active high-resolution Ξ-absorber target

The main purpose of the active high-resolution Ξ-absorber target is the tracking and stopping of the produced cascade hyperons and their decay products. The active part

Fig. 1 Scheme of the hypernuclear setup inside $\overline{\text{P}}$ANDA. In the backward region of the spectrometer a dedicated synthetic diamond target will be installed inside a beam pipe of reduced diameter (*1*). A mechanical support structure surrounds this target and holds the active Ξ-absorber target (*2*). A compact γ-array with maximized solid-angle acceptance will be used to detect radiative de-excitations (*3*)

of the secondary target will be made from silicon strip sensors. The slowing down of the Ξ^- proceeds by energy-loss during the passage through additional absorber layers. If decelerated to rest before decaying, the Ξ^- can be captured inside a nucleus, eventually releasing two Λ hyperons and forming a double hypernuclei. The geometry of the target is essentially determined by the life-time of the hyperons and their stopping time in silicon and the absorber materials.

In analogy to the germanium detectors array, the silicon detector has to be able to operate in extreme conditions such as a large hadronic environment since it is close to the interaction point. Furthermore, the material budget on the detector volume must be kept low. The feasibility of such a device has recently been studied in Mainz with a single sided microstrip sensor with a strip pitch of 50 μm and $2{\times}2\,\text{cm}^2$ size. In addition, studies on ultra-thin cables based on adhesive-less aluminium-polyimide foiled dielectrics, the 128-channel APV-25-S1 front-end chip [4], the modified electronics board, and the mechanics of the target are being performed.

5 The challenge for a compact γ-array inside the $\overline{\text{P}}$ANDA setup

High purity germanium (HPGe) detectors are key instruments in nuclear structure physics for detecting the radiative de-excitation of excited nuclei. Typically, the crystals are cooled with liquid nitrogen and operated in the temperature range of 77–115 K. The requirement of minimal distance of the secondary target from the interaction point combined with the necessary support structures nested inside the $\overline{\text{P}}$ANDA target spectrometer leaves very restricted space for the installations of the hypernuclear physics setup. The situation is displayed in Fig. 1, showing the anti-proton beam pipe surrounded by the targets and an γ-array of 15 n-type HPGe triple cluster detectors.

For an effective integration of the array into the $\overline{\text{P}}$ANDA spectrometer an electro-mechanical cooling device will be used instead of liquid nitrogen with its bulky dewars. Composite detectors made of three large volume encapsulated Ge crystals and cooled by the electromechanical cooling device X-Cooler II by ORTEC are

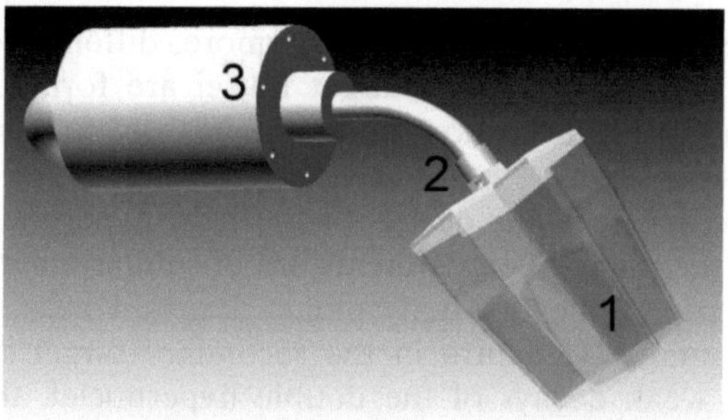

Fig. 2 Drawing of one HPGe detector system assembled to an X-Cooler II device. Three encapsulated coaxial HPGe crystals (*1*) are arranged in one capsule. The flexible section of the thick cold finger (*2*) enables the placement of the cluster at the restricted space inside the $\overline{\text{P}}$ANDA spectrometer. The X-Cooler II (*3*) replaces the standard liquid nitrogen cooling devices

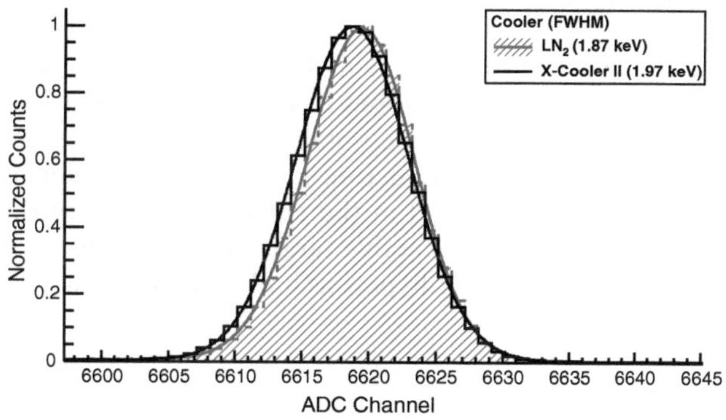

Fig. 3 Measured energy spectra of the 1.332 keV line of a ^{60}Co calibration source taken with two different cooling devices. For the *dashed spectrum* with a line width of FWHM = 1.97 keV the HPGe crystal was cooled electromechanically, for the *solid spectrum* with a line width of FWHM = 1.87 keV a liquid nitrogen cooling system was used [5]

being considered. Fig. 2 shows the drawing of a triple HPGe cluster assembled to an X-Cooler II device. The individual Ge crystals are sealed in an aluminium can and installed in a common vacuum cryostat. An intermediate thermal shield may be applied in order to act as a heat reflector thus reducing the heating of the encapsulated Ge crystals.

The energy resolution of such a system with electromechanical cooling has been determined with a standard ^{60}Co calibration source and a line width of 1.97 keV for the 1.332 keV γ-line was found, see Fig. 3. In comparison to a cooling device based on liquid nitrogen where a line width of 1.86 keV was measured, the electromechanical cooling seems to have no negative impact on the performance [5].

Another major challenge at $\overline{\text{P}}$ANDA is the operation of the germanium detectors close to a strong magnetic field over long periods. It was demonstrated that a good energy resolution can be preserved up to 1 T [6].

6 Performance of the proposed hypernuclear experiment

The production of excited states in $\Lambda\Lambda$ hypernuclei was studied following the micro-canonical break-up of an initially excited double hypernucleus created by the absorption and conversion of a stopped Ξ^- hyperon [7, 8]. In these calculations the formation of excited states dominates. Furthermore, different double hypernuclei isotopes which depend on the initial target nuclei are formed. Thus, the ability to assign the observable γ-transitions in a unique way to a specific double $\Lambda\Lambda$ hypernucleus seems possible.

The non-mesonic and mesonic decays of the light hypernuclei to be studied in the initial phase of the planned experiments are of similar magnitude. The analysis strategy will make use of the two-body pionic decays, where the mono-energetic pions will leave a unique signature in the secondary target. In the case of two sequential mesonic weak decays of the double hypernuclei, the momenta of the two pions are strongly correlated. Thus, a coincidence measurement in the active Ξ-absorber target will provide an effective method to tag the production of a double hypernucleus.

7 Conclusions

At the $\overline{\text{P}}$ANDA experiment at FAIR it will become possible to explore the level scheme of low-mass isotopes of double hypernuclei for the first time. The spectroscopic information will be obtained via γ-ray detection using an array of 15 n-type HPGe triple cluster detectors located near a dedicated arrangement of targets: a primary diamond target at the entrance to the central tracking detector of $\overline{\text{P}}$ANDA, and a small secondary active target composed of silicon detectors and absorbers to decelerate and stop Ξ-hyperons and to identify the weak decay products.

References

1. Pochodzalla, J.: Nucl. Instrum. Methods Phys. Res. B **214**, 149 (2004)
2. $\overline{\text{P}}$ANDA Collaboration: Physics performance report for $\overline{\text{P}}$ANDA: strong interaction studies with antiprotons. arXiv:0903.3905 (2009)
3. Iazzi, F.: Double hypernuclei: internal target report (2011). PANDA Meeting, GSI 5 Sept 2011
4. Dalla Betta, G.F., et al.: Nucl. Instrum. Methods Phys. Res. A **460**, 306 (2001)
5. Steinen, M.: Optimierung von elektrisch gekühlten Germanium-Detektoren für das $\overline{\text{P}}$ANDA-Experiment. Diploma thesis, Johannes Gutenberg-Universität Mainz (2011)
6. Sanchez Lorente, A., et al.: Nucl. Instrum. Methods Phys. Res. A **573**, 410 (2007)
7. Sanchez Lorente, A., Pochodzalla, J., Botvina, A.: Int. J. Mod. Phys. E **19**, 2644 (2010)
8. Sanchez Lorente, A., Botvina, A.S., Pochodzalla, J.: Phys. Lett. B **697**, 222 (2011)

Hyperfine Interact (2012) 209:105–110
DOI 10.1007/s10751-012-0581-z

Studying hadrons in matter with PANDA

P. Bühler on behalf of the PANDA collaboration

Published online: 23 February 2012

Abstract With the PANDA experiment at the FAIR facility in Darmstadt, Germany it will be possible to investigate antiproton-nucleus reactions in an energy range not explored so far. This provides opportunities for unique measurements of which some are outlined in this article. Possible modifications of hadron properties in nuclear matter is subject of extensive theoretical and experimental studies. With PANDA it will be possible to extend this kind of studies to the charm sector. A study of particular interest will be to measure the J/Ψ-nucleon dissociation cross-section. This cross-section is relevant for the interpretation of the J/Ψ suppression observed in high energy heavy ion reactions. Further topics include the study of antibaryons in nuclei and short-range nucleon-nucleon correlations.

Keywords Antiproton-nucleus reactions · PANDA

1 Introduction

The Facility for Antiproton and Ion Research (FAIR) is currently constructed on the site of the GSI Helmholtzzentrum für Schwerionenforschung, close to Darmstadt in Germany. In a few years from now, when the facility becomes operational it will be host of various experiments exploring different aspects of nuclear matter with ions and antiprotons in a medium energy range.

One of these experiments is PANDA (Proton-Antiproton Annihilation at Darmstadt) [1]. With PANDA collisions of antiprotons with protons and nuclei will be investigated at center-of-mass energies between 2.7 and 16 GeV. One part of the

P. Bühler (✉)
Stefan Meyer Institute for subatomic Physics, Austrian Academy of Sciences,
Boltzmanngasse 3, 1090 Vienna, Austria
e-mail: paul.buehler@oeaw.ac.at

scientific program will use antiproton-proton collisions to e.g. perform charmonium spectroscopy below and above open charm threshold, investigate the nucleon structure, and hypernuclei [2]. Alternatively nuclear targets can be used to investigate antiproton-nucleon reactions in an energy range which has not been explored so far.

2 Antiproton-nucleon reactions with PANDA

Nuclear targets can be realized by e.g replacing the hydrogen gas in the cluster jet target by heavier gases or by using fiber or wire targets. It however has to be considered, that the achievable luminosity is reduced for higher mass targets and correspondingly the observation time needed to reach a given number of interactions is longer. In Fig. 1a the maximum luminosity is shown as function of the nuclear charge Z of the target material for different antiproton momenta (three lines from bottom to top: $1.5, 9, 15$ GeV/c) [2]. The reduction of the luminosity relative to $Z = 1$ is most pronounced at lower energies but in any case is considerable.

2.1 In medium mass modifications

At finite nuclear density chiral symmetry is expected to be partially restored leading to modifications of mass and lifetimes of particles (see e.g. [3]). So far studies of this phenomena were mainly restricted to the light quark sector [4]. With $\overline{\text{p}}$A reactions in PANDA these studies can be extended into the charm sector. Predicted mass shifts and related references of selected mesons are shown in Table 1. In many cases the predicted shifts are large compared to the natural width of the particles (Γ in third column of Table 1). However, due to the Fermi motion of the nucleons the measured widths are broadened by typically 250 MeV and thus shifts of less than 100 MeV will not be easily observable. Relative shifts of neighboring charmonium states could lead to the opening or closing of decay channels and thus to observable changes of particle widths. It was e.g. considered by Hayashigaki [5] that the combined shift of the D and $\overline{\text{D}}$ mesons could lead to a lowering of the D$\overline{\text{D}}$ threshold below the ϕ' of χ_{c2} charmonium states, thus opening the possibility of these mesons to decay into D$\overline{\text{D}}$. Recent coupled channel calculations reveal the momentum dependence of in medium particle property modifications [6, 7], suggesting the modifications to be largest at small momenta.

Experimentally there are three techniques available to observe the in-medium modifications: measurement of line shapes, transparency ratios, and excitation functions. In any case most significant results can be expected if the investigated particles are produced in the center of the nucleus (high density) at low momentum. To produce e.g. the lowest charmonium state η_c in a $\overline{\text{p}}$p reaction an antiproton momentum of approximately 2.6 GeV/c is needed. Thus charmed hadrons with low momentum are hardly produced with this reaction. Lower momenta are achieved in multi-nucleon reactions, for which however the cross-section is highly reduced. The total cross section $\sigma_{\overline{\text{p}}\text{p}}$ is approximately 60 mb at $p_{\overline{\text{p}}} \approx 2$ GeV/c and normal nuclear density ([8], section 41). Thus antiprotons with this energy can be expected to penetrate typically a few fm into the nucleus and only a small fraction reaches the center. The production volume is thus rather spread over one hemisphere of the

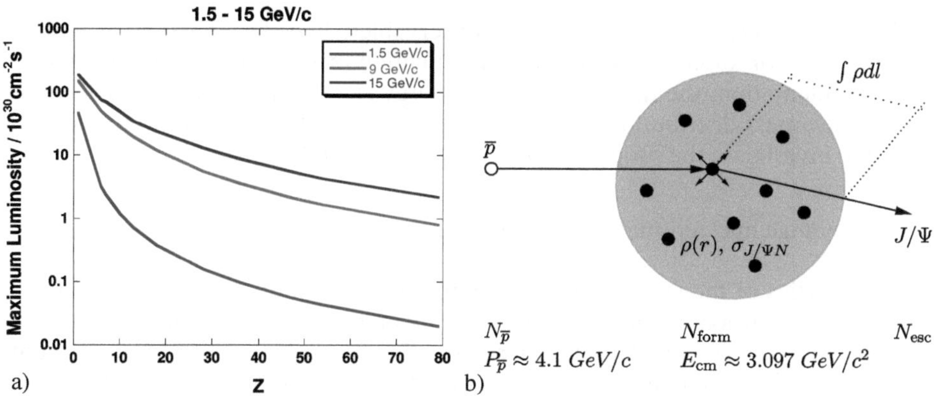

Fig. 1 **a** Luminosity of the HESR with a nuclear target as function of the nuclear charge Z of the target material [2]. **b** Illustration of the $\overline{p}$A reaction which is used to measure the J/Ψ-nucleon dissociation cross section $\sigma_{J/\Psi N}$

Table 1 Predicted in-medium mass shifts of selected charmed mesons at normal nuclear density

	Mass shift [MeV]	Refs.	Γ [MeV]	References
η_c	−5	[1], [5]	30	[1] Klingl, et al., PRL **82**, 3396 (1999)
J/Ψ	−4 to −24	[2], [3], [4], [5]	0.09	[2] Hayashigaki, PLB **487**, 96 (2000)
$\chi_{c0,1,2}$	−40	[5]	<10	[3] Lee and Ko, PRC **67**, 038202 (2003)
$\Psi(3686)$	−100	[3], [5]	0.3	[4] Krein, et al., PLB **697**, 136
$\Psi(3770)$	−100	[3], [5]	30	[5] Lee, AIP Proc. Conf. **717**, 780 (2004)
D-mesons	−50	[2]	<1	[6] Tolos, et al., PRC **77**, 015207 (2008)
$\overline{\Lambda}_c$, $\overline{\Sigma}_c$	−40	[6]	<3	

The third column contains the vavuum width for reference

nucleus (over range of nuclear densities) than concentrated at the center which will reduce the effects and make the data interpretation more challenging.

2.2 Antibaryons in nuclei

In $\overline{p}$A reactions antibaryon—baryon pairs can be produced, e.g. $\overline{p}$p, $\overline{\Lambda}\Lambda$, D^+D^- opening interesting opportunities to study the potential of these particles in nuclei.

Pochodzalla [9] has argued, that in exclusive production reactions the relative transverse momentum of antibaryon and baryon contains information on the difference between their respective potentials. The reaction cross sections are large such that a measurement of the transverse momentum asymmetry in case of $\overline{\Lambda}\Lambda$ could be performed in less than an hour and less than a day for $\overline{\Psi}\Psi$.

In events in which the baryon flies in forward direction—taking the majority of the momentum—antibaryons are produced in the nucleus with very small momentum. It was suggested [10, 11] that the presence of an antibaryon can drastically modify the arrangement of nucleons in a nucleus and lead to a significant increase of the binding energy and local compression. Due to the low reaction Q value in medium the annihilation rate of the antibaryon is reduced resulting in a relatively long-lived antibaryon—nucleus system. Different observables, like the emission of a single photon, pion, or kaon due to the final annihilation of the antibaryon, an explosive multifragmentation, the hardening of the energy spectrum of emitted nucleons, or the missing mass distribution of the emitted baryon [12] have been proposed to be suited to investigate the properties of such states with PANDA.

2.3 Short range nucleon-nucleon correlations

The properties of the strong force determine the motion of the nucleons in nuclei. The strong repulsion at small distances results in a phenomenon know as nucleon-nucleon short-range correlation. A large fraction of nucleons with large momenta (above the fermi momentum) are found to form pairs, with roughly equal, back-to-back momenta [13] (short-range correlated pairs, SRC pairs). Most SRC pairs consist of a proton and a neutron but also pairs of two correlated protons or two neutrons exist. The properties of short-range correlations are traditionally explored using the reactions $A(e, e'NN)$ and $A(p, p'NN)$, in which a nucleon-pair is knocked-out of the nucleus by a high energetic projectile [14]. With PANDA the reaction $A(\overline{\mathrm{p}}, \overline{\mathrm{p}}'NN)$ can be exploited, and due to the ability of the detector to measure protons as well as neutrons, all possible combinations of nucleon-pairs can be investigated simultaneously.

2.4 J/Psi-nucleon dissociation cross section

The inelastic J/Ψ-nucleon cross section $\sigma_{J/\Psi\mathrm{N}}$ is important in understanding the role of the formation of a Quark-Gluon-Plasma, QGP in the J/Ψ suppression observed in high energy heavy-ion collisions [15]. For an interpretation of these data the understanding of the effects which are not related to QGP formation, but also affect J/Ψ production in nucleus-nucleus reactions is crucial. The inelastic scattering of the J/Ψ state in cold nuclear matter is expected to be the dominant contribution of this kind. Its strength is described by the J/Ψ-nucleon inelastic cross section, $\sigma_{J/\Psi\mathrm{N}}$.

$\sigma_{J/\Psi\mathrm{N}}$ can be determined with reactions, in which J/Ψ mesons are produced in nuclear medium. It is from the comparison of the expected number of produced J/Ψs and the actually measured number of J/Ψs escaping from the nuclear medium that the dissociation cross section is derived. In $\overline{\mathrm{p}}$A reactions the momentum of the incident $\overline{\mathrm{p}}$ can be tuned such that J/Ψs are produced exclusively at a relatively well defined momentum.

The idea for this experiment for measuring $\sigma_{J/\Psi\mathrm{N}}$ can be traced back to a paper by Brodsky & Mueller [16]. The principle of the measurement is illustrated in Fig. 1b. The number of J/Ψs formed in the nucleus $\mathrm{N_{form}}$ and the number of J/Ψs escaping the Nucleus $\mathrm{N_{esc}}$ are related through

$$\mathrm{N_{esc}} = \mathrm{N_{form}} \left(1 - \sigma_{J/\Psi\mathrm{N}} \left\langle \int \rho \cdot dl \right\rangle \right) \quad (1)$$

$\langle \int \rho \cdot dl \rangle$ is the average integrated nucleon density along the path of the J/Ψ through the nucleus, where the average is to be taken over all formations.

Equation 1 represents a recipe for the measurement of $\sigma_{J/\Psi\mathrm{N}}$. It contains a few quantities which either need to be determined experimentally ($\mathrm{N_{esc}}$) or by simula-

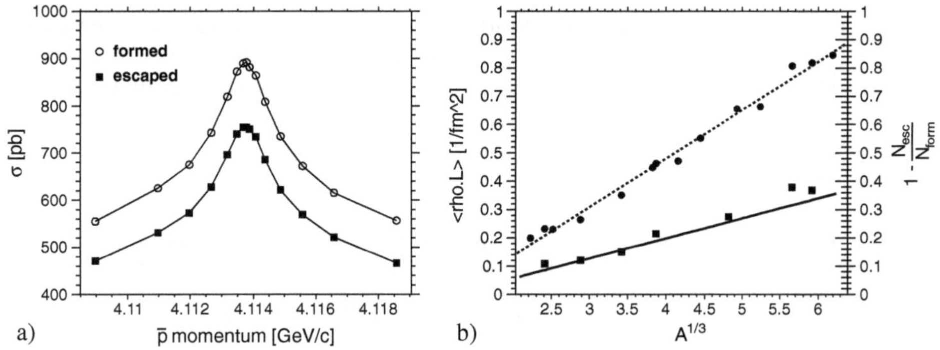

Fig. 2 **a** $\langle\sigma_{\text{form}}\rangle$ and $\langle\sigma_{esc}\rangle$ as function of the $\overline{\text{p}}$ momentum in a ^{40}Ca target **b** $(1-\text{N}_{\text{esc}}/\text{N}_{\text{form}})$ (*square symbols*) and $\left\langle\int\rho\cdot dl\right\rangle$ as function of nuclear mass of the target material

tion/calculation (N_{form}, $\left\langle\int\rho\cdot dl\right\rangle$). The success of the measurement hence depends on the exact determination of the absolute values of these quantities.

J/Ψs are experimentally best identified by their decay into lepton pairs $e^+e^-/\mu^+\mu^-$. The simple topology in this reaction allows to efficiently identify J/Ψs with a detector like PANDA and also can be exploited to reduce the background to an acceptable level. The cross section for formation of J/Ψ in $\overline{\text{p}}$A is by a factor of a few times 10^9 smaller than the total inelastic cross section of typically 1 b. Thus a background suppression of at least 10^{-10} has to be and can be achieved with PANDA (see [2], section 4.4.3 for further details).

The number of formed J/Ψs, N_{form} is given by $\text{N}_{\text{form}} = L\cdot\langle\sigma_{\text{form}}\rangle$. L is the luminosity integrated over the measurement time. $\langle\sigma_{\text{form}}\rangle$ is the average formation cross section. The formation probability in a single $\overline{\text{p}}$p reaction is well described by the Breit-Wigner formula. The formation cross section at resonance in a $\overline{\text{p}}$p and final decay into a e^+e^- pair is 275.7 nb (relevant parameters have been determined by [17]). The effective formation cross section $\langle\sigma_{\text{form}}\rangle$ in $\overline{\text{p}}$A reactions however is considerably smaller than this peak value due to the Fermi motion of the nucleons in the nucleus.

The effective value of $\langle\sigma_{\text{form}}\rangle$ was determined by Monte Carlo simulations using a Glauber type model. The nucleus is described by a radial nucleon density profile $\rho(r)$ which is taken from [18, 19]. The distribution of the Fermi-motion is modeled using a local Thomas-Fermi approximation in which the Fermi momentum P_{F} depends on the local nucleon density $\rho(r)$. Figure 2 shows results of these simulations for a ^{40}Ca target.

In the left panel $\langle\sigma_{\text{form}}\rangle$ and the cross section for escaping and decaying into an electron-positron pair $\langle\sigma_{esc}\rangle$ is plotted as function of the beam momentum. The peak value of $\langle\sigma_{esc}\rangle$ is in the order of 700 pb in this case. With a total reconstruction efficiency of 50% and an average luminosity of 10^{31} cm^{-2}s^{-1} the number of J/Ψs per day would be around 300.

According to (1) $\sigma_{J/\Psi\text{N}}$ is obtained by the ratio of $(1-\text{N}_{\text{esc}}/\text{N}_{\text{form}})$ and $\left\langle\int\rho\cdot dl\right\rangle$. For consistency checks, measurements of N_{esc} should be carried out with different target materials. The square symbols plotted in the right panel of Fig. 2 represent the quantity $(1-\text{N}_{\text{esc}}/\text{N}_{\text{form}})$ for $\sigma_{J/\Psi\text{N}} = 4.2$ mb, and the circular symbols show the quantity $\left\langle\int\rho\cdot dl\right\rangle$ for different target materials. The dotted line is the best linear fit to the $\langle \int \rho \cdot dl \rangle$ points, whereas the bold line is the dotted line scaled to match the $(1 - N_{esc}/N_{form})$ data points best. The scaling factor is the measurement of $\sigma_{J/\Psi N}$. Here its value is found to agree with the value used as input to the simulations.

References

1. PANDA experiment website. http://www.panda.gsi.de/. Accessed 20 February 2012
2. The PANDA collaboration: Physics performance report for PANDA (2009). arXiv:0903.3905v1 [hep-ex]. Accessed 20 February 2012
3. Saito, K., Tsushima, K., Thomas, A.W.: Prog. Part. Nucl. Phys. **58**, 1 (2007)
4. Metag, V.: AIP Conf. Proc. **1322**, 73 (2010)
5. Hayashigaki, A.: Phys. Lett. B **487**, 96 (2000)
6. Lutz, M.F.M., Korpa, C.L.: Phys. Lett. B **633**, 43 (2006)
7. Mitzutani, T., Ramos, A.: Phys. Rev. C **74**, 065201 (2006)
8. Nakamura, K., et al. (Particle Data Group): J. Phys. G **37**, 075021 (2010). http://pdg.lbl.gov
9. Pochodzalla, J.: Phys. Lett. B **669**, 306 (2008)
10. Mishutsin, I.N., et al.: Phys. Rev. C **71**, 035201 (2005)
11. Larionov, A.B., et al.: Phys. Rev. C **78**, 014604 (2008)
12. Aslanides, E., et al.: Nucl. Phys. A **470**, 445 (1987)
13. Subedi, R., et al.: Science **320**, 1476 (2008)
14. Arrington, J., et al. (2011). arXiv:1104.1196v2 [nucl-ex]. Accessed 20 February 2012
15. NA50 collaboration: Eur. Phys. J. C **48**, 329 (2006)
16. Brodsky, S.J., Mueller, A.H.: Phys. Lett. B **206**, 685 (1988)
17. Armstrong, T.A., et al.: Phys. Rev. D **47**,772 (1993)
18. De Vries, H. et al.: At. Data Nucl. Data Tables **36**, 495 (1987)
19. Reuter, W., et al.: Phys. Rev. C **26**, 806 (1982)

Hyperfine Interact (2012) 209:111–115
DOI 10.1007/s10751-012-0565-z

Determination of the $D_{s0}^*(2317)$ width with the PANDA detector

Marius Christian Mertens for the PANDA collaboration

Published online: 24 January 2012

Abstract The $D_{s0}^*(2317)$ meson which was discovered at BaBar in 2003 has the interesting properties of a surprisingly narrow width and a mass just below the DK threshold. Different theoretical models try to explain the nature of its properties. A precise knowledge of the width is an important criterion to evaluate these models. However, only an upper limit of 3.8 MeV is known so far. A suitable method to determine the width of particles which are significantly narrower than the experimental mass resolution is to measure the production cross section as a function of the center of mass energy. The shape of this excitation function allows to deduce the width. At PANDA, the measurement of the production cross section will be possible in antiproton-proton collisions. The PANDA experiment at the future FAIR facility is designed to combine precisely adjustable beam momenta and high luminosities which make it an excellent tool for this kind of measurement. In the following we will describe the experimental procedure to carry out this measurement with the PANDA detector in order to achieve a resolution in the order of 0.1 MeV for the width of the $D_{s0}^*(2317)$.

Keywords PANDA · FAIR · $D_{s0}^*(2317)$ · Width · Excitation function

1 Introduction

Currently, various theoretical models exist which try to explain the nature of the $D_{s0}^*(2317)$ (e.g., [1, 2]). They feature different predictions for the width, making it an important experimental parameter to distinguish between the models. A direct width measurement by simply measuring the invariant mass distribution of the decay products only works where the experimental resolution is better than the width to

M. C. Mertens (✉)
Forschungszentrum Jülich GmbH, Wilhelm-Johnen-Strasse, 52428 Jülich, Germany
e-mail: m.mertens@fz-juelich.de

be measured. With a width less than 3.8 MeV [3], the $D_{s0}^*(2317)$ is too narrow for a direct width measurement. However, a resonance or threshold scan can be carried out instead which does not depend on the experimental mass resolution.

2 Experimental method

During an energy scan, the cross section is measured as a function of the center of mass energy. The shape of this excitation function depends on the particle width, thus the measurement of the shape can be used to deduce the particle width. With PANDA it is possible to study the reaction $\bar{p}p \to D_s^{\pm} D_{s0}^*(2317)^{\mp}$.

It has been shown in [4] (Section 4.2.4) that an inclusive reconstruction of the signal via $D_s^{\pm} \to \phi\pi^{\pm}, \quad \phi \to K^+K^-$ using the sum mass spectrum of the reconstructed D_s and the missing mass provides a promising combination of a relatively high signal count and good separation from the background which can be described by an Argus function. Therefore the following analysis is based on this reconstruction strategy.

Throughout the energy scan, the production rate of the $D_{s0}^*(2317)$ is measured for different beam momenta, i.e., center of mass energies. The experimental data then yield the mass and width of the $D_{s0}^*(2317)$ by fitting the excitation function ([4] Section 4.2.4, see (1)) to the scanpoints.

$$\frac{\sigma_\Gamma\left(\sqrt{s}\right)}{|M|^2} = \frac{\Gamma}{4\pi \cdot \sqrt{s}} \cdot \int_{-\infty}^{\sqrt{s}-m_{D_s}} \frac{\sqrt{\left(s - \left(m + m_{D_s}\right)^2\right) \cdot \left(s - \left(m - m_{D_s}\right)^2\right)}}{\left(m - m_{D_{s0}^*}\right)^2 + \left(\frac{\Gamma}{2}\right)^2} dm \tag{1}$$

Note that (1) can be very well approximated by (2) below (Christoph Hanhart, priv. comm.). Here, m_{D_s} and $m_{D_{s0}^*}$ appear only as sum, which indicates that any deviation of m_{D_s} from the true D_s mass only causes a corresponding shift of the fitted $D_{s0}^*(2317)$ mass but has no influence on the reconstructed width.

$$\sigma(\lambda) \approx \sqrt{\left(m_{D_s} + m_{D_{s0}^*}\right) \cdot \Gamma} \cdot |M|^2 \cdot \frac{1}{2\pi} \cdot \int_{-\infty}^{\lambda} \frac{\sqrt{\lambda - x}}{x^2 + 1} dx \tag{2}$$

$$\lambda = \frac{2}{\Gamma} \cdot \left(\sqrt{s} - \left(m_{D_s} + m_{D_{s0}^*}\right)\right) \tag{3}$$

Beam momentum spread The individual antiprotons which form the beam follow a finite distribution around the nominal beam momentum, resulting in a finite distribution of the center of mass energy around the nominal value. The expected beam momentum spread of the HESR running in high luminosity mode is in the order of 10^{-4} [4] (Section 2.3.2.2) and results in a center of mass energy spread of 192 keV at the threshold energy (4286 MeV, 8802 MeV/c beam momentum). This effect is included in the simulations by convoluting the excitation function with a Gaussian bell curve which describes the momentum distribution of the antiproton beam.

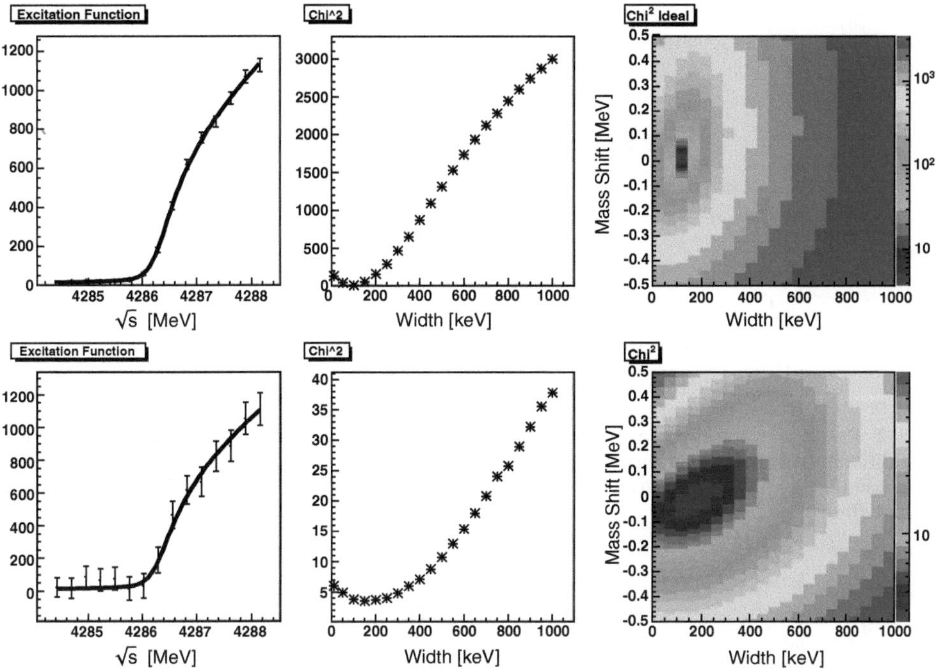

Fig. 1 Fit of the excitation function to simulated scan data for a 100 keV wide $D_{s0}^*(2317)$ and 60 days of total measurement time. The *upper row* shows the results without any background taken into account, the *lower row* includes an Argus background for a signal to background ratio of 1. *Left diagram:* simulated scan data drawn with an excitation function based on the input parameters. *Center:* χ^2 distribution for excitation functions with varying width hypotheses. *Right:* 2D χ^2 distribution for the excitation functions with varying mass and width hypotheses

3 Simulation overview

The following parameter space of the simulations is presented here: Relative beam momentum spread of 10^{-4}, 15 scanpoints in two energy ranges (± 1 MeV, ± 2 MeV) around the threshold, optionally two additional scanpoints located 4 MeV above the threshold, signal to background ratio of 1, 60 days total measurement time with a daily integrated luminosity of 9000 nb^{-1}, reconstruction efficiency of 0.25 (compare [4], Section 4.2.4).

While the specific background channels containing D_s mesons can be suppressed to half the signal size, the suppression of the generic background is more challenging. In [4] Section 4.2.4, a large suppression to $2 \cdot 10^{-5}$ has already been achieved, however there are still some parameters left for optimization (e.g. vertex and momentum constraints, pion identification). Since the optimization of kaon identification alone changed the background suppression by a factor of 10, an improvement of another order of magnitude to reach a total signal to background ratio of 1 seems feasible.

For historical reasons, the production cross section σ_0 has been normalized to the cross section of a 1 MeV wide $D_{s0}^*(2317)$ at threshold energy which corresponds to roughly 5 nb cross section 5 MeV above the threshold energy for $\sigma_0 = 1$.

Figure 1 shows the simulated signal counts for a 100 keV wide $D_{s0}^*(2317)$ together with a fit of the excitation function, where the parameters mass and width are found

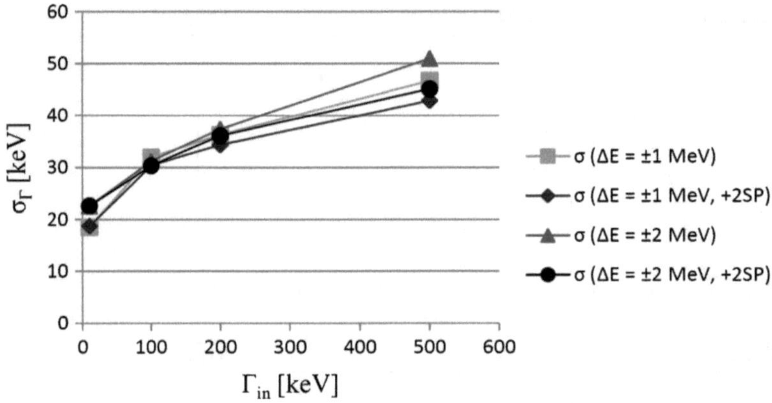

Fig. 2 Resulting resolutions for different $D^*_{s0}(2317)$ input widths. The four curves show the results for different scanpoint distributions around the threshold. The curves marked with +2SP feature two additional scanpoints located at 4 MeV above the threshold with otherwise identical scanpoint distributions

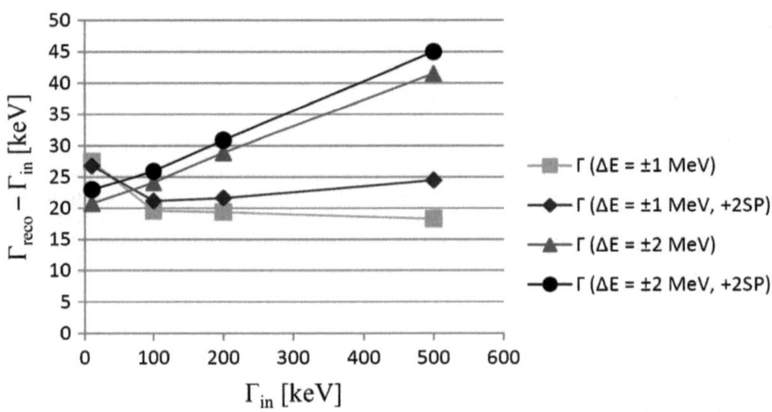

Fig. 3 Resulting deviations from the expectation value for different $D^*_{s0}(2317)$ input widths. The four curves show the results for different scanpoint distributions around the threshold. The curves marked with +2SP feature two additional scanpoints located at 4 MeV above the threshold with otherwise identical scanpoint distributions

by minimizing χ^2. The correlations which are observed when background is present might be related to the observation of an increased systematic deviation of the width from the expectation value with increasing background (compare Section 4). These correlations and in particular finding a strategy to eliminate their effect are subject to further study.

4 Results

Repeating the simulation and fitting procedures for different width hypotheses yields the expected resolutions (Fig. 2) and deviations from the expectation value (Fig. 3).

The statistical errors returned from the fit procedures were in all cases less than 3 keV (width resolution) and 6 keV (deviation from the expectation value). It becomes clear that the narrower energy range should be the location within which the final distribution of scanpoints should be optimized. The two additional scanpoints 4 MeV above the threshold do not improve the results but actually worsen it due to the increased background. The systematic deviations appear only with background, therefore it seems a reasonable assumption that this is a fit artifact which should be compensated in further studies. However, a conservative linear sum of the 50 keV resolution and 50 keV currently present systematic shift yields a very promising value of (100 ± 10) keV possible resolution.

For comparison: Leaving the other parameters unchanged, a resolution of 150 keV is obtained for a signal to background ratio of 0.1. The systematic deviation remains the same, indicating an overall resolution of (200 ± 10) keV for a 10 times larger background.

5 Conclusion

The resulting numbers for the systematic shift and the resolution indicate that a $D_{s0}^*(2317)$ width resolution in the order of 0.1 MeV is feasible. While for most experimental parameters the PANDA design values have been included into the simulations, the remaining main source of uncertainty is the as of yet unknown production cross section.

References

1. Faessler, A., et al.: Strong and radiative decays of the Ds(2317) meson in the DK-molecule picture. Phys. Rev. D **76**, 014005 (2007)
2. Guo, F.-K., et al.: More kaonic bound states and a comprehensive interpretation of the DsJ states. Phys. Rev. D **84**, 014013 (2011)
3. Nakamura, K., et al.: Particle data group. J. Phys. G **37**, 075021 (2010)
4. PANDA Collaboration: Physics Performance Report for PANDA. ArXiv:0903.3905v1[hep-ex] (2009)

Hyperfine Interact (2012) 209:117–120
DOI 10.1007/s10751-011-0554-7

Pion production on exotic nuclei by antiproton annihilation

Stefanie Lourenço · Horst Lenske · Slawomir Wycech

Published online: 26 February 2012

Abstract Pion production by antiproton annihilation on neutron- or proton-rich nuclei is studied in a fully quantum mechanical description of the reaction process in terms of a distorted wave approach. The elementary elastic $\overline{N}N$ and the $\overline{N}N \to n\pi$ annihilation vertices are obtained from a t-matrix accounting for coupled channels effects. The theoretical amplitudes are used to derive a microscopic $\overline{N}A$ optical model potential and the meson production vertices. Nuclear structure effects are taken into account microscopically. Results for meson production cross sections are presented.

Keywords Antiproton · Exotic nuclei · Pion production

1 Introduction

Despite the decades-long experimental and theoretical efforts many open questions are persisting on the interaction of antinucleons with matter. The antinucleon-nucleon data [1–3] are showing clearly that reactions are dominated by annihilation processes, finally producing a broad distributions of pions [4].

Our interest in $\overline{p}A$ and $\overline{n}A$ interactions is equally directed towards investigations of the basic $\overline{N}N$ and $\overline{N}A$ interactions and the use of antinucleons as probes for nuclear studies. Especially antiproton physics is currently becoming of renewed

Supported in part by DFG, contract Le439/8-1.

S. Lourenço (✉) · H. Lenske
Institut fuer Theoretische Physik, JLU Giessen, Heinrich-Buff-Ring 16,
35392 Giessen, Germany
e-mail: Stefanie.Lourenco@physik.uni-giessen.de

S. Wycech
National Center for Nuclear Studies, Hoza 69, 00-681, Warsaw, Poland

interest by the new opportunities provided by the upcoming FAIR@GSI facility. The PANDA experiment is one example, but in addition there are other experiment proposals. Among them, the Antiproton-Ion-Collider (AIC) proposal [5] is intended for making use of in-beam $\overline{p}A$ interactions for spectral studies of exotic nuclei.

In previous works [6, 7], we showed that total $\overline{p}A$ annihilation cross sections are well suited for investigations of neutron skins in heavy exotic nuclei. Here, our interest is in the mesonic annihilation channels, exploring their potential use for investigations of in-medium $\overline{p}N$ interactions and for more detailed nuclear studies. For that aim, we study theoretically coherent annihilation reactions into two mesons $m_{1,2}$ and a residual nucleus B. A fully quantum mechanical approach is used, deriving theoretically the $\overline{N}A$ interactions and the elementary $\overline{N}+N \to m_1 + m_2$ vertices from free-space interactions. After a brief discussion of the theoretical background we present first results on $\overline{p}A$ elastic scattering and absorption cross sections and meson production on ^{78}Ni.

2 Theory of $\overline{N}A$ and πA interactions

The invariant cross section for the 2-to-3 body reaction $\overline{N}+A \to m_1 + m_2 + B$ is given by

$$d^9\sigma_{\alpha\beta} = N_{\alpha\beta}\left(\frac{\hbar c}{2\pi}\right)^9 \frac{d^3k_1}{E_1}\frac{d^3k_2}{E_2}\frac{d^3k_B}{E_B}\left|M_{\alpha\beta}\left(\mathbf{k_1},\mathbf{k_2},\mathbf{k_B};\mathbf{k}_\alpha\right)\right|^2$$
$$\times\,\delta\left(\mathbf{k_1}+\mathbf{k_2}+\mathbf{k_B}\right)\delta\left(E_1+E_2+E_B-\sqrt{s}\right). \quad (1)$$

The various Dirac-delta functions are accounting for energy-momentum conservation, essentially defining the kinematical phase space of the reaction. The dynamics are contained in the matrix element

$$M_{\alpha\beta} = \left\langle \Phi^{(-)}_{m_1 B}\Phi^{(-)}_{m_2 B}|\varphi_{BA} t_{\overline{N}N,m_1m_2}|\chi^{(+)}_{\overline{N}A}\right\rangle. \quad (2)$$

Initial and final state interactions are contained in the distorted waves $\chi^{(+)}_{\overline{N}A}$ in the initial channel and $\Phi^{(-)}_{m_i B}$ for the outgoing mesons, respectively. The wave function of the initially bound nucleon N on which the incoming antinucleon is annihilated is given by the overlap matrix element $\varphi_{BA} \equiv \langle B|\Psi_N|A\rangle$ where Ψ_N is a nucleon field operator. The remaining operator parts of the annihilation vertex are included into the elementary production vertex $t_{\overline{N}N,m_1m_2}$.

We consider first the initial state interaction (ISI). Our present understanding of the $\overline{N}N$ interactions is that they contain an elastic, diffractive part, given by the G-parity transformed NN interaction, and a strong dispersive component resulting from the coupling to annihilation channels which, in fact, is the dominant part. Theoretically, these effects are taken into account in various approaches [8–10]. These approaches describe the interactions in terms of meson-exchange and annihilation diagrams. A simple but realistic model is to use the optical theorem and derive the on-shell $\overline{N}N$ t-matrix from the total elastic cross section. This can be done using either the amplitudes derived from measured cross sections [1] or directly the theoretically calculated amplitudes $t_{\overline{N}N}(\sqrt{s}, q)$ where the latter are also known at the off-shell momentum q. Presently, we are using the $\overline{N}N$ t-matrices of [11]. In either case, we derive the $\overline{N}A$ optical model by a folding approach

$$U_{opt}(\mathbf{r}) = \sum_{N=p,n} \int \frac{d^3q}{(2\pi)^3} \rho_N(q) t_{\bar{p}N}\left(\sqrt{s}, q^2\right) e^{i\mathbf{q}\cdot\mathbf{r}} \tag{3}$$

corresponding to the known t-ρ approximation, i.e. neglecting at this stage the medium dependence of the elementary scattering amplitude. The target ground state Hartree-Fock-Bogoliubov (HFB) densities as used in [12, 13] define the nuclear mass form factors $\rho_N(q)$.

With the choice of an optical potential model we solve in partial wave representation the wave equation describing the $\overline{p}A$ relative motion. We use a Schrödinger equation but taking into account relativistic kinematics.

The final state interactions (FSI) requires knowledge of the meson-nucleus interactions. In the case considered here the FSI is given by the interaction of the two outgoing pions with the residual nucleus B. At present, we neglect the pion-pion interactions assuming that the πA optical potential dominates. Since the two pions share the available center-of-mass energy either of both may emerge over a wide range of kinetic energies, from close to the threshold up to rather high energies in the GeV-region. Hence, we need a pion-nucleus potential over a correspondingly wide range of kinetic energies. As a starting point we use the pion-nucleus potential of Kisslinger-type from [14]. At present, we extend the approach to higher energies. This requires to include also higher nucleon resonances beyond the $\Delta_{33}(1232)$ state.

3 Meson production

The central quantity for the meson production process, (1) is the transition t-matrix $t_{\overline{N}N,m_1m_2}$. In principle, there are two processes contributing: An s-channel process where first in a $\overline{N}N$ s-wave reaction a 1^{--} vector meson is formed, followed by the decay into two mesons; A t-channel radiative process proceeding through the exchange of a nucleon, a nucleon resonance or, if also kaon production is considered, any other octet baryon. The second process is part of the Juelich $\overline{N}N$-model [10, 15, 16] and implicitly also contained in other models. As a first step we consider the production of a ρ-meson, which then decays into two pions accordingly to the study of $\rho \to \pi\pi$ dynamics in [17]. In Fig. 1 first results are shown. For a ^{78}Ni target the annihilation on the full set of bound neutrons was considered, using single particle binding energies and wave functions from HFB calculations [12, 13]. The angular distributions are calculated quantum mechanically including Coulomb interactions and partial waves with $\ell \leq 90$. The cross section is seen to be of a sizeable magnitude, well within the limits of present experimental possibilities.

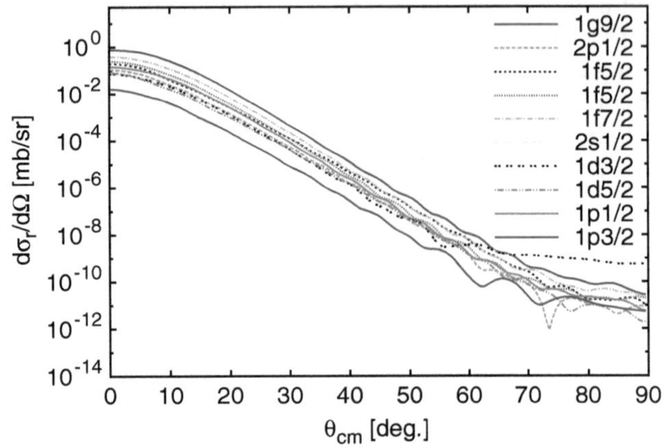

Fig. 1 As a first step we calculate the cross section of one meson production, namely $\rho^{0,-}$, which then decays into two pions. The relative motion of the two pions has been integrated out. The differential cross sections are shown for meson production on a ^{78}Ni target and $T_{lab} = 800$ MeV

4 Summary and outlook

Interactions of antinucleons with nucleons and nuclei were discussed for elastic and meson production reactions. Our approach is based on a fully microscopical and quantum mechanical description of the reaction mechanism and nuclear structure aspect. This allows to investigate $\overline{N}A$ reactions also in target mass and incident energy regions where data are not available. A First result for meson production on ^{78}Ni was presented. The approach indicates a new way for studying $\overline{N}N$ in-medium interactions and particle production processes. Obviously, the formalism can be applied correspondingly also to $\overline{B}B$ baryon production channels. The calculated cross sections are of a magnitude making them accessible for projects like PANDA and AIC@FAIR.

Acknowledgements Discussions and collaboration with J. Haidenbauer (FZ Jülich) and B. Loiseau (Orsay) are gratefully acknowledged.

References

1. Nakamura, K., et al. (Particle Data Group): J. Phys. G **37**, 075021 (2010)
2. Amsler, C.: Rev. Mod. Phys. **70**, 1293 (1998)
3. Klempt, E., Batty, C., Richard, J.-M.: Phys. Rep. **413**, 197 (2005)
4. Clover, M.R., et al.: Phys. Rev. C **26**, 2138 (1982)
5. Krücken, R., Kienle, P., et al.:Proposal for an Antiproton-Ion Collider at FAIR, (AIC), GSI, Darmstadt (2005)
6. Lenske, H., Kienle, P.: Phys. Lett. B **647**, 82 (2007)
7. Lenske, H., et al.: Prog. Part. Nucl. Phys. **59**, 114 (2007)
8. Cote, J., et al.: Phys. Rev. Lett. **48**, 1319 (1982)
9. Timmers, P.H., et al.: Phys. Rev. D **29**, 1928 (1984)
10. Hippchen, T., Holinde, K., Plessas, W.: Phys. Rev. C **39**, 761 (1989)
11. El-Bennich, B., Lacombe, M., Loiseau, B., Wycech, S.: Phys. Rev. C **79**, 054001 (2009)
12. Hofmann, F., Lenske, H.: Phys. Rev. C **57**, 2281 (1998)
13. Tsoneva, N., Lenske, H.: Phys. Rev. C **77**, 024321 (2008)
14. Johnson, M.B., Satchler, G.R.: Ann. Phys. **248**, 134 (1996)
15. Mull, V., Holinde, K.: Phys. Rev. C **51**, 2360 (1995)
16. Mull, V., Haidenbauer, J., Hippchen, T., Holinde, K.: Phys. Rev. C **44**, 1337 (1991)
17. Peters, W., Post, M., Lenske, H., Leupold, S., Mosel, U.: Nucl. Phys. A **632**, 109 (1998)

Hyperfine Interact (2012) 209:121–126
DOI 10.1007/s10751-011-0541-z

Results from the kaonic hydrogen X-ray measurement at DAFNE and outlook to future experiments

Michael Cargnelli · M. Bazzi · G. Beer · C. Berucci · L. Bombelli · A. M. Bragadireanu · A. Clozza · G. Corradi · C. Curceanu (Petrascu) · A. d'Uffizi · C. Fiorini · F. Ghio · B. Girolami · C. Guaraldo · R. S. Hayano · M. Iliescu · T. Ishiwatari · M. Iwasaki · P. Kienle · P. Levi Sandri · V. Lucherini · J. Marton · S. Okada · D. Pietreanu · K. Piscicchia · M. Poli Lener · T. Ponta · R. Quaglia · A. Rizzo · A. Romero Vidal · E. Sbardella · A. Scordo · H. Shi · D. L. Sirghi · F. Sirghi · H. Tatsuno · A. Tudorache · V. Tudorache · O. Vazquez Doce · E. Widmann · J. Zmeskal

Published online: 30 December 2011

Abstract The $\overline{K}N$ system at rest plays a key role for the understanding of strong interaction of hadrons with strangeness involved. The experiment SIDDHARTA used X-ray spectroscopy of kaonic atoms to measure the strong interaction induced

M. Cargnelli (✉) · T. Ishiwatari · J. Marton · E. Widmann · J. Zmeskal
Stefan-Meyer-Institut für subatomare Physik, Vienna, Austria
e-mail: michael.cargnelli@oeaw.ac.at

M. Bazzi · C. Berucci · A. M. Bragadireanu · A. Clozza · G. Corradi · C. Curceanu (Petrascu) · A. d'Uffizi · C. Guaraldo · M. Iliescu · P. Levi Sandri · V. Lucherini · S. Okada · K. Piscicchia · M. Poli Lener · A. Rizzo · A. Romero Vidal · E. Sbardella · A. Scordo · D. L. Sirghi · F. Sirghi · H. Tatsuno · O. Vazquez Doce
INFN, Laboratori Nazionali di Frascati, Frascati, Roma, Italy

G. Beer
Department of Physics and Astro., University of Victoria, Victoria, BC, Canada

L. Bombelli · C. Fiorini · R. Quaglia
Politechno di Milano, Sez. di Elettronica, Milano, Italy

A. M. Bragadireanu · M. Iliescu · D. Pietreanu · T. Ponta · D. L. Sirghi · F. Sirghi · A. Tudorache · V. Tudorache
IFIN-HH, Magurele, Bucharest, Romania

F. Ghio · B. Girolami
INFN Sez. di Roma I and Inst. Superiore di Sanita, Roma, Italy

R. S. Hayano · H. Shi
University of Tokyo, Tokyo, Japan

M. Iwasaki
RIKEN, The Institute of Phys. and Chem. Research, Saitama, Japan

P. Kienle
Physik Dep., Tech. Univ. München, Garching, Germany

shift and width of the ground state. It was the first experiment on kaonic He3 and deuterium ever, kaonic hydrogen was measured with improved precision resulting in $\epsilon_{1s} = -283 \pm 36(\text{stat}) \pm 6(\text{syst})$ eV and $\Gamma_{1s} = 541 \pm 89(\text{stat}) \pm 22(\text{syst})$ eV. Additionally a scheme for an improved future experiment on kaonic deuterium is introduced in this contribution.

Keywords Kaonic hydrogen · Antikaon-nucleon physics · Silicon drift detectors

1 Introduction

The strong interaction between hadrons at low energies can not be described in terms of quarks and gluons. Instead, effective field theories are used which rely on experimental input. The data come from kaon scattering experiments, atomic X-ray measurements and the energy and width of known resonances.

The kaonic atom experiments measure the effect of the strong interaction on the X-ray transitions, namely the shift and width of the ground state. These quantities are key input for theory and permit testing the ability of theoretical models to accommodate the different types of experimental information. For current theory see [1, 2] and references therein.

The $\overline{K}N$ interaction at rest is normally described in terms of complex scattering lengths. To extract the isospin dependant $\overline{K}N$ scattering lengths a_0 and a_1 a first measurement of kaonic deuterium is necessary in addition to kaonic hydrogen, no experiments were done before now.

Kaonic atoms are bound electromagnetically, the energy levels with strong interaction "switched off" are known precisely from QED. The strong interaction modifies the binding and causes the absorption of the kaon by the nucleus. This leads to a significant shift and broadening of the 1 s state of kaonic $Z = 1$ atoms, higher states are only marginally affected. By measuring the K-series X-ray transitions the strong-interaction induced shift and width are directly observable. Via Deser-type formulae [3] the shift and width are connected with the real and imaginary part of the K^{-}p and K^{-}d scattering length, which themselves are linear combinations of the isospin scattering lengths.

The best previous experimental results for kaonic hydrogen (KpX [5] and DEAR [6]) suffer from limited precision, the improvement reported herein is of high relevance for testing and advancing the theory in this field. Furthermore, by using only DEAR results, most of the theoretical models had difficulties in explaining all experimental evidence in a consistent way [4].

2 Experiment

Figure 1 illustrates the principle of our experiment: Both kaons of a charged kaon pair are detected and define the trigger. Hits in the X-ray detectors coincident with the trigger contain kaonic hydrogen signals along with background from kaon decay and kaon absorption in hydrogen and structure materials. Due to the limited timing resolution, some accidental (not kaon-correlated) background remains.

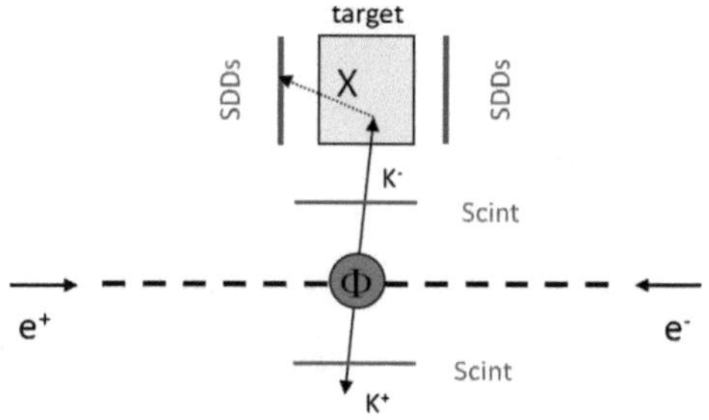

Fig. 1 Schematic view of the SIDDHARTA setup at DAΦNE

The e^+ e^- collider DAΦNE of Laboratori Nazionali di Frascati produced a peak luminosity of 3 10^{32} cm^{-2} s^{-1} and an integrated usable[1] luminosity of up to 8 pb^{-1} per day. The 2.2 liter cryogenic gas target was operated at 1.5–1.8% of liquid hydrogen density. The entrance window was 20 cm above the interaction point (IP). The kaon trigger consisted of two 1.5 mm thick plastic scintillators situated 6 cm above and below the IP.

A kaonic hydrogen atom is formed when a negative kaon enters the target cell, looses its remaining kinetic energy through ionization and excitation and finally is captured, replacing the electron, in an excited orbit ($n \simeq 25$). Via different cascade processes the kaonic atom deexcites to lower states. When the kaon reaches a low-n state with small angular momentum, it is absorbed by the nucleus. A small fraction of kaonic hydrogen atoms survive long enough to make one of the (np-1s) transitions (K-series).

The experimental challenge is the very small kaonic hydrogen X-ray yield, and the difficulty in doing X-ray spectroscopy in the radiation environment of an accelerator. Therefore new X-ray detectors were developed and built in the framework of the SIDDHARTA[2] project.

For each event, the data acquisition system stored the signal amplitudes of hits on the 144 SDDs and, if a kaontrigger was present, the time correlation between X-ray and kaon.

The basic energy calibration for each of the SDDs was obtained from periodic measurements of fluorescence lines from Ti and Cu foils excited by an X-ray tube during beam. This procedure delivered the information to compensate the individual gains, making it possible to sum the 144 spectra. The response function of the overlayed detector-system was found to deviate slightly from the pure Gaussian shape. Various approximations to account for this effect were tested, e.g. 2 exponential tail-functions convoluted in each fit evaluation. The rate dependance was studied and finally the parameters of the response *width* were derived from calibration data taken with very low X-ray tube current, which resulted in SDD rates similar to those in real data taking.

[1]During injection of e^+ and e^- into the rings, high beam-background prohibited X-ray measurements.

[2]Silicon Drift Detectors for Hadronic Atom Research by Timing Application.

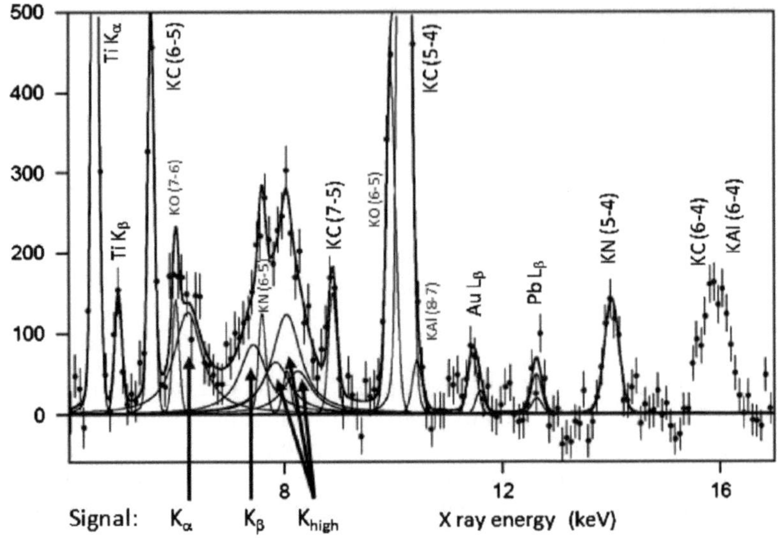

Fig. 2 X-ray spectrum of kaonic hydrogen measured in SIDDHARTA. The continuous background fit-component is already subtracted. The *kaonic lines* are from stops in the kapton foils of the target walls

Using the Cu and Au fluorescence lines in the untriggered spectra with X-ray tube off—they are excited by background—we derived a small gain correction[3] of a few eV which results from rate difference between data with and without the X-ray tube. Additionally we checked the calibration using the triggered spectra and the kaonic lines from wallstops (Fig. 2).

For the fit the kaonic hydrogen K-series transitions were represented by Voigt functions with the (energy dependant) Gauss sigma fixed to the values obtained by interpolating the values from the calibration lines. The pattern of the transition energies is given by the electro-magnetic values. The background lines (kaonic carbon etc. from stops in foils of the target windows) were Gaussians, the continuous background, a quadratic polynomial. The total spectrum was convoluted with the tail functions.[4] Since the individual lines from the higher transitions are high in total intensity, but completely overlapped, they tend to spoil the determination of shift and width of the K-series and a fixing of the intensity ratio would introduce a model-dependance. Therefore we concentrated our analysis on the K_α and K_β lines while the K_{high} lines were treated as "background". The procedure is as follows: (1) fit with all line intensities free, free common shift of all transitions, one Lorentz width for all transitions. (2) fix the line position of the (4–1) and higher transitions, fit free common shift of K_α, K_β. (3) modify the fixed line positions by the last shift result

[3]Corresponds to the line positions.

[4]Various models were tested, their deviation contributed to the systematic error.

then resume at step (2). After few iterations we obtained a consistent shift and width for all transitions. The result was:

$$\epsilon_{1s} = -283 \pm 36(\text{stat}) \pm 6(\text{syst})\ \text{eV} \quad \text{and} \quad \Gamma_{1s} = 541 \pm 89(\text{stat}) \pm 22(\text{syst})\ \text{eV}.$$

3 Towards SIDDHARTA-2

The analysis of the kaonic deuterium (KD) data in SIDDHARTA showed a 2 sigma hint on the signal, establishing only an upper limit for the K-series yield. A quantitative measurement of KD X-ray shift and width requires a substantial improvement of the applied experimental technique. Since the yield is expected to be about 1/10 of the KH yield and the width about twice the KH width, we need a reduction in background of about 20 and an increase of detection efficiency of 2–3 anlong with an integrated luminosity of around 800 pb^{-1}. Given such a situation, our Monte Carlo studies approximate a reachable precision of about 70 eV for the shift and 170 eV for the width.

The changes in experimental technique go into four directions:

(1) Trigger geometry and target density: By placing the upper kaon-trigger detector closely in front of the target entry window, the probability that a triggered kaon really enters the gas and is stopped there is much enlarged. Making the detector smaller than the entry area gives away some signal, but suppresses efficiently the kaonic lines from "wallstops" (kaons entering the gas volume, but passing from inside to the cylindrical target walls). The "signal per trigger" number goes up, which reduces also the accidental background coming along with every trigger. We intend to double the gas density which enhances the gasstops and further reduces the wallstops.

(2) K^+ discrimination by two complementary methods: A "kaonstopper" scintillator is placed directly below the lower kaon trigger. In case a K^- is stopped there, only one (large) signal from pileup of stopping and kaon-absorption secondaries is seen, however in case of a K^+, the kaon-decay particles are seen after the signal from the stopping (mean K^+ lifetime 12.8 ns), with a flash-ADC we will be able to distinguish the two cases with high efficiency. In addition we will use scintillators surrounding the target to measure K^- absorption secondaries (pions). The time-window for gasstops is about 4 ns wide. By this condition we also suppress stops in the entry window.

(3) Active shielding: The scintillators surrounding the target will also be used as a prompt anticoincidence if the spatial correlation of SDDs hit and scintillator indicates that the hit has originated from a pion ("charged particle veto"). An anticoincidence covering the SDD time window of about 600 ns (with an exception of the 4 ns of the stopping time in the gas) will reduce the accidental background. Although the scintillators have only low efficiency for gammas, the high rate of secondaries from the electromagnetic showers allows a relevant reduction of accidental ("beam") background. Remarkably, the upper trigger scintillator has two functions, it is also used as anticoincidence counter: after the passing of the kaon and eventual prompt kaon-absorption secondaries, it vetos beam-background.

(4) Operate SDDs at lower temperature: tests indicate that an improvement of the timing resolution from 900 ns to 600 ns is feasible by more cooling. The energy resolution decreases somewhat, but will still be sufficient for the broad KD transitions.

The signal enhancement is due to moving the target cell closer to the IP, by changing its shape and by better solid angle of the SDDs and by the higher gas-density.

4 Summary and conclusion

Within the SIDDHARTA project we measured X-ray transitions of light kaonic atoms. The experimental part covering kaonic hydrogen (KH) delivered the most precise measurement of the K-series X-rays in KH [7]. Our determination of the shift and width provides new constraints on theories, having reached a quality which will demand refined calculations of the low-energy $\overline{K}N$ interaction.

We worked out an improved experimental technique [8] which—using the existing SDD detectors—promises to allow a measurement of the shift and width in kaonic deuterium, long awaited data for the description of strong interaction at lowest energies in hadronic systems with strangeness.

Acknowledgements Part of this work was supported by the European Community-Research Infrastructure Integrating Activity "Study of Strongly Interacting Matter" (acronym HadronPhysics2, Grant Agreement no. 227431) under the Seventh Framework Programme of EU; HadronPhysics I3 FP6 European Community program, Contract No. RII3-CT-2004-506078; Austrian Federal Ministry of Science and Research BMBWK 650962/0001 VI/2/2009; Romanian National Authority for Scientific Research, Contract No. 2-CeX 06-11-11/2006; and the Grant-in-Aid for Specially Promoted Research (20002003), MEXT, Japan.

References

1. Döring, M., Meißner, U.-G.: Phys. Lett. B **704**, 663 (2011)
2. Ikeda, Y., Hyodo, T., Weise, W.: Phys. Lett. B **706**, 63 (2011)
3. Meißner, U.-G., Raha, U., Rusetsky, A.: Eur. Phys. J. C **35**, 349 (2004)
4. Weise, W.: Nucl. Phys. A **835**, 51 (2010)
5. Iwasaki, M., et al.: Phys. Rev. Lett. **78**, 3067 (1997)
6. Beer, G., et al., DEAR Collaboration: Phys. Rev. Lett. **94**, 212303 (2005)
7. Bazzi, M., et al., SIDDHARTA Collaboration: Phys. Lett. B **704**, 113 (2011)
8. SIDDHARTA-2 Collaboration: Proposal of Laboratori Nazionali di Frascati of INFN. The upgrade of the SIDDHARTA apparatus for an enriched scientific case (2010)

Hyperfine Interact (2012) 209:127–132
DOI 10.1007/s10751-011-0515-1

Future directions in kaonic atom physics

Eli Friedman

Published online: 9 December 2011

Abstract Recent progress and open problems in kaonic atom physics are presented. A connection between phenomenological deep potentials and the underlying K^-N interaction is established as well as the need for a theory for multinucleon absorption of kaons. K^- absorption at rest to specific Λ hypernuclei states is briefly discussed.

Keywords K^-N interaction · Kaonic atoms

1 Introduction and background

Attempting to foresee directions of further research into the field of kaonic atoms one naturally focuses on open problems on the one hand and on recent progress on the other. All the data on strong interaction effects in medium-weight and heavy kaonic atoms come from experiments in the 1970's [1, 2], using solid targets and a single GeLi X-ray detector. Analyses of the data show good consistency between the various experiments, and sub-sets of the data do not lead, in general, to conclusions which differ from what is found from analyses of the whole set of data. A new generation of experiments characterized by the use of many detectors and of tracking facilities for reducing background [3] and with excellent timing capabilities [4] addressed 'puzzles' with kaonic atoms of hydrogen and ^{4}He.

Starting with medium-weight and heavy elements, a long standing problem with these kaonic atoms has been the question of the depth of the real optical potential. Deep potentials, in the range of Re $V_{K^-}(\rho_0) \sim -(150\text{–}200)$ MeV are obtained in comprehensive global fits to K^--atom strong-interaction shift and width data by introducing empirical density dependent (DD) effective K^-N amplitudes [5]. In contrast, considerably shallower potentials, in the range of Re $V_{K^-}(\rho_0) \sim -(40\text{–}60)$ MeV, are obtained for zero kinetic-energy kaons when *threshold* chiral scattering

E. Friedman (✉)
Racah Institute of Physics, Hebrew University, Jerusalem, Israel
e-mail: elifried@cc.huji.ac.il

amplitudes are used and self energy (SE) contributions are included in the in-medium corrections [6]. The large difference between the two results has been known as the 'deep *vs.* shallow puzzle' and is of interest not only because of the credibility of the two approaches but also because the depth of the potential may have far-reaching consequences for the possible existence of kaon-nucleon clusters. Very recently [7, 8] this problem was solved by using in-medium sub-threshold K^-N scattering amplitudes which reproduced the empirical characterization of deep potentials [5], namely, that the real potentials are compressed relative to the nuclear charge distribution. These results demonstrated the need to supplement the theory with multi-nucleon absorption terms.

For the very light kaonic atoms new experiments [3, 4] removed the long-standing puzzle of kaonic hydrogen where now the results are fully consistent with the rest of the available data on antikaon-nucleon interactions near threshold. For ^{4}He the new results [9] removed the conflict of more than an order of magnitude between any credible theory and experiment.

The future directions outlined below result from this background.

2 Sub-threshold amplitudes

Introducing empirical density dependent scattering amplitudes into global optical-model fits to data over the whole of the periodic table, deep real potentials were obtained that produced superior fits compared to fixed-amplitude approaches [5]. Another feature of these fits was a 'compression' of the real potential relative to the nuclear charge distribution. This result means that an underlying antikaon-nucleon amplitude must increase with the nuclear density such that it overshadows any finite-range effect that might be present. This is a general result, independent of details. We therefore focus attention on the density dependence of the K^-N scattering amplitude, where its energy dependence near threshold is strongly affected by the proximity of the Λ1405 resonance. Figure 1 shows chirally leading-order Tomozawa-Weinberg K^-N scattering amplitudes marked TW1 solution, in free space and in symmetric nuclear medium for half the nuclear density, where medium corrections with self-energy and without are included, see [7, 8] for details.

The in-medium amplitudes are defined by the Mandelstam parameter $s = (E_{K^-} + E_N)^2 - \left(\mathbf{p}_{K^-}^{cm} + \mathbf{p}_N^{cm}\right)^2$ where the momenta are those implied by a kaon bound in the atom and by a nucleon which is part of the nucleus. Averaging over directions, using the Fermi gas model for the nucleus and taking the local kinetic energy for the kaon, one ends up with

$$\sqrt{s} \approx E_{th} - B_N - \xi_N B_K - 15.1(\rho/\rho_0)^{2/3} + \xi_K(\mathrm{Re}\, V_{\mathrm{opt}} + V_c). \ (\mathrm{MeV}) \qquad (1)$$

in obvious notation, with B_N an average binding energy for a nucleon and $\xi_N = m_N/(m_N + m_K)$, $\xi_K = m_K/(m_N + m_K)$, $E_{th} = m_K + m_N$. We note that the potential is proportional to the amplitude, but the amplitude depends on the potential through the argument $\sqrt{s}$. Therefore a self consistent iterative procedure was applied at every radial point for each nucleus in the data. That transforms the energy dependence into density dependence.

Figure 2 shows an example for the density dependence of the scattering amplitude obtained for the K^-N interaction in Ni. Near the nuclear surface the real part

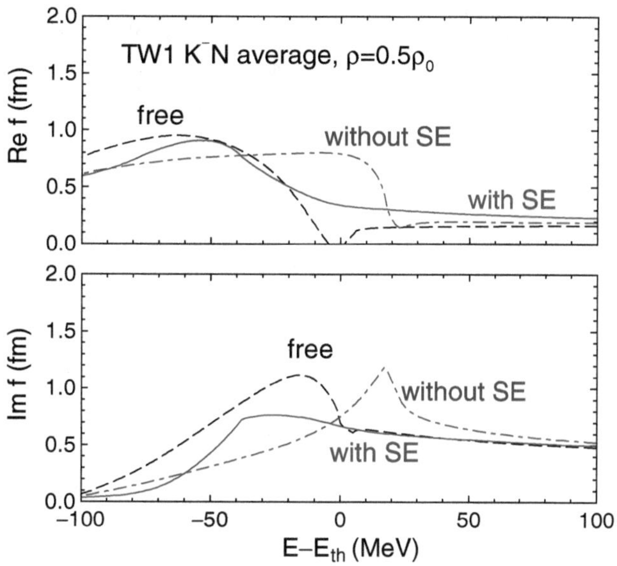

Fig. 1 Examples for average scattering amplitudes for a leading-order Tomozawa-Weinberg formulation. *Dashed curves:* free space, *dot-dashed:* Pauli blocked amplitude (without SE) at $\rho = 0.5\rho_0$, *solid curves:* including meson and baryon self energies (with SE) at $\rho = 0.5\rho_0$

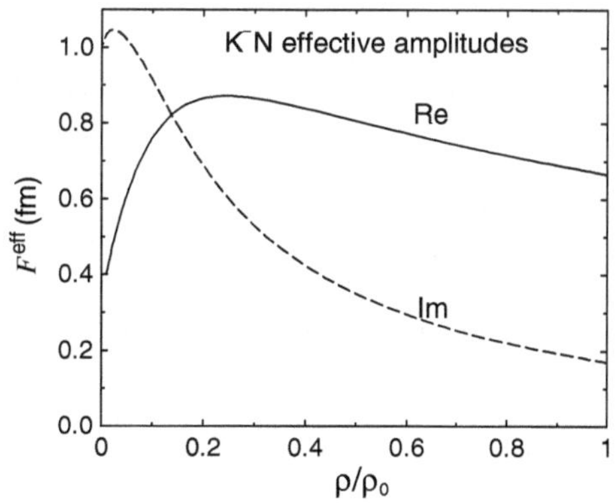

Fig. 2 Effective amplitudes for K^- Ni interaction

increases with density whereas the imaginary part decreases, in line with the compression and inflation of the real and imaginary potentials, respectively, observed in [5]. The small values of the imaginary part at large densities is due to the sharp drop when energies go well below threshold, Fig. 1, towards the $\pi\Sigma$ threshold at $E - E_{th} \simeq -100$ MeV.

3 Kaonic atom potentials

3.1 Medium-weight and heavy nuclei

Comparisons between data and predictions made with optical potentials based on the above model lead to inadequate agreement with experiment, with χ^2 per point of 10. Adding a phenomenological potential with four adjustable parameters (for 65

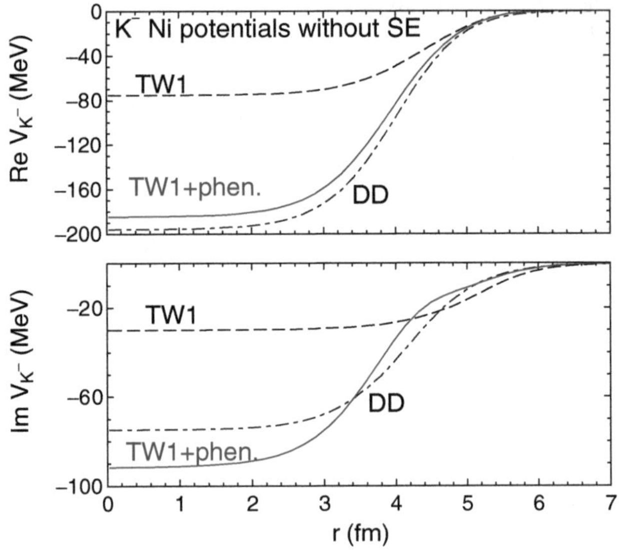

Fig. 3 K^- nuclear potentials for K^- atoms of Ni

data points) leads to χ^2 per point of 2. This phenomenological term is dominated by ρ^2 dependence, suggesting indeed two-nucleon absorption which is not included in the present K^-N amplitudes. A real dispersive term is to be expected.

Figure 3 shows examples for the K^-Ni potential. The real potential based on the TW1 amplitudes [8] is 80 MeV deep compared to 40–50 MeV deep in [6]. The cause for the latter is that [6] used the amplitudes at threshold whereas Fig. 1 shows that at threshold the real part of the SE version is $\approx 50\%$ of typical subthreshold values. For the full nuclear density the corresponding energy is as low as 50 MeV below threshold. Also shown is the best fit potential when a phenomenological term is added, and for reference, also the purely empirical DD potential. It is tempting to reject the phenomenological additional term as being unacceptably large. However, the values at the nuclear center are not the meaningful quantities as kaonic atoms are sensitive mostly to the potential close to the nuclear surface [1, 2]. Considering that strong interaction effects in kaonic atoms are dominated by absorption, we note that the shape of the imaginary potential in the surface region ($\approx$ 4 fm for Ni) is modified by the phenomenological term by $\pm 30\%$, close to estimates of multi-nucleon absorptions obtained from old emulsion and bubble-chamber experiments [10]. Replacing the additional phenomenological term by a multinucleon extension of the present 'microscopic' approach is a high priority line of research.

3.2 Light nuclei

Formation rates of hypernuclear states following the ${}^AZ(K^-_{\rm stop}, \pi^-)^A_\Lambda Z$ reaction have been another probe for studying the interaction of K^- mesons with nuclei at zero kinetic energy, where experimental results have been limited mostly to light p-shell nuclei. The analysis of these reactions is the other extreme compared to global analysis of kaonic atoms, by dealing separately not only with each nuclear species but also by handling explicitly specific kaonic atom states. Recent analysis of

Table 1 Shifts and widths for the $2p$ state in ^{3,4}He

	^{3}He	^{4}He
Shift (eV)	0.3	−0.2
Width (eV)	2.1	1.6

experimental results from FINUDA [11] showed [12] that with the sub-threshold approach the deep K^--nucleus potential is favored. This conclusion is based on *relative* formation rates but the absolute scale of these is not reproduced by the calculations. Understanding the absolute scale is another direction of work in the field of kaonic atom physics.

Finally we turn to kaonic atoms of He. Table 1 summarizes predictions made for ^{3,4}He with a global optical potential based on the TW1 amplitude plus a phenomenological term. Other variations such as the removal of the phenomenological term or the inclusion of a p-wave term change these values by up to 0.3 eV. However, future high precision experiment, well beyond the present capabilities, could bring interesting results.

4 Conclusions

Having established a connection between *deep* real optical potentials and underlying chiral-motivated K^-N scattering amplitudes, the next step will be to implement a 'microscopic' multinucleon absorption terms to supplement the one-nucleon sub-threshold approach to kaonic atom potentials. Another near-future direction could be studies of the absolute scale of formation rates in stopped kaon absorption experiments where the current predictions underestimate experimental results by as much as a factor six. Refinements of global analyses by studies of different levels in a given kaonic atoms could bring interesting results. We note also that the role of a p-wave term in kaonic atoms has not been established [8]. Last but not least, any new high quality experiment will be most welcome.

Acknowledgements I wish to thank Aleš Cieplý, Avraham Gal, Daniel Gazda and Jiří Mareš who contributed much to the emerging sub-threshold approach.

References

1. Batty, C.J., Friedman, E., Gal, A.: Strong interaction physics from hadronic atoms. Phys. Rep. **287**, 385–445 (1997); for reviews
2. Friedman, E., Gal, A.: In-medium nuclear interactions of low-energy hadrons. Phys. Rep. **452**, 89–153 (2007)
3. Iwasaki, M., et al.: Observation of kaonic hydrogen K_α X Rays. Phys. Rev. Lett. **78**, 3067–3069 (1997)
4. Bazzi, M., et al.: A new measurement of kaonic hydrogen X rays. Phys. Lett. B **704**, 113–117 (2011)
5. Friedman, E., Gal, A., Batty, C.J.: Density-dependent K^- nuclear optical potential from kaonic atoms. Nucl. Phys. A **579**, 518–538 (1994)
6. Ramos, A., Oset, E.: Properties of (K)over-bar in the nuclear medium. Nucl. Phys. A **671**, 481–502 (2000)
7. Cieplý, A., et al.: Chirally motivated K^- nuclear potentials. Phys. Lett. B **702**, 402–407 (2011)

8. Cieplý, A., et al.: K^- nuclear potentials from in-medium chirally motivated models. Phys. Rev. C **84**, 045206 (2011)
9. Okada, S., et al.: Precision measurements of the 3d→2p X-ray energy in kaonic ^{4}He. Phys. Lett. B **653**, 387–391 (2007)
10. Vander Velde-Wilquet, C., et al.: Determination of branching fractions for K-mesons absorption at rest in carbon nuclei. Nuovo Cim. **39A**, 538–547 (1977); and references therein
11. Agnello, M., et al.: Hypernuclear spectroscopy with K^- at rest on ^{7}Li, ^{9}Be, ^{13}C and ^{16}O. Phys. Lett. B **698**, 219–225 (2011)
12. Cieplý, A., et al.: Constraints on the threshold K^- nuclear potential from FINUDA $^AZ(K^-_{\rm stop},\pi^-)^A_\Lambda Z$ spectra. Phys. Lett. B **698**, 226–230 (2011)

Hyperfine Interact (2012) 209:133–138
DOI 10.1007/s10751-011-0539-6

Kaonic atoms and $\Lambda(1405)$

Slawomir Wycech

Published online: 9 February 2012

Abstract Studies of $\Lambda(1405)$ in $\Sigma\pi$ decay channels are briefly presented and the related uncertainties indicated. The advantages of measurements in the $\overline{K}$N channel are stressed. Two methods: studies of upper levels in K-mesic atoms and radiative decays from the hydrogen atom to $\Lambda(1405)$ are discussed.

Keywords Nuclear states of kaons · Exotic atoms

1 Introduction

Understanding of $\Lambda(1405)$ is significant for several reasons:

- On the theory side the still unsettled question is the strength of pure quark components that possibly admix to the $\overline{K}$N quasi-bound state [1].
- On the phenomenological level the open question is whether one can generate $\Lambda(1405)$ with a short range attraction in the $\overline{K}$N channel or is there an additional support from a resonating $\Sigma\pi$ channel. The first option has a long history, the attraction according to the early idea of Dalitz is due to vector meson exchanges. The second option is described by $SU(3)$ chiral effective models and presents a crucial test of the applicability of such models [2–4] in the strange sector.
- The interest in nuclear physics concerns energies and widths of $\overline{K}$ nuclear states [5]. These properties are sensitive to the spectrum of $\Lambda(1405)$ in the $\bar{K}$ N channel.

Supported by European Hadron 2, LEANNIS project.

S. Wycech (✉)
National Centre for Nuclear Studies, Hoza 69, 00-681, Warsaw, Poland
e-mail: wycech@fuw.edu.pl

1.1 Profiles of $\Lambda(1405)$ in $\Sigma\,\pi$ channels

The $\Lambda(1405)$ has isospin 0 but may be detected in states of mixed isospin. As a consequence the positions of resonant peaks generated by $\Lambda(1405)$ are different in different reactions. In the simplest version of theory there might be at least six different maxima of cross sections or other measurable quantities. These are indicated below.

The elastic scattering amplitudes in the $\bar{K}N$ system are superimposed by two isospin amplitudes A_0, A_1. These are related to channel amplitudes

$$A(K^- p \rightarrow K^- p) = \frac{1}{2} A_0 + \frac{1}{2} A_1$$

$$A(K^- n \rightarrow K^- n) = A_1$$

where for the simplicity of argument isospin symmetry violations are neglected. One can notice that for a single pole in the A_0 amplitude the modulus of $K^- p$ amplitude displays maximum at energy shifted with respect to the pole energy. This shift is generated by the background isospin 1 amplitude A_1.

The inelastic cross sections are given by the transition amplitudes T

$$T(K^- p \rightarrow \Sigma^+ \pi^-) = \frac{1}{\sqrt{6}} T_0 + \frac{1}{2} T_1$$

$$T(K^- p \rightarrow \Sigma^- \pi^+) = \frac{1}{\sqrt{6}} T_0 - \frac{1}{2} T_1$$

$$T(K^- p \rightarrow \Sigma^0 \pi^0) = \frac{1}{\sqrt{6}} T_0$$

and elastic scattering in decay channels is given by three amplitudes corresponding to isospin 0,1,2,

$$F(\Sigma^+ \pi^- \rightarrow \Sigma^+ \pi^-) = \frac{1}{3} F_0 + F_1 \frac{1}{2} + F_2 \frac{1}{6}$$

$$F(\Sigma^- \pi^+ \rightarrow \Sigma^- \pi^+) = \frac{1}{3} F_0 + F_1 \frac{1}{2} + F_2 \frac{1}{6}$$

$$F(\Sigma^0 \pi^0 \rightarrow \Sigma^0 \pi^0) = \frac{1}{3} F_0 + \frac{2}{3} F_2$$

Unitarity relates amplitudes A,T and F for each isospin and this relation may be checked for amplitudes A and T against experiments performed above the $\bar{K}$N threshold. On the other hand, it is known that multichannel potential and/or K-matrix models are unstable against subthreshold extrapolation to the $\Lambda(1405)$ region. Hence, the importance of experimental checks in the Σ, π channels. This may be done via studies of final state interactions. Very recent experiments allow for a systematization but so far have not been analyzed in terms of models. In a crude way these are presented in Table 1 and discussed below.

The $\Lambda(1405)$ is located close to the $\bar{K}$ N threshold an thus the $\Lambda(1405)$ profile differs from the lorentzian one. Moreover $\Lambda(1405)$ is not an S-matrix eigen-resonance in the multichannel space and may differ in each channel. On top of that the experimental data refers to interfering resonant and background isospin 1,2

Table 1 Maxima in the $\Sigma\pi$ spectra, recent experiments

Channel	$\gamma p \rightarrow K^+ \Sigma \pi$, [6]	$pp \rightarrow p K^+ \Sigma \pi$ [7, 8]
	E_{max}	E_{max}
Σ^+, π^-	1,420	1,400
Σ^0, π^0	1,415	1,390
Σ^-, π^+	1,410	1,400

amplitudes. Assuming a single pole nature of isospin 0 amplitude one expects three different maxima in three different Σ, π charge channels if the $\Lambda(1405)$ formation is initiated in $\bar{K}$ N channel. This apparently happens in $\gamma p \rightarrow K^+ \Sigma \pi$ reactions and the shifted maxima are indicated in Table 1. On the other hand, if the $\Lambda(1405)$ formation is initiated in the Σ, π channel one expects two different maxima. This apparently happens in $pp \rightarrow p\ K^+\ \Sigma\ \pi$ reactions and the shifted maxima are indicated in Table 1.

Qualitatively one obtains consistency of the various positions of the resonant maxima in different charge states with the expectations coming from the amplitudes given above. At this moment a comparison of specific models with the fresh data is not available. Two points are clear however: (1) the differences in the maxima positions depend on poorly known isospin 1 and unchecked isospin 2 amplitudes and (2) the position of maxima observed in final states—and the shape of $\Lambda(1405)$—depend on the formation mechanism. Moreover, the $\Sigma\ \pi$ pair is formed with K^+—another strongly interacting particle. Thus the final state interactions have to be corrected for K, π interactions which resonate at K^* state present in some regions of the available final state phase space.

These uncertainties indicate usefulness of studying $\Lambda(1405)$ in the basic $\bar{K}$N channel.

2 Profile of $\Lambda(1405)$ in $\overline{K}$N channel

To observe the $\Lambda(1405)$ in the main $\overline{K}$N channel one has to bind either meson or nucleon or perhaps both particles with the binding energy in the 0–50 MeV region. Another required condition is that the $\overline{K}$N pair should be quasi-free. Such conditions are met at distant nuclear surfaces and may be materialized in high angular momentum states of K-mesic atoms. Another possibility is the reaction $\overline{K} + N \rightarrow \Lambda(1405) + \gamma$. These two suggestions are discussed below.

- ***The highest accessible levels in hadronic atoms***

Collisions of atomic $\bar{\text{K}}$ mesons with bound nucleons are described by the elastic scattering amplitude

$$a(E) = A_{\bar{K}N}\left(E^{CM} = -E_S - E_A - E_R\right), \tag{1}$$

and the energy in $\bar{\text{K}}$N center of mass E^{CM} is given by a separation energy of the hit nucleon E_S, binding energy of the atomic meson E_A and a recoil energy of the $\bar{K}N$ pair relative to the nucleus E_R. The latter is calculable in terms of nucleon and atomic wave functions. It amounts to about 10 MeV. Thus the region of energies in atomic states which involve predominantly valence nucleons cover the range of −5 MeV (Deuteron) to −37 MeV (Helium). In these states the atomic level shift ΔE

Fig. 1 The *Im* parts of effective $\overline{K}$N length as a function of valence nucleon separation energy

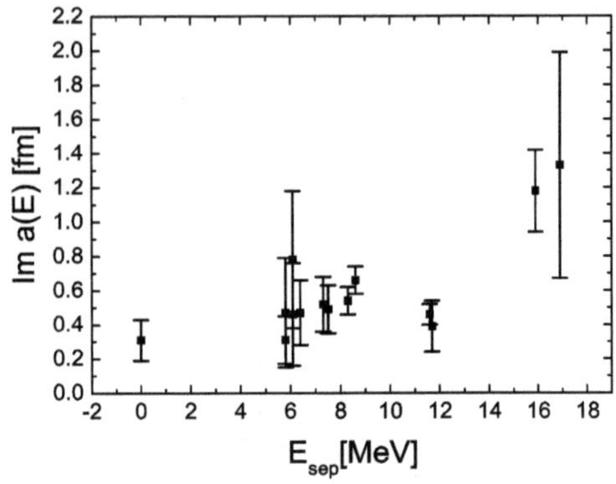

and width Γ are conveniently calculated as a sum of meson multiple scattering terms. One has [9]

$$\Delta E - i\Gamma/2 = < T > \left[1 - \frac{< TDT >}{< T >} + \frac{< TDT >^2}{< T >^2} - \frac{< TDTDT >}{< T >}..\right]^{-1} \tag{2}$$

where

$$T(r, r') = -\frac{2\pi}{\mu} a \cdot \Sigma_i\ \rho(r)_i\ \delta(\mathbf{r} - \mathbf{r}'), \qquad D = -\frac{m}{2\pi|\mathbf{r} - \mathbf{r}'|}. \tag{3}$$

and μ is the reduced K-N and m is the K meson mass, $\rho(r)_i$ is the single nucleon density, the summation extends over all nucleons and the averaging is understood as $< T > = \int\int d\mathbf{r}d\mathbf{r}'\psi(r)^* T(r, r')\psi(r)$ where ψ is the atomic wave function.

Result (2) is equivalent to that generated by the optical potential of strength $V = -\frac{2\pi}{\mu} a \cdot \Sigma_i \rho(r)_i$ provided the convergence in the denominator is fast enough. In the known "upper levels" ≈80% of the complex shift is given just by $< T >$ and in most cases only $< TDT > / < T >$ is necessary to reach a few % precision. The advantage of (2) is clear in light systems as it may avoid unnecessary double collisions on the same nucleon.

Measured upper level widths in H,Cd,Ag,In,Cu,Cl,P,Sn,Al,S,Si,Mg,C,Be, are ordered here with increasing E_S. These data handled with (2) allow to extract the Im $a(E)$ parameters, which are plotted in Fig. 1. Calculations were done with Re $a = -0.5\ fm$ [10] corresponding to the central Re $V_{\rm optical} = \simeq -60$ MeV. (A change to Re $a = -0.7\ fm$ would reduce these numbers by ~5%. Im a increases for the largest available separation energies corresponding to subthreshold energy of about 26 MeV. It could indicate deeply bound Λ(1405), but the argument is weak at this moment, as the experimental errors are large. Precise measurement of upper widths in Be,C,O and complex level shifts in He, De would allow to separate $A_{K^-p}(E)$ and $A_{K^-n}(E)$. Such a program was performed with precise measurements of levels in antiprotonic H^1,H^2,He^3,He^4 and indicate S and P waves quasi-bound states in the N$\overline{N}$ system [11]. The same may be obtained with the K^- mesons.

•• *Atom → Nucleus radiative transitions*

The simplest γ-ray radiative transitions from kaonic hydrogen are given in (4)

$$p\ K^- \rightarrow \gamma\ \Lambda. \tag{4}$$

Table 2 Partial widths of the $2P$ states in K-mesic hydrogen

$E_B(\Lambda(1405))$ [MeV]	Γ_{abs} [meV]	Γ_γ [meV]	$\Gamma_\gamma/\Gamma_{abs}$ (%)
27	$1.5 \cdot 10^{-2}$	3	0.5
12	$3.9 \cdot 10^{-2}$	3	1.3

The first experiment performed in liquid hydrogen [12] gave an upper limit for the rate $R_{\Lambda\gamma} < 4 \cdot 10^{-4}$ per stopped meson. Another experiment in liquid [16] came with $R_{\Lambda\gamma} = (0.86 \pm 0.07^{+}0.10\_0.08) \cdot 10^{-3}$. The physics behind these experiments was based on reaction $p\,K^- \to \Lambda(1405) \to \gamma\,\Lambda$. However, another radiative transition is possible

$$p\,K^- \to \gamma\,\Lambda(1405). \tag{5}$$

which resembles an ordinary atomic radiation. Transition rates are calculable for some states. Thus the radiative transition from an $|2, P>$ atomic state to the state of $\Lambda(1405)$ is found under the assumption that the latter is the $N\overline{K}$ quasi-bound state described by a wave function ψ_Λ. For the electric radiative transition one finds the widths given in Table 2. As K^-p state is a mixture of isospin 0 and isospin 1 states an additional factor of 1/2 was included in Γ_γ of Table 2, as well as the Fried Martin factor.

This γ-transition rate may be compared to the "nuclear" absorption rate Γ_{abs} which is obtained from the $\overline{K}N\Sigma(1385)$ coupling [13]. The latter gives the basic width of this state. The values in Table 2 are shown for two binding energies of $\Lambda(1405)$ characteristic for potential models and chiral effective models. In both cases the wave function of $\Lambda(1405)$ was obtained with the Yamaguchi separable model of inverse range $\kappa = 3.8/fm$ and a strength parameter chosen to fit the binding energy. Once the widths of $|2P>$ states are known the widths of $|nP>$ states are obtained by simple re-scaling of the atomic wave function at nuclear distances. This consists of the barrier factor and normalization $\psi \sim r^L\,N(n, L)$. Normalization factors N scale both widths in the same way and one obtains $\Gamma(n, P) = \Gamma(2, P) \cdot 32/3 \cdot (n-1)(n+1)n^{-5}$.

The atomic cascade of $\overline{K}$-meson involves several types of transitions: Auger electron emissions, X-ray emissions, collision induced transitions, the γ-ray emissions of (4) and the nuclear capture dominated in S waves by the $\Lambda(1405)$ and in P-waves by $\Sigma(1385)$ formation. The cascade populates many atomic states for some time. From the $|nP>$ states the γ-ray transitions offer noticeable branching ratios. Thus to estimate the chance for the γ-ray transitions one needs to know the total probability of nuclear absorption from all P states. It is commonly assumed that most of mesons reach high $|n, S>$ or $|n, P>$ states and are absorbed by the nucleus. In this way the occupation probability of $|n, P>$ states is apparently sizable. Some cascade models [14, 15] estimate these at a few percent level at least in dilute gases. All together one could expect the emission of one γ ray per $10^3 - 10^4$ mesons stopped in gas hydrogen.

Measurements would meet a heavy background due to pionic decays. Old experiments looking for $K^-p \to \gamma + X$ [16, 17] devised ingenious techniques to reduce it which was successful with high energy γ, but in the region of reaction (5) this background was difficult to resolve and removed from the published data (Horvath, private communication). An improvement requires a cleaner beam, tagging of the decay products and theory of the line shape. The last point is briefly discussed below.

The γ line shape may be determined with the realization that the real final state in (5) consists of the photon and the Σ, π pair. The full amplitude for the reaction is a product of three factors: the γ emission vertex, the intermediate propagation of $K^- p$ and the transition matrix $T(K^- p \rightarrow \Sigma, \pi)$. The probability of decay involves a modulus square of the latter summed over channels and the phase space of the final pair. By the unitarity condition this yields the absorptive part of the elastic $K^- p$ scattering amplitude Im $A(K^- p \rightarrow K^- p)$ extrapolated to the $\Lambda(1405)$ region. The other ingredients: contribution from the photon phase space and effects of $K^- p$ propagation in the intermediate states are standard. These factors strongly deform the Lorentzian shape expected for a quasi-bound state. The detailed calculation will be published elsewhere (Wycech, in preparation), here only an approximate formula for the line shape is given. One has $\Gamma_\gamma = \int dk \ S(k)$ where the spectral density is given by

$$S(k) = Im \ A(K^- p \rightarrow K^- p)(E_{\text{atom}} - k) \frac{k \cdot const}{[(k + B_{\text{atom}} + k^2(1/2\mu_{KN} + 1/2M_{KN})]^2}. \tag{6}$$

In this expression, k is the photon energy, μ_{KN} and M_{KM} are the reduced and total masses of the KN system, B_{atom} is the atomic binding and $E_{\text{atom}} = M_K + M_N - B_{\text{atom}}$. The quantity of interest is $Im \ T_{KN,KN}(E_{\text{atom}} - k)$ for energies extending below the KN threshold. The resonant peak is strongly deformed on the low k side but the related factor given by (6) may be calculated with a good accuracy. Experiments would resolve the position of $\Lambda(1405)$ and distinguish the two approaches: the phenomenological one from the chiral one.

References

1. Martin, A.D.: Nucl. Phys. B **179**, 33 (1981)
2. Hyodo, T., Weise, W.: Phys. Rev. C **77**, 035204 (2008)
3. Borasoy, B., Meißner, U.-G., Nißler, R.: Phys. Rev. C **74**, 055201 (2006)
4. Oset, E., Ramos, A., Bennhold, C.: Phys. Lett. B **527**, 99 (2002)
5. Zmeskal, J.: Prog. Part. Nucl. Phys. **61**, 512(2008)
6. Moriya, K., Schumacher, R.: (CLAS) arXiv:0911.2705v1
7. Zychor, I., et al.: (ANKE) Phys. Lett. B **660**, 167 (2008)
8. Siebenson, J.: (HADES) arXiv:1009.0946v1
9. Wycech, S., Green, A.M.: Zeit. Phys. A **344**, 117 (1992)
10. Friedman, E., Gal, A.: Phys. Rep. **452**, 89 (2007)
11. Wycech, S., et al.: Phys. Rev. C **76**, 034316 (2007)
12. Davies, J., et al.: Nucl. Phys. B **160**, 492 (1979)
13. Braun, O., et al.: Nucl. Phys. B **129**, 1 (1979)
14. Raeisi, M., Kalantari, S.Z.: Phys. Rev. A **79**, 012510 (2009)
15. Faifman, M.P.: Frascati Series XVI, vol. 637 (1999)
16. Whitehouse, D.A., et al.: Phys. Rev. Lett. **63**, 1352 (1989)
17. Larsen, K.D., et al.: Phys. Rev. D **47**, 799 (1993)

Hyperfine Interact (2012) 209:139–143
DOI 10.1007/s10751-011-0552-9

Kaonic ^{3}He and ^{4}He X-ray measurements in SIDDHARTA

Tomoichi Ishiwatari · M. Bazzi · G. Beer · C. Berucci · L. Bombelli · A. M. Bragadireanu · M. Cargnelli · A. Clozza · G. Corradi · C. Curceanu (Petrascu) · A. d'Uffizi · C. Fiorini · F. Ghio · B. Girolami · C. Guaraldo · R. S. Hayano · M. Iliescu · M. Iwasaki · P. Kienle · P. Levi Sandri · V. Lucherini · J. Marton · S. Okada · D. Pietreanu · K. Piscicchia · M. Poli Lener · T. Ponta · R. Quaglia · A. Rizzo · A. Romero Vidal · E. Sbardella · A. Scordo · H. Shi · D. L. Sirghi · F. Sirghi · H. Tatsuno · A. Tudorache · V. Tudorache · O. Vazquez Doce · E. Widmann · B. Wünschek · J. Zmeskal

Published online: 13 January 2012

Abstract The SIDDHARTA experiment measured the kaonic ^{3}He and ^{4}He $3d \rightarrow 2p$ X-ray transitions using gaseous targets for the first time. The strong-interaction shift both of the kaonic ^{3}He and ^{4}He $2p$ states was determined with a precision of

T. Ishiwatari (✉) · M. Cargnelli · J. Marton · E. Widmann · B. Wünschek · J. Zmeskal
Stefan-Meyer-Institut für subatomare Physik, Vienna, Austria
e-mail: tomoichi.ishiwatari@assoc.oeaw.ac.at

M. Bazzi · C. Berucci · A. M. Bragadireanu · A. Clozza · G. Corradi · C. Curceanu (Petrascu) · A. d'Uffizi · C. Guaraldo · M. Iliescu · P. Levi Sandri · V. Lucherini · S. Okada · K. Piscicchia · M. Poli Lener · A. Rizzo · A. Romero Vidal · E. Sbardella · A. Scordo · D. L. Sirghi · F. Sirghi · H. Tatsuno · O. Vazquez Doce
INFN, Laboratori Nazionali di Frascati, Frascati (Roma), Italy

G. Beer
Department of Phys. and Astro., University of Victoria, Victoria B.C., Canada

L. Bombelli · C. Fiorini · R. Quaglia
Sez. di Elettronica, Politechno di Milano, Milano, Italy

A. M. Bragadireanu · M. Iliescu · D. Pietreanu · T. Ponta · D. L. Sirghi · F. Sirghi · A. Tudorache · V. Tudorache
IFIN-HH, Magurele, Bucharest, Romania

F. Ghio · B. Girolami
INFN Sez. di Roma I and Inst. Superiore di Sanita, Roma, Italy

R. S. Hayano · H. Shi
Univ. of Tokyo, Tokyo, Japan

M. Iwasaki
RIKEN, The Inst. of Phys. and Chem. Research, Saitama, Japan

P. Kienle
Physik Dep., Tech. Univ. München, Garching, Germany

a few eV. The shift of kaonic ^{4}He is much smaller than the values obtained in the experiments performed in 70's and 80's, while it is consistent with the recent result of the KEK E570 experiment, as well as the theoretical calculated values. Therefore, the problem on kaonic helium (the "kaonic helium puzzle") was definitely solved. The first observation of the kaonic ^{3}He X-rays was also achieved in the SIDDHARTA experiment. The shift both of kaonic ^{3}He and ^{4}He was found to be as small as a few eV.

Keywords Kaonic helium · Silicon drift detectors · X-ray spectroscopy

1 Introduction

There was a serious problem in the strong-interaction shift of the kaonic ^{4}He $2p$ state (the "kaonic helium puzzle"). The shift determined by the X-ray spectroscopy of kaonic ^{4}He was large (about -40 eV) [1], while a value deduced from the kaonic atom data of $Z \geq 3$ was about 0 eV [2, 3]. This discrepancy remained for more than 20 years.

This abnormal $2p$ level shift was focused in recent theoretical studies related to kaonic nuclear states. The largest shift of 10 eV was estimated in either the kaonic ^{3}He or ^{4}He $2p$ states [4]. However, the abnormal shift could not be explained even using the new theories. More precise and accurate measurements both of kaonic ^{3}He and ^{4}He were needed to understand this unclear situation.

The KEK E570 collaboration measured the kaonic ^{4}He $3d \rightarrow 2p$ transition with better detector resolution and a precise calibration method. The shift was determined to be $+2 \pm 2$ (stat.) ± 2 (syst.) eV [5], which disagreed with the previously found average value of -43 ± 8 eV [1]. The new result is consistent with the values according to theory [2, 3], but further confirmations by experiment were needed.

To confirm the result obtained by the E570 experiment, the SIDDHARTA experiment measured the kaonic ^{4}He X-rays for the first time using a gaseous target [6]. In addition, the kaonic ^{3}He X-rays were measured using the same setup, by replacing the target gas [7]. It is note that the world first's observation of the kaonic ^{3}He X-rays was achieved by the SIDDHARTA experiment. This article gives a summary of the kaonic He X-ray measurement in SIDDHARTA.

2 Experiment

The kaonic He X-rays were measured in the framework of the SIDDHARTA experiment in the DAΦNE e^+e^- collider at LNF (Italy). Figure 1 shows the schematic view of the experimental setup. Charged kaon pairs produced by the annihilation of e^+e^- beams were detected by two scintillators mounted in the interaction point (kaon detector). A cryogenic gas target was installed above the interaction point. Large area silicon drift detectors (SDDs) with a total area of 144 cm^2 [8] were installed to surround the target cell. The advantage of the gas target is a negligible effect of Compton scattering in helium. A contribution of the Compton tail was one of the difficulties in the analysis of the previous experiments [1, 5].

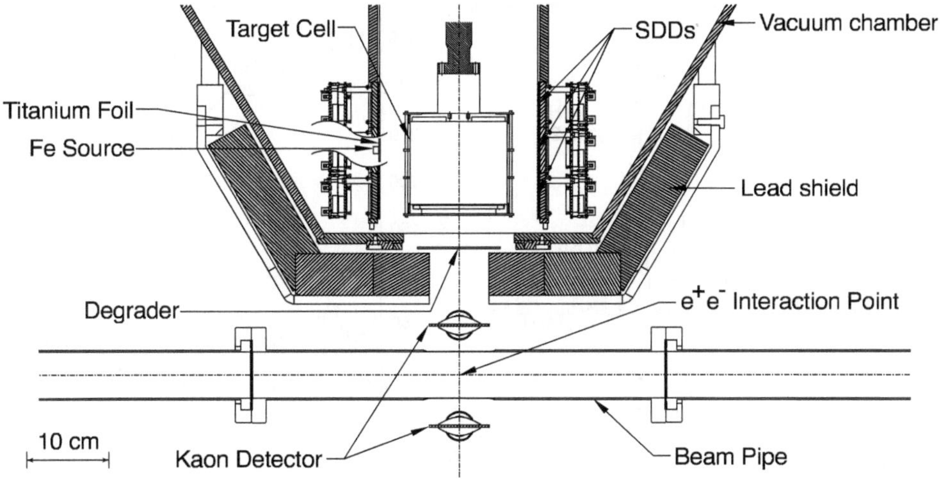

Fig. 1 Schematic view of the SIDDHARTA experimental setup

The data of kaonic helium were taken in the SIDDHARTA beam time, together with the kaonic hydrogen/deuterium X-ray measurement [9, 10]. The measurement of kaonic helium is important not only for the determination of the shift, but also for the performance check of the experimental apparatus. Because the X-ray yields of the kaonic helium $3d \rightarrow 2p$ transition are about 10 times higher than the yield of the kaonic hydrogen $2p \rightarrow 1s$ transition, the kaonic helium X-rays can be observed with excellent statistics in a short period of time.

The energy scale of each SDD was calibrated using the fluorescence X-rays from the Ti and Cu foils. An energy resolution of about 150 eV (FWHM) at 6 keV was obtained. Background events associated with the accelerator were rejected using the timing resolution of the SDDs. The main contribution of the background in DAΦNE is charged particles scattered from the beams. These background events are uncorrelated to the timing of the kaon production. The selection of the events correlated to the timing of K^+K^- pairs reduced background by several order of magnitudes.

Figure 2a shows the X-ray energy spectrum of kaonic ^{4}He [6]. The X-ray peak at 6.4 keV corresponds to the kaonic ^{4}He $3d \rightarrow 2p$ transition. Together with the kaonic helium X-rays, the Mn Kα (5.9 keV) and Kβ (6.4 keV) lines were observed because of the installation of an ^{55}Fe source intended for calibration and stability check of the SDDs [6, 8]. The shift of kaonic ^{4}He was determined to be

$$\Delta E_{2p}^{^4\mathrm{He}} = 0 \pm 6\,(\mathrm{stat.}) \pm 2\,(\mathrm{syst.})\ \mathrm{eV}. \quad (1)$$

After the successful measurement of kaonic ^{4}He with the ^{55}Fe calibration source, kaonic ^{4}He and ^{3}He X-rays were measured without the source. Figure 2b shows the X-ray energy spectrum of kaonic ^{3}He [7]. The X-ray peak at 6.2 keV corresponds to the kaonic ^{3}He $3d \rightarrow 2p$ transition. In addition to the kaonic helium X-rays, other kaonic atom X-ray peaks (kaonic carbon, oxygen, and nitrogen) were observed, which were originated from the material of the target window (Kapton Polyimide).

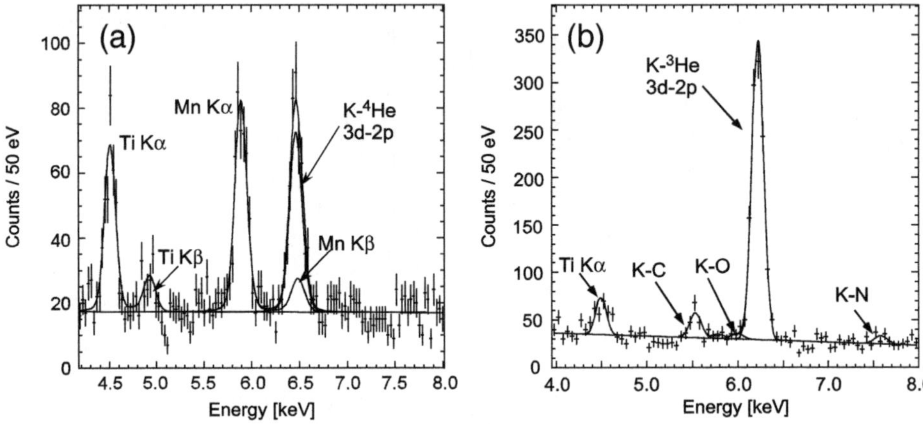

Fig. 2 X-ray energy spectra of the $3d \rightarrow 2p$ transition of kaonic ^{4}He (**a**) and ^{3}He (**b**). The energy of these transitions were determined within a precision of a few eV

The shift of kaonic ^{3}He was determined to be

$$\Delta E_{2p}^{^3\mathrm{He}} = -2 \pm 2\ (\mathrm{stat.}) \pm 4\ (\mathrm{syst.})\ \mathrm{eV}. \tag{2}$$

By replacing the target gas, the kaonic ^{4}He X-rays were remeasured. The fit result gave a shift of

$$\Delta E_{2p}^{^4\mathrm{He}} = +5 \pm 3\ (\mathrm{stat.}) \pm 4\ (\mathrm{syst.})\ \mathrm{eV}. \tag{3}$$

3 Conclusions

The SIDDHARTA experiment measured the strong-interaction shift both of the kaonic ^{3}He and ^{4}He $2p$ levels with a precision of a few eV. They were measured using the gaseous targets for the first time, and the world first's observation of kaonic ^{3}He X-rays was achieved.

The determined values are summarized in Fig. 3, together with the results of the previous experiments. A large shift of the order of −40 eV determined by the experiments performed in 70's and 80's was not observed neither in kaonic ^{3}He nor kaonic ^{4}He. Both shifts of kaonic ^{3}He and ^{4}He are consistent with 0 eV within the errors. These results agree with theoretical values determined from other kaonic atoms with $Z \geq 3$ using optical model approaches [2, 3]. However, the theoretical prediction of a possible shift of more than 1 eV [4] cannot be ruled out within our precision.

A small isotope difference between kaonic ^{3}He and kaonic ^{4}He might exist. A much precise measurement, such as the J-PARC E17 experiment [11], will help for better understanding of the kaon-helium interaction.

The determination of the strong-interaction width of the kaonic helium $2p$ state is also important for the study of the kaon-helium interaction. However, the width

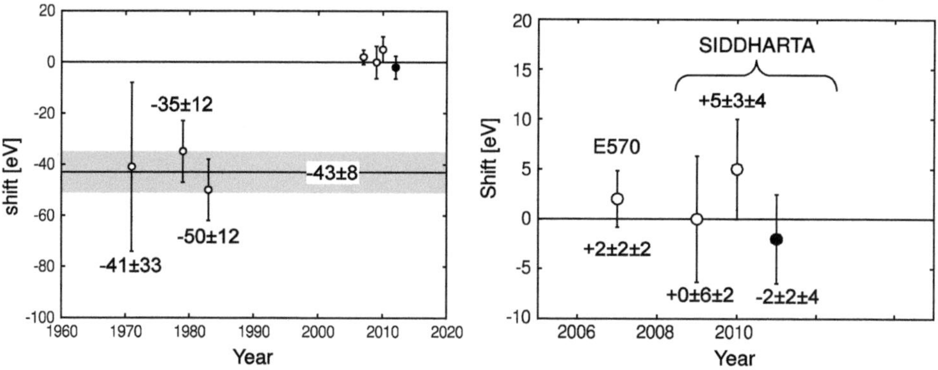

Fig. 3 Comparison of experimental results. *Open circle:* K-^{4}He $2p$ state, *Closed circle:* K-^{3}He $2p$ state. A negative (positive) value of the shift shows a repulsive (attractive) shift

is much smaller than the detector resolution. The upper limit of the $2p$ level width both of kaonic ^{3}He and ^{4}He will be reported in future.

Acknowledgements We thank C. Capoccia, B. Dulach, and D. Tagnani from LNF-INFN; and H. Schneider, L. Stohwasser, and D. Stückler from Stefan-Meyer-Institut, for their fundamental contribution in designing and building the SIDDHARTA setup. We thank as well the DAΦNE staff for the excellent working conditions and permanent support. Part of this work was supported by HadronPhysics I3 FP6 European Community program, Contract No. RII3-CT-2004-506078; the European Community Research Infrastructure Integrating Activity "Study of Strongly Interacting Matter" (HadronPhysics2, Grant Agreement No. 227431) under the Seventh Framework Programme of EU; Austrian Federal Ministry of Science and Research BMBWK 650962/0001 VI/2/2009; Romanian National Authority for Scientific Research, Contract No. 2-CeX 06-11-11/2006; Grant-in-Aid for Specially Promoted Research (20002003), MEXT, Japan; and the Austrian Science Fund (FWF): [P20651-N20].

References

1. Baird, S., et al.: Nucl. Phys. A **392**, 297 (1983)
2. Batty, C.J.: Nucl. Phys. A **508**, 89c (1990)
3. Friedman, E.: Hyperfine Interact (2011). doi:10.1007/s10751-011-0515-1, arXiv:1111.7194v1 [nucl-th]
4. Akaishi, Y.: In: Proc. Inter. Conf. on Exotic Atoms (EXA05), Austrian Academy of Sciences Press, Vienna, p. 45 (2005). doi:10.1553/exa05s45
5. Okada, S., et al.: Phys. Lett. B **653**, 387 (2007)
6. SIDDHARTA Collaboration: Phys. Lett. B **681**, 310 (2009)
7. SIDDHARTA Collaboration: Phys. Lett. B **697**, 199 (2011)
8. Bazzi, M., et al.: Nucl. Instrum. Methods A **628**, 264 (2011)
9. SIDDHARTA Collaboration: Phys. Lett. B **704**, 113 (2011)
10. Cargnelli, M., et al.: Hyperfine Interact (2011). doi:10.1007/s10751-011-0541-z
11. Hayano, R.S., et al.: Proposal of J-PARC 50-GeV PS. Precision spectroscopy of Kaonic Helium 3 $3d \to 2p$ X-rays (2006)

Hyperfine Interact (2012) 209:145–149
DOI 10.1007/s10751-011-0537-8

Calculation of cascade processes rates and simulation of the transitions in Kaonic ^{4}He atom

Seyed Zafarollah Kalantari · Shahin Sanaye Hajari · Mohsen Dayyani Kelisani

Published online: 29 December 2011

Abstract Interactions of K^- in a ^{4}He target and the cascade processes rates in K^4He atoms are investigated by a full-quantum mechanical approach. Then the cascade processes and x-ray yields of kaonic helium atoms in liquid helium are simulated by Monte-Carlo method. For this purpose the rates of scattering, capturing and internal Auger effect are calculated, after that the initial distribution of n, l states in $K\alpha^+$ ion are determined, then the cascade processes rates in $K\alpha^+$ ion are calculated by quantum mechanical methods. The relative x-ray L series are compared with the experimental results.

Keywords Kaonic atoms · Cascade processes of K^{-4}He atoms · Capturing of K^- · Monte – Carlo simulation

1 Introduction

When a K^- enters in a target, initially stopping of kaon and kaonic atom formation in a highly excited state occur. It is followed by a complicated interplay of competitive collisions and radiative deexcitation processes. Radiative transitions produce x-ray. Strong interactions between the nucleus and kaon lead to a shift and finite absorption width of the low lying levels of kaonic atoms, which are determined by x-ray measurement [1, 2]. The absorption during the cascade, also affect the x-ray yields.

The cascade dynamics for K^4He atoms are not fully understood yet. Fortunately, precise experiments have been carried out to detect x-ray yields of K^4He atoms recently [1, 2].

In this paper, first we have calculated the interaction rates of K^- in a ^{4}He target. Internal Auger effect leads to produce a $K\alpha^+$ ion which starts cascade processes until either K^- is absorbed by the nucleus or decays. $K\alpha^+$ deexcites to lower states

S. Z. Kalantari (✉) · Sh. S. Hajari · M. D. Kelisani
Department of Physics, Isfahan University of Technology, Isfahan 8415683111, Iran
e-mail: zafar@cc.iut.ac.ir

by competitive cascade processes: Stark mixing, external Auger effect, Coulomb deexcitation and radiative transitions [3–6]. We have also calculated the cascade processes rates in the $K\alpha^+$ ion.

For this purpose we have used perturbation theory to calculate the rates of K^- scattering, $K\alpha e$ atom formation, internal Auger effect and the cascade processes. Actually, our calculations are done without any semi classical approximations or any free parameters. For the full quantum mechanical calculations, detailed analytical and numerical calculations should be done for many transitions.

The initial probability distribution for n, l states in $K\alpha e$ atom and $K\alpha^+$ ion are determined. Finally the cascade processes and x-ray yields of kaonic helium atoms in liquid helium targets are obtained by Monte-Carlo method and compared with the existing measurements.

2 Quantum mechanical calculations of the processes rates

After decreasing the kinetic energy of the entering K^- due to scattering, deexcitation and ionization of helium atoms, the kaon is captured in a highly excited ^{4}He atomic orbit, replacing one of the electrons, which is ejected from the atom and a $K\alpha e$ atom is produced. Then the $K\alpha e$ atom is decexcited by internal Auger effect and the other electron is ejected. After that we have a $K\alpha^+$ ion which starts cascade processes until either the K^- is absorbed by the nucleus or decays.

In this paper all the interaction rates of a K^- with a ^{4}He atom and the cascade processes after kaonic atom formation, are calculated quantum mechanically. For example the time dependent perturbation theory for calculation of the K^- capture rate by the ^{4}He atom is presented:

During collision of kaon on helium atom, the Hamiltonian is given:

$$H = H_{He} + H_{k\alpha} + \frac{1}{2|\vec{r}_{k\alpha} - \vec{r}_1|} + \frac{1}{2|\vec{r}_{k\alpha} - \vec{r}_2|} \tag{1}$$

where H_{He} and $H_{k\alpha}$ are Hamiltonians of helium atom and $k\alpha$, respectively and $\vec{r}_{k\alpha}$, $\vec{r}_1$ and $\vec{r}_2$ are displacement vector of kaon and two electrons relative to α particle, respectively. We consider, $\frac{1}{2|\vec{r}_{k\alpha}-\vec{r}_1|} + \frac{1}{2|\vec{r}_{k\alpha}-\vec{r}_2|}$, as the perturbation potential.

The Hamiltonian of the $K\alpha e$ atom after capturing is:

$$H_{k\alpha e} = H_{k\alpha} + H_{e\alpha} + \frac{\lambda}{2|\vec{r}_{k\alpha} - \vec{r}|} \tag{2}$$

Time independent perturbation theory is use to calculate the eigenket of the $K\alpha e$ atom, where λ is a dimensionless perturbation parameter and eventually, we may set $\lambda \rightarrow 1$.

Finally, we have used time dependent perturbation theory and Born approximation to determine the amplitude of the transition probabilities. For this purpose many detailed analytical and numerical calculations have been done. The capturing rate of K^- by ^{4}He atom is given by:

$$\Gamma_C^{n\ell m} = \int N_{He} V \left|V_{fi}\right|^2 \frac{2\pi}{k} \delta\left(k - k'\right) d^3k, \tag{3}$$

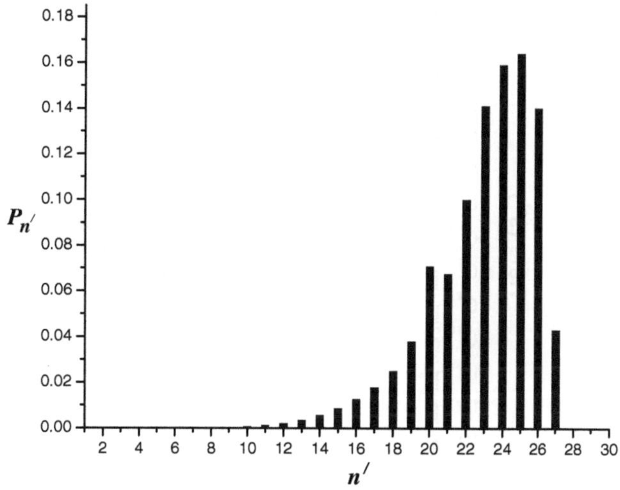

Fig. 1 Initial distribution of the orbital n state in $K\alpha^+$ ion before cascade processes

where $\vec{k}$ is the momentum of the ejected electron relative to $K\alpha e$ atom, whose allowed value $\left(\vec{k}'\right)$ is determined by the conservation of energy, N_{He} is the atomic density and V is the volume of the target and V_{fi} is matrix elements of the potential. Now we can calculate the probability distribution (P_n) for the n state, in which K^- is captured and produces the $K\alpha e$ atom.

By the same procedure for our full quantum mechanical approach and Fermi golden rule, the amplitudes of the internal Auger transition probabilities are also calculated. Internal Auger effect leads to a new distribution of n', ℓ' states for $K\alpha^+$ ion rapidly. Figure 1 shows the initial distribution of the orbital n' state in $K\alpha^+$ ion before cascade processes.

3 Monte-Carlo simulation of the cascade processes of $K\alpha^+$ ions

By the same procedure for the mentioned full quantum mechanical approach to calculate the capture rate of K^- by the ^4He atom, and time dependent perturbation theory, the rates of the Coulomb deexcitation, Stark mixing and external Auger effect are calculated for the $K\alpha^+$ ion. The radiative transition rate is calculated same as hydrogen atom. The calculated cascade processes rates in a liquid density of helium target are compared in Fig. 2.

Now we can simulate the cascade processes in $K\alpha^+$ ion. For this purpose, the Monte-Carlo method is used. Simulation is started from the determination of the n state of the produced $K\alpha^+$ ion (Fig. 1). By our simulation many quantities such as the number of any kind of events and the x-ray yields can be calculated. Figure 3 shows that our results for the relative x-ray L series are comparable with the measurement [1]. This is an approximated comparison, because we have considered the efficiency of the detectors approximately as a same value in the three transitions. It should be noted that in contrast reference [6] we do not use any tuned parameter in our calculations.

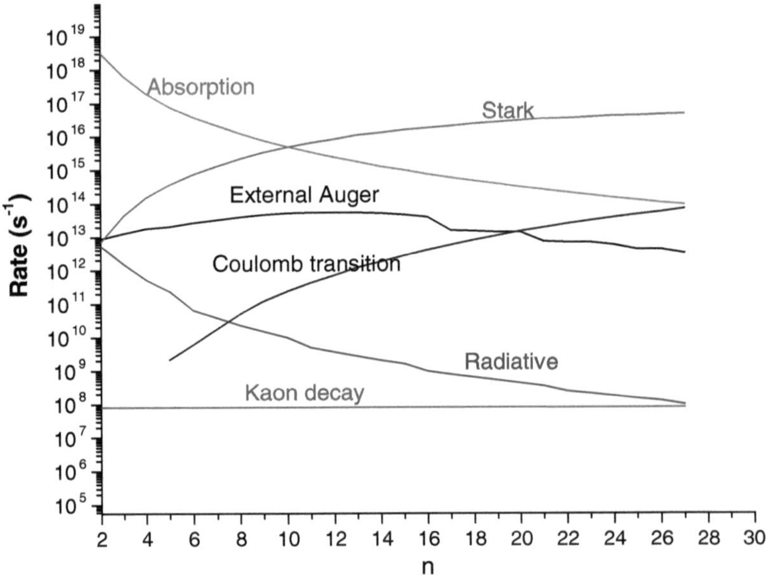

Fig. 2 Comparison of the cascade processes rates in $K\alpha^+$ ion in a liquid helium target

Fig. 3 Comparison of the relative L series x-ray yields with measurement [1] and other work [6]

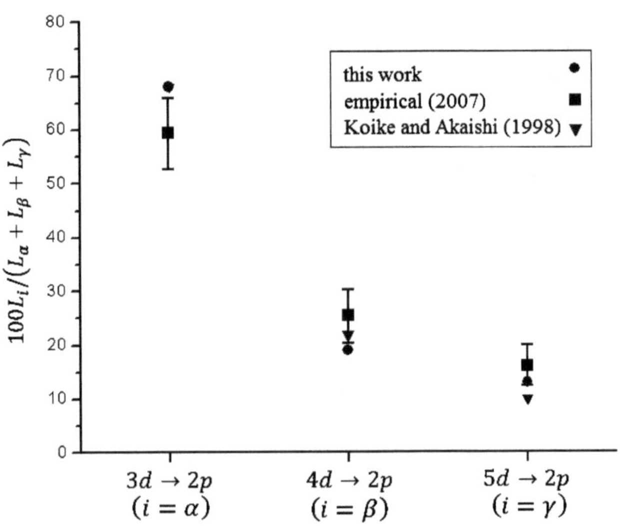

4 Conclusions

In this paper we have presented a full quantum mechanical approach to calculate the rates of kaonic atom formation and internal Auger effect in ^{4}He target to determine the distribution of the orbital n levels before cascade processes in $K\alpha^+$ ion. We have also calculated the cascade process rates by the same full quantum mechanical calculations which will be presented in the forthcoming paper. This results have been used to simulate the cascade dynamics of $K\alpha^+$ ion by the Monte-Carlo method and calculate the x-ray yields. Some of our results are presented here. Figure 1 shows that the distribution of the n states is important to simulate the cascade processes and using the semi-classical result ($n = 25$) is not a good approximation. Figure 2 shows the Stark mixing rate is higher than the rates of the other cascade processes and it leads kaons to the lower l states where absorption of kaon is high. This effect is very important to affect the x-ray yields. It shows also the Coulomb deexcitation is not very important to cascade dynamics. Finally the cascade rates are higher than the rate of kaon decay. It leads to more x-ray yields in ^{4}He target with respect to a hydrogen or deuterium target.

References

1. Okada, S., et al.: Phys. Lett. B. **653**, 387 (2007)
2. Bazzi, M., et al.: Phys. Lett. B. **681**, 310 (2009)
3. Raeisi, G.M., Kalantari, S.Z.: Phys. Rev. A **79**, 012510 (2009)
4. Kalantari, S.Z., Raeisi, G.M.: Phys. Rev. C **81**, 014608 (2010)
5. Raeisi, G.M., Kalantari, S.Z.: Phys. Rev. A **82**, 042501 (2010)
6. Koike, T., Akaishi, Y.: Nucl. Phys. A **639**, 521c (1998)

calculate the [illegible] that the [illegible] and [illegible] [illegible]

References

[illegible]

Hyperfine Interact (2012) 210:1–6
DOI 10.1007/s10751-011-0518-y

g-2 of the Muon

After 10 years still a puzzle for the now consistent theory – The Brookhaven experiment moves to Fermilab

Klaus P. Jungmann

Published online: 21 December 2011

Abstract The experimental value a_μ^{exp} for the muon magnetic anomaly measured at the Brookhaven National Laboratory (BNL), Upton, USA, and the latest theoretical value a_μ^{theo} based on a number of calculations and auxiliary experiments differ today by 3.3 standard deviations. Discrepancies between different independent approaches towards the theoretical value could recently be removed and had yielded a consistent value for a_μ^{theo}. At the Fermi National Laboratory (Fermilab), Batavia, USA, a new experiment has been approved which aims to improve the present experimental uncertainty by a factor of about five. At this level the muon magnetic anomaly is superior in sensitivity to, e.g., LHC concerning tests of several speculative models beyond standard theory. The new experiment relies in the essential parts on concepts proven at BNL such as a muon storage ring at 1.45 T field to store muons at 3.1 GeV/c momentum and field magnetometry based on NMR in water. At Fermilab predominantly a significantly higher number of muons can be exploited.

Keywords Muon magnetic anomaly · Standard model test · New physics search

1 Introduction

The magnetic anomaly of leptons is the relative deviation of the leptons magnetic moment from the Dirac value two. It can be very accurately calculated [1–3]. The by far dominating contribution arises from electromagnetic interactions and can be calculated within the framework of Quantum Electrodynamics (QED). The main differences between the leptons arise from their mass differences. The sensitivity of

On behalf of the muon g-2 collaboration E989 at Fermilab.

K. P. Jungmann (✉)
Kernfysisch Versneller Instituut, University of Groningen, Zernikelaan 25, 9747 AA Groningen, The Netherlands
e-mail: jungmann@kvi.nl

the magnetic anomaly to other than electromagnetic interactions increases with the square of the mass of the lepton. Whereas for the electron other interactions are no reason of concern at the present level of accuracy in theory and experiment, the muon, on the contrary, is very sensitive to other interactions at this level. It is even sensitive to potential new, yet unknown interactions, which could show up in vacuum polarization loops through new particles characteristic for these new interactions. Therefore the muon magnetic anomaly is a calibration point for our theoretical understanding of the fundamental interactions. Precise measurements can provide limits on potential new interactions or hints to new physics [4–7].

The theoretical value a_μ^{theo} for the muon magnetic anomaly is composed of a value a_μ^{SM} that can be calculated within the present Standard Theory and a potential contribution a_μ^{NP} from new interactions outside of the Standard Model in particle physics:

$$a_\mu^{\text{theo}} = a_\mu^{\text{SM}} + a_\mu^{\text{NP}} \qquad \text{, where} \tag{1}$$

$$a_\mu^{\text{SM}} = a_\mu^{\text{QED}} + a_\mu^{\text{weak}} + a_\mu^{\text{strong}}. \tag{2}$$

a_μ^{SM} consists of three parts which can presently be determined to sufficient accuracy within the Standard Model. They arise from electromagnetic interactions a_μ^{QED}, from weak interactions a_μ^{weak} and strong interactions a_μ^{strong}.

A manifest deviation of the experimental value a_μ^{exp} from the Standard Theory value a_μ^{SM} would indicate the presence of physics beyond the Standard Model, which is searched for with various approaches in high energy physics, such as, e.g., experiments at LHC or in dedicated precision experiments at lower energies, such as searches for permanent electric dipole moments [8], or violations of fundamental symmetries [9–12].

The present Standard Model is an excellent description of all known and confirmed physical processes, however, it lacks in various aspects a deeper explanation of physical phenomena beyond their successful and bare description. A number of speculative theories were invented in order to provide deeper explanations for features in nature not fully explained in the Standard Model like, e.g., the mass hierarchy of fundamental fermions or the number of particle generations. Those speculative theories which in themselves are fully consistent, however, share the lack of experimental verification or even any realistic evidence, yet. We have as examples of such models Supersymmetry, LeftRight Symmetry, Technicolor, Universal extra Dimensions, Littlest higgs with T-parity, two Higgs doublets and shadow Higgs, which all could give rise to a small contribution $\Delta a_\mu^{\text{new}}$ to the muon magnetic anomaly a_μ. Larger values of $\Delta a_\mu^{\text{new}}$ could arise from, e.g., Randall Sundrum models and Models with additional light bosons, which could affect electromagnetic interactions and which are difficult to study at LHC (see e.g. [1]).

2 Situation of theory

The hadronic contributions to a_μ^{theo} have been carefully investigated by different theoretical groups using input from a number of experiments (see for details [1] and

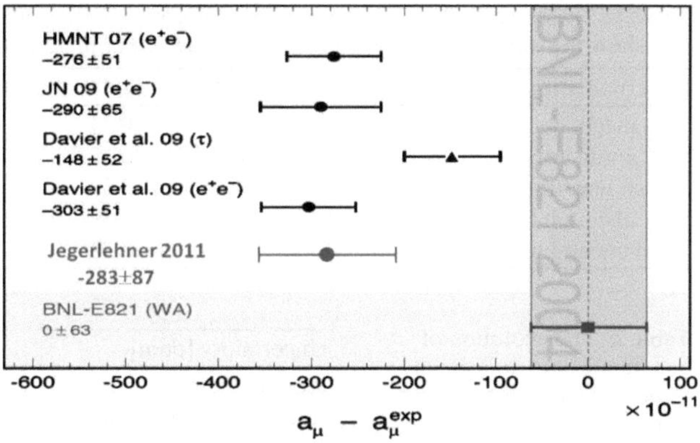

Fig. 1 The calculations of hadronic corrections to a_μ agree now within their uncertainties [13]. The value of Davier et al. 09 (τ) does not yet include all neccessary terms

references therein). The most severe problem which arose in the past decade was the fact that the contributions from strong interactions to a_μ^{theo} were calculated and gave two different results, depending on the chosen route and experimental input. Calculations exploiting electron and positron annihilation into hadrons produced different results from those which were using hadronic τ-decays. It was a major step forward when it was shown by Jegerlehner that terms had been omitted in evaluations involving τ-decays (see Fig. 1). These terms relate to isospin breaking when calculating the hadronic contributions using experimental data from hadronic τ-decays in the region $\sqrt{s}$ around 1 GeV. At this point both values agree satisfactorily and as a consequence there exists one single reliable theoretical value [13]

$$a_\mu^{\text{theo}} = 11\,659\,179.7(6.0) \cdot 10^{-10}. \tag{3}$$

3 Muon magnetic anomaly at Brookhaven National Laboratory

The experiment at the Brookhaven National Laboratory (BNL) [4–7] employed a superferric 7.112 m diameter magnetic storage ring at 1.45 T field to store muons at 3.1 GeV/c momentum, which were injected with longitudinal polarization into the storage volume. The experiment has measured the muon magnetic anomalies for both possible signs of charge. Electrons (positrons) from the decay $\mu^- \rightarrow e^- + \nu_\mu + \overline{\nu_e}$ ($\mu^+ \rightarrow e^+ + \overline{\nu_\mu} + \nu_e$) were detected in calorimeters consisting of scintillating fibers embedded in lead as a function of time. The experiment determined the muon spin precession frequency ω_a in a homogeneous magnetic field measured and mapped through the NMR frequency of protons in water ω_p [14]. Together with the ratio of muon magnetic moment and proton magnetic moment (as determined from muonium spectroscopy [15]), this provided for a determination of the respective magnetic anomalies. They are

$$a_\mu^+ = 11\,659\,204(7)(5) \cdot 10^{-10} \;\; (0.7\text{ppm}) \quad \text{and} \tag{4}$$

$$a_\mu^- = 11\,659\,214(8)(3) \cdot 10^{-10} \;\; (0.7\text{ppm}), \tag{5}$$

Table 1 The error budget was continuously improved in the course of experiment E821 at BNL. We also show the estimated uncertainties for the new experiment P989 at Fermilab

Uncertainty [ppm]	1998	1999	2000	2001	E821 final	P989 goal
Syst. magn. field ω_p	0.5	0.4	0.24	0.17		0.07
Syst. anom. precession ω_a	0.8	0.3	0.31	0.21		0.07
Statist. uncertainty	4.9	1.3	0.62	0.66	0.46	0.10
Syst. uncertainty	0.9	0.5	0.39	0.28	0.28	0.10
Total uncertainty	5.0	1.3	0.73	0.72	0.54	0.14

Table 2 The evolution of systematic uncertainties of ω_a. Expected improvements due to Fermilab beam structure and improved detectors and electronics are indicated by $^{(*)}$

Uncertainty [ppm]	1999	2000	2001	P989 goal
Pile-up	0.13	0.13	0.08	$0.04^{(*)}$
Accelerator background	0.10	0.10	0.015	
Lost muons	0.10	0.10	0.09	$0.02^{(*)}$
Timing shifts	0.10	0.02	0.02	
E-field, pitch	0.08	0.03	0.06	0.03
Fitting, binning	0.07	0.06	0.06	
Coherent betatron oscillations	0.05	0.21	0.07	0.04
Beam debunching	0.04	0.04	0.04	
Gain change	0.02	0.13	0.13	$0.02^{(*)}$
Total	0.3	0.31	0.21	0.07

and they appear to agree very well for both signs of charge. Here the first uncertainty is statistical and the second systematic. With the assumption of CPT being a good symmetry we can combine them to receive

$$a_\mu = 11\,659\,208(6) \cdot 10^{-10}(0.5\text{ppm}). \tag{6}$$

This experimental value differs from the latest theory value by 3.3 standard deviations. The uncertainties in the experiment (see also Table 1) are mostly statistical. Therefore, a new experiment will mostly need to record significantly more muon decays.

In addition to the magnetic anomaly, the experiment has provided a new limit on the muon electric dipole moment at $d_e < 1.8 \cdot 10^{-19} e\,\text{cm}$ (95% C.L.) [16]. Further, bounds on potential CPT and Lorentz invariance violating terms [17] in a Standard Model extension [18, 19], which surpass previous bounds for muons [20], could also be extracted.

4 Muon magnetic anomaly at Fermilab

The muon g-2 experiment will now move from BNL to Fermilab [21]. The new experiment aims for a 5-fold improvement over the BNL result. The ring magnet, the central device, is presently being dismantled at BNL and it will be reinstalled in a new building at FNAL. In particular the some 15 m diameter special manufactured magnet coils need to be shipped from the Long Island Sound to Lake Michigan and airlifted at both ends of the journey between the nearest port and the respective national laboratories.

In the new experiment the positron(electron) detectors and the electronics will take advantage of new technology which became available since the BNL

Table 3 The evolution of systematic uncertainties of ω_p

Uncertainty [ppm]	1999	2000	2001	P989 goal
Absolute calibration	0.05	0.05	0.05	0.05
Calibration of trolley	0.20	0.15	0.09	0.06
Trolley field measurement	0.10	0.10	0.05	0.02
Interpolation with fixed probes	0.15	0.10	0.07	0.06
Inflector fringe field	0.20			
Muon distribution	0.12	0.03	0.03	0.02
Other	0.15	0.10	0.10	0.05
Total	0.4	0.24	0.17	0.11

experiment. A major advantage over the BNL experiment will be the new beamline providing a much cleaner muon beam to the experiment due to its significantly larger length. This essentially reduces hadronic beam contamination and will enable starting the measurement cycles much earlier than at BNL, where a delayed start of positron recording was needed to circumvent nonlinearities in the detectors which were gated to avoid the flashes originating from the hadronic beam contamination. The expected improvements on the systematic errors on the anomaly frequency ω_a are given in Table 2.

The magnetic field measurement system has been designed for BNL with sufficient base accuracy to stand also the challenges of the new experiment. This narrow band pulsed NMR system [14] only needs some modifications in the operating procedures such as more frequent measurements and some fixes of broken parts to achieve the prospected accuracy (see Table 3). The basic concept will be maintained and all the crucial parts of the equipment will be refurbished. Additional measures will be taken for cross checking and calibration. As an example, some of the 360 fixed probes distributed around the ring will be relocated to strategically better suited locations. A second absolute calibration method is being considered in which nuclear magnetic resonance in optically pumped ^{3}He gas is used.

The collaboration and the laboratory are confident, that an improved value for a_μ can be available available within about half a decade. There is a robust potential [22] to either substantiate or disfavour the present 3.3 standard deviation difference between the Standard Model and the experimental value.

Acknowledgements The author is indebted to B.L. Roberts, Boston University, co-spokesman of the new muon g-2 experiment at Fermilab, for providing actual and essential slides for the presentation of this paper. We thank C.J.G. Onderwater for reading the manuscript. The organizers of EXA2011 provided for the participants a stimulating atmosphere and we appreciate the hospitality which we received during the conference. This work was impart supported by the Dutch FOM Programmes 48 and 114 (TriμP).

References

1. Roberts, B.L., Marciano, W.J. (eds.): Lepton Dipole Moments. World Scientific, Singapore (2010)
2. Czarnecki, A., Marciano, W.J.: Electromagnetic Dipole Moments and New Physics. In: Roberts, B.L., Marciano, W.J. (eds.) Lepton Dipole Moments, pp. 119–156. World Scientific, Singapore (2010)

3. Miller, J.P., Roberts, B.L., Jungmann, K.: Measurement of the Muon (g-2) Value. In: Roberts, B.L., Marciano, W.J. (eds.) Lepton Dipole Moments, pp. 333–392. World Scientific, Singapore (2010)
4. Bennett G.W., et al.: Final report of the E821 muon anomalous magnetic moment measurement at BNL. Phys. Rev. D **73**, 072003 (2006)
5. Bennett, G.W., et al.: Measurement of the negative muon anomalous magnetic moment to 0.7 ppm. Phys. Rev. Lett. **92**, 161802 (2004)
6. Bennett, G.W., et al.: Measurement of the positive muon anomalous magnetic moment to 0.7 ppm. Phys. Rev. Lett. **89**, 101804 (2002)
7. Brown, H.N., et al.: Precise measurement of the positive muon anomalous magnetic moment. Phys. Rev. Lett. **86**, 2227 (2001)
8. Onderwater, C.J.G.: Search for electric dipole moments at storage rings. EXA2011 Conference (2011, to be published in Hyperfine Interact.)
9. Willmann, L.: The TRIμP facility. EXA2011 Conference (2011, to be published in Hyperfine Interact.)
10. Abele, H.: New experiments in neutron physics. EXA2011 Conference (2011, to be published in Hyperfine Interact.)
11. Lehnert, R.: CPT-symmetry studies with Antihydrogen. EXA2011 Conference (2011, to be published in Hyperfine Interact.)
12. Müller, S.: Lorentz invariance on trial in the weak decay of rubidium atoms. EXA2011 Conference (2011, to be published in Hyperfine Interact.)
13. Jegerlehner, F.: ρ^0-γ mixing in the neutral channel pion form factor $F_\pi^{(e)}(s)$ and its role in comparing e^+e^- with τ spectral functions. Eur. Phys. J. C **71**, 1632 (2011)
14. Prigl, R., et al.: A high precision magnetometer based on pulsed NMR. Nucl. Instrum. Methods A **374**, 118 (1996)
15. Liu, W., et al.: High precision measurements of the ground state hyperfine structure interval of muonium and of the muon magnetic moment. Phys. Rev. Lett. **82**, 711 (1999)
16. Bennett, G.W., et al.: Improved limit on the muon electric dipole moment. Phys. Rev. D **80**, 052008 (2009)
17. Bennett, G.W., et al.: Search for Lorentz and CPT violation effects in muon spin precession. Phys. Rev. Lett **100**, 091602 (2008)
18. Bluhm, R., Kostelecki, V.A., Lane, C.D.: CPT and Lorentz tests with muons. Phys. Rev. Lett. **84**, 1098 (2000)
19. Kostelecky, V.A., Russell, N.: Data tables for Lorentz and CPT violation. Rev. Mod. Phys. **83**, 1 (2011)
20. Hughes, V.W., et al.: Test of CPT and Lorentz Invariance from muonium spectroscopy. Phys. Rev. Lett. **87**, 111804 (2001)
21. Roberts, B.L.: Status of the Fermilab muon (g-2) experiment. Chinese Physics C **34**, 741 (2011)
22. Bandyopadhyay, A., et al.: Physics at a future Neutrino Factory and Super Beam Facility. Reports on Progress in Physics **72**, 106201 (2009)

Hyperfine Interact (2012) 210:7–12
DOI 10.1007/s10751-011-0517-z

Quantum vacuum magnetic birefringence

Paul Berceau · Rémy Battesti · Mathilde Fouché · Paul Frings · Marc Nardone · Oliver Portugall · Geert L. J. A. Rikken · Carlo Rizzo

Published online: 22 December 2011

Abstract In this contribution to EXA2011 congress, we present the status of the BMV (*Biréfringence Magnétique du Vide*) experiment which is based on the use of a state-of-the-art optical resonant cavity and high pulsed magnetic fields, and it is hosted by the *Laboratoire National des Champs Magnétiques Intenses* in Toulouse, France.

Keywords QED · Birefringence · Magnetic field · High finesse cavity

1 Introduction

In 1934 [1] Dirac made the assumption that vacuum was filled by negative energy levels fully occupied by electrons. This charge distribution is not accessible to observation because of its uniformity all over the space. Negative energy electrons have to overcome a $2m_ec^2$ gap, where m_e is the electron mass, to jump to positive energy levels. The corresponding non occupied negative energy level appears as a hole in the original negative charge distribution which can be assumed to be a positron. This vacuum level structure looks like the one of a crystal meaning that non linear optical phenomena should exist in vacuum.

Starting from the Dirac model, Euler and Kochel in 1935, and Heisenberg and Euler in 1936 established the effective lagrangian of the interaction between electromagnetic fields [2, 3]. Thanks to the Heisenberg-Euler lagrangian, it has been shown in the seventies [4] that in the presence of external fields the velocity of light depends on the light polarization. Because of this linear magnetic birefringence, a linearly polarized light passing through a region where a magnetic field B perpendicular to

P. Berceau · R. Battesti · M. Fouché · P. Frings · M. Nardone ·
O. Portugall · G. L. J. A. Rikken · C. Rizzo (✉)
Laboratoire National des Champs Magnétiques Intenses,
(UPR 3228, CNRS-UPS-UJF-INSA), 31400 Toulouse, France
e-mail: carlo.rizzo@lncmi.cnrs.fr

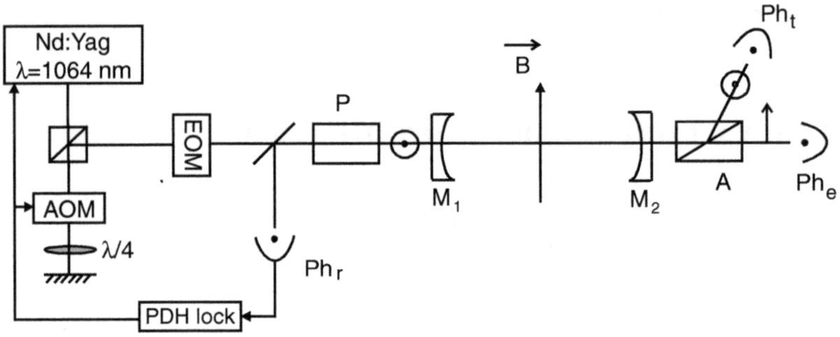

Fig. 1 Scheme of experimental setup

the direction of propagation is present will therefore acquire an ellipticity Ψ. Ψ can be written as $\Psi = \frac{\pi}{\lambda}\Delta n B^2 L \sin(2\theta)$, where λ is the light wavelength in vacuum, Δn is the difference between the index of refraction of the light polarized parallel to the magnetic field and the index of refraction of the light polarized perpendicular to the magnetic field, L is the length of the magnetic field region, and θ the angle between light polarization and the magnetic field. This kind of magnetic effect is also known as the Cotton-Mouton effect. Such an effect is usually very small especially in dilute matter [5]. It is clear looking to previous equation that a critical parameter for experiments looking for vacuum magnetic birefringence is B^2L.

Following [4] and [6], and using the CODATA recommended values [7] for the fundamental constants, at the lower orders in α, the fine structure constant, the vacuum magnetic birefringence value is $\Delta n = (4.031699 \pm 0.000002) \times 10^{-24}\, B^2$ with B in Tesla. The measurement of Δn up to a precision of a few ppm remains a pure QED test. It is worth mentioning that QED predictions have been widely and successfully tested in bound systems like hydrogen, muonic hydrogen, muonium, and in the case of isolated charged particles for example as far as the anomalous magnetic moment of electron or muon is concerned. Tests of QED prediction for the photon behavior are still lacking.

All recent experiments, done [8] or under way (see e.g. [9, 10]), concerning vacuum magnetic birefringence follow the 1979 proposal of Iacopini and Zavattini. In 1979, Iacopini and Zavattini [11] proposed to measure the ellipticity induced in a linearly polarized laser beam by the presence of a transverse magnetic field using an optical cavity in order to increase the optical path in the field, and to modulate the effect in order to increase the signal to noise ratio.

2 BMV experiment

The BMV experiment is detailed in [12]. Briefly, as shown in Fig. 1, a 30 mW linearly polarized Nd:Yag laser beam ($\lambda = 1064$ nm) is injected into a Fabry-Perot cavity made of mirrors M_1 and M_2. The beam reflected by the cavity is then analyzed on a photodiode Ph_r. This signal is used to drive the acousto-optic modulator (AOM) to lock the laser frequency to the cavity resonance frequency using the Pound-Drever-Hall method [13]. To this end, the laser is phase-modulated at 10 MHz with an electro-optic modulator (EOM).

Our birefringence measurement is based on an ellipticity measurement following [11]. Light is polarized just before going into the cavity by a polarizer P. The beam transmitted by the cavity is then analyzed by an analyzer A crossed at maximum extinction and collected by a low noise photodiode (intensity of the extraordinary beam I_e). The analyzer also has an escape window which allows us to extract the ordinary beam (intensity I_t) which corresponds to the polarization parallel to P. This beam is also collected by a photodiode.

A strong magnetic field is provided by pulsed magnets, which is a novelty as far as linear magnetic birefringence is concerned. In principle, pulsed magnetic fields can be as high as several tens of Tesla, which increases the signal, and they are rapidly modulated which increases the signal to noise ratio. B^2 can be so high that a relative short L is sufficient to obtain a B^2L parameter of few hundreds of T^2m. We have been therefore able to mount a table top experiment which in our opinion is the best choice to get high ellipticity sensitivity.

Taking into account the gain in ellipticity due to the Fabry-Perot cavity, the ellipticity acquired by the linearly polarized beam going through the magnet region can be finally written as $\Psi = \frac{2F}{\lambda}(B^2L)\Delta n \sin(2\theta)$, where F is the Fabry-Perot cavity finesse. For a cavity constituted by two identical mirrors of reflectivity R the finesse can be written as $F = \pi/(1-R)$. It is clear that cavity finesse F is an experimental parameter as important as B^2L. The ellipticity Ψ can be calculated from the measurements of I_t and I_e since $I_e = I_t\sigma^2 + I_t(\Gamma + \Psi(t))^2$ where σ^2 is the polarizer extinction, and Γ is the ellipticity due to the intrinsic birefringence of cavity mirrors. σ^2 is measured when no cavity is present, and Γ is measured for each magnetic pulse just before switching on the magnetic field, i.e., when $\Psi(t) = 0$.

$$\Psi(t) = \sqrt{\frac{I_e - I_t\sigma^2}{I_t}} - \Gamma \tag{1}$$

$\Psi(t)$ is proportional to the square of the magnetic field, $\Psi(t) = \kappa B(t)^2$. A correlation between $\Psi(t)$ measured and the $B(t)^2$ delivered by our magnets gives the κ factor for each pulse. A final statistical analysis gives the average value of κ and its standard deviation. Knowing the experimental parameters like F, L, λ and θ, one can finally calculate the measured Δn.

BMV experimental final goals are a cavity finesse approaching 1 000 000 thanks to very high reflectivity mirrors provided by the *Laboratoire Matériaux Avancés* of IN2P3, France, and a B^2L in excess of 600 T^2m produced by optimized pulsed field coils designed and built at the *Laboratoire National des Champs Magnétiques Intenses* of CNRS, France, which also hosts the experiment on its Toulouse site [14]. Thanks to these experimental parameters the ellipticity to be measured should be as high as 5×10^{-9} rad.

The only method to obtain a high magnetic field is to have a strong current circulating into a coil. This creates two major problems. The first one is heating. Our solution is to pulse the field for a duration short enough that the coil, starting at liquid nitrogen temperature, remains always at a safe level, i.e., below room temperature. The disadvantage in this case is that one has to wait for the coil to reach liquid nitrogen temperature before pulsing again reducing the duty cycle. Once a high field is reached another problem appears, i.e., the magnetic pressure proportional to B^2/μ_0 which obliges us to use strong conductors and to have a very efficient external reinforcement.

Our present apparatus consists of two magnets, called Xcoils because of their X-geometry, that have been designed at the LNCMI in Toulouse. The principle of these magnets and their properties are described in details in [15]. The basic idea was to get the current creating the transverse magnetic field as close as possible to the light path over a length as long as possible to maximize the integral of the square of the field over the magnet length L. Outer dimensions are 250 mm × 100 × 100 mm. The internal hole available for optical measurements is 17 mm. Like for conventional pulsed magnets, the coil is placed in a liquid nitrogen cryostat. The maximum field obtained has been 14.3 T corresponding to a B^2L of about 25 $\mathrm{T}^2\mathrm{m}$.

Long terms perspectives depend on the possibility to have higher magnetic fields. We have designed a new pulsed coil, called XXL-coil, which has already reached a field higher than 30 T when a current higher than 27 000 A is injected. This corresponds to more than 300 $\mathrm{T}^2\mathrm{m}$.

The other key point of our experiment is to accumulate the effect due to the magnetic field by trapping the light between two interferometric mirrors constituting a Fabry-Perot cavity. Its length has to be large enough to leave a wide space so as to insert our magnets. Its length is $L_c = 2.27\,\mathrm{m}$. In order to increase the induced signal, a finesse as high as possible is essential. Our best finesse has been obtained using mirrors provided by the LMA of IN2P3, France. We have got a finesse of $F = 529\,000$ and a linewidth of the cavity resonance of $\Delta\nu = c/2nL_cF = 124\,\mathrm{Hz}$. As far as we know, nowadays our interferometer is the sharpest in the world.

Following the feasibility study published in [12], the apparatus has been designed in order to minimize the cross talk between the optics and the magnetic field and the spurious effect that could be originated by it. The whole apparatus has been mounted in a clean room since high reflectivity mirrors are very sensitive to pollution.

All the optical components from the polarizer P to the analyzer A are placed in a ultra high vacuum chamber. Mirror and polarizer mounts can be tilted and rotated from outside the vacuum chamber and we can therefore align optics under high vacuum.

For security reasons, once the optics is ready, in particular the laser is locked to the cavity, and the magnets operational, physicists must leave the clean room before pulsing the magnet. This is done from a control room which is contiguous to the clean room.

Figure 2 presents raw signals obtained with about 32×10^{-3} atm of molecular nitrogen. The intensity of the ordinary beam I_t (top) remains constant while the intensity of the extraordinary beam I_e (middle) varies when the magnetic field (bottom) is applied. The magnetic field reaches its maximum of 5.2 T within less than 2 ms. The laser beam remains locked to the Fabry-Perot cavity, despite of the acoustic and mechanical vibrations caused by the pulsed magnetic field.

We see that the minimum of I_e does not coincide with the maximum of B^2. This phenomenon is due to the cavity filtering as explained in detail in [16].

In Fig. 3, we plot the square of the magnetic field filtered by the cavity and the ellipticity calculated with (1) as a function of time. We note that both quantities reach their extremum at the same time and their variation can be perfectly superimposed, providing a very precise measurement of magnetic linear birefringence of nitrogen

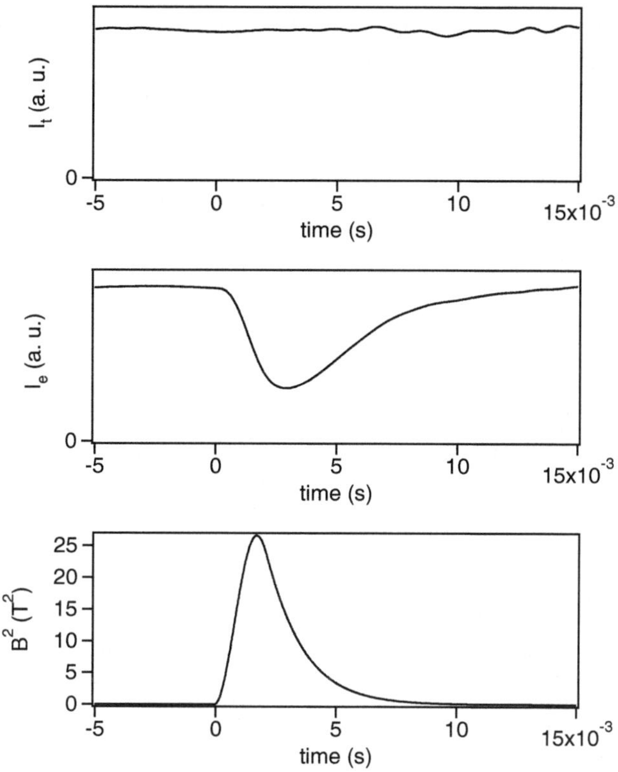

Fig. 2 Cotton-Mouton effect measurement on about 32×10^{-3} atm of molecular nitrogen. (*top*) Intensity of the ordinary beam as a function of time. (*middle*) Intensity of the extraordinary beam as a function of time. (*bottom*) Square of the magnetic field as a function of time

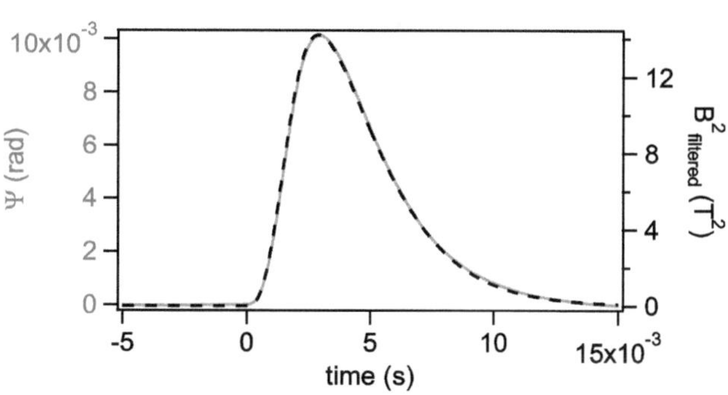

Fig. 3 Cotton-Mouton effect measurement on about 32×10^{-3} atm of molecular nitrogen. *Gray line:* Total ellipticity as a function of time. *Dashed line:* Square of the magnetic field filtered by a first-order low pass filter corresponding to the cavity filtering

gas. Our experimental value for nitrogen magnetic birefringence at $B = 1$ T and $P = 1$ atm is $\Delta n_{\mathrm{u}} \simeq -2.00 \times 10^{-13}\,\mathrm{atm}^{-1}\mathrm{T}^{-2}$. Our value agrees perfectly well with other measurements. It therefore provides a successful calibration of the whole apparatus.

Our present sensitivity corresponds to a record sensitivity of $\Delta n_{\mathrm{sens}} = 5 \times 10^{-20}\,\mathrm{T}^{-2}\,\frac{1}{\sqrt{\mathrm{Hz}}}$. This proves that modulation of the effect naturally induced by the pulsed field allows to get a sensitivity that with continuous magnets can only be reached with thousand of seconds of integration time (see e.g. [9]).

If we want to observe vacuum magnetic birefringence in a thousand pulses, i.e., about 20 working days, we need a sensitivity per pulse of about $\Delta n = 1 \times 10^{-22}\,\mathrm{T}^{-2}$.

A factor of the order of one hundred should be gained by the upgrade of X-coils into XXL-coils. We are very confident that eventually the vacuum magnetic birefringence will be observed thanks to our original set-up based on pulsed magnets coupled to a very high finesse Fabry-Perot cavity.

Acknowledgements We thank L. Pinard, R. Flaminio, and the technical staff of LMA of Lyon for providing very high finesse mirrors. We thank all the technical staff of LNCMI, in particular J. Béard, J. Billette and J. Mauchain. We acknowledge the support of the *ANR-Programme non thématique* (ANR-BLAN06-3-139634), and of *Fondation pour la recherche IXCORE.*

References

1. Dirac, P.A.M.: Rapport du 7e Conseil Solvay de Physique. Structure et Proprietés des Noyaux Atomiques, 203 (1934)
2. Euler, H., Kochel, B.: Naturwiss **23**, 246 (1935)
3. Heisenberg, W., Euler, H.: Z. Phys. **38**, 714 (1936)
4. Bialynicka-Birula, Z., Bialynicki-Birula, I.: Phys. Rev. D **2**, 2341 (1970)
5. Rizzo, C., Rizzo, A., Bishop, D.M.: Int. Rev. Phys. Chem. **16**, 81 (1997)
6. Ritus, V.I.: Sov. Phys. JETP **42**, 774 (1975)
7. http://physics.nist.gov/cuu/Constants/index.html
8. Cameron, R., et al.: Phys. Rev. D **47**, 3707 (1993)
9. Zavattini, E., et al.: Phys. Rev. D **77**, 032006 (2008)
10. Chen, S.-J., Mei, H.-H., Ni, W.-T.: Mod. Phys. Lett. A **22**, 2815 (2007)
11. Iacopini, E., Zavattini, E.: Phys. Lett. B, **85**, 151 (1979)
12. Battesti, R., et al.: Eur. Phys. J. D **46**, 323 (2008)
13. Drever, R.W.P., et al.: Appl. Phys. B **31**, 97 (1983)
14. http://www.toulouse.lncmi.cnrs.fr/?lang=en
15. Batut, S., et al.: IEEE Trans. Appl. Supercond. **18**, 600 (2008)
16. Berceau, P., et al.: Appl. Phys. B **100**, 803 (2010)

Hyperfine Interact (2012) 210:13–17
DOI 10.1007/s10751-011-0494-2

Neutron bound beta-decay: BOB

Josephine McAndrew · Stephan Paul · Ralf Emmerich · Ralf Engels · Peter Fierlinger · Mirko Gabriel · Erwin Gutsmiedl · Johannes Mellenthin · Johannes Schön · Wolfgang Schott · Andreas Ulrich · Florian Grüenauer · Anton Röhrmoser

Published online: 19 April 2012

Abstract An experiment to observe the bound beta-decay (BOB) of the free neutron into a hydrogen atom and an electron anti-neutrino is described. The hyperfine spin state population of the monoenergetic hydrogen atom yields the neutrino left-handedness or possible right-handed admixture as well as possible small scalar and tensor contributions to the weak force. The BOB H(2s) hyperfine states can be separated with a Lamb-Shift Spin Filter. These monoenergetic H(2s) atoms are ionised into H^- by charge exchanging within an argon cell. These ions are then separated using an adaptation of a MAC-E Filter. A first experiment is proposed at the FRMII high thermal-neutron flux beam reactor SR6 through-going beam tube, where we will seek to observe this rare neutron decay-mode for the first time and determine the branching ratio. After successful completion, the hyperfine spin state population will be determined, possibly at the ILL high-flux beam reactor through-going beam tube H6–H7, where the thermal neutron flux is a factor of four larger.

Keywords Neutron · Weak interaction · Scalar coupling constant · Tensor coupling constant · Neutrino helicity

J. McAndrew (✉) · S. Paul · R. Emmerich · M. Gabriel · E. Gutsmiedl · J. Mellenthin · J. Schön · W. Schott · A. Ulrich · F. Grüenauer
Physik-Department, TUM, 85748 Garching, Germany
e-mail: j.mcandrew@tum.de

R. Engels
Institut Für Kernphysik, Forschungszentrum Jülich, 52425 Jülich, Germany

P. Fierlinger
Excellence Cluster Universe, TUM, 85748 Garching, Germany

A. Röhrmoser
FRMII, 85748 Garching, Germany

Table 1 The six possible spin configurations, i, in the neutron bound beta-decay

Configuration, i	$\bar{\nu}$	n	p	e^-	W_i (%)	F	m_F	$\mid m_s m_I\rangle$
1	$\leftarrow$	$\leftarrow$	$\leftarrow$	$\rightarrow$	44.14 ± 0.05	0,1	0	$\mid +-\rangle$
2	$\leftarrow$	$\leftarrow$	$\rightarrow$	$\leftarrow$	55.25 ± 0.04	0,1	0	$\mid -+\rangle$
3	$\leftarrow$	$\rightarrow$	$\rightarrow$	$\rightarrow$	0.622 ± 0.011	1	1	$\mid ++\rangle$
4	$\rightarrow$	$\leftarrow$	$\leftarrow$	$\leftarrow$	0.0	1	−1	$\mid --\rangle$
2’	$\rightarrow$	$\rightarrow$	$\rightarrow$	$\leftarrow$	0.0	0,1	0	$\mid -+\rangle$
1’	$\rightarrow$	$\rightarrow$	$\leftarrow$	$\leftarrow$	0.0	0,1	0	$\mid +-\rangle$

By convention the neutron, n, is in the rest frame, the anti-neutrino, $\bar{\nu}$, goes to the left and the hydrogen atom, H, goes to the right. The spins of each of the particles, including the electron, e^-, are indicated by arrows in the table. The population probablility of each of the spin configurations according to V-A theory, W_i, is also given. F is the total spin with hyperfine interactions, with its projection given by m_F. $\mid m_s m_I\rangle$ are the Paschen-Back states, where m_s and m_I represent the e^- and p quantum numbers respectively

1 Introduction

The bound beta-decay of the free neutron is a rare neutron decay-mode first predicted by Nemenov [1], in which the neutron decays to a hydrogen atom and an electron anti-neutrino:

$$n \rightarrow H + \bar{\nu} \tag{1}$$

So far unobserved, the main challenges in observing this decay lie in the small branching ratio predicted to be $\sim$4x10^{-6} [1, 2] of the three-body decay mode and in the initial and final states containing electrically neutral particles.

Since this is a two-body decay, the kinetic energy of the emerging H atom is well-defined as $T_H = 325.7$ eV. Due to the short range of the weak interaction, it is expected that only atomic states with zero angular momentum will be populated; 83.2% of the H atoms will occupy the 1s state and 10.4% the 2s state. Through investigation of the hyperfine spin configurations of the metastable H(2s) states, the properties of this exotic decay mode can be investigated. The H(2s) is of particular interest as it is long-lived and it can be quenched selectively to the H(1s) state.

Table 1 shows the six possible spin configurations of the neutron, hydrogen atom and electron anti-neutrino involved in the bound beta-decay process along with their respective population probablilities, W_i, according to the V-A theory. In this theory, Configurations 4, 2’ and 1’ cannot be populated as the emission of a left-handed anti-neutrino would be required. A non-zero value of W_4 to $W_{1'}$ would therefore imply a left-right symmetric V+A theory.

2 Physics motivation

By measuring the hyperfine spin state populations of the emerging H(2s) atoms, new limits on the scalar and tensor coupling constants of the weak force may be determined. For example, assuming a purely left-handed V-A interaction, the population probabilities of configurations 1 to 3 are given by [3]:

$$W_1 = \frac{(\chi - 1)^2}{2(\chi^2 + 3)} \quad W_2 = \frac{2}{\chi^2 + 3} \quad W_3 = \frac{(\chi + 1)^2}{2(\chi^2 + 3)} \tag{2}$$

Table 2 The population probability, W_i (%), for various small admixutres of g_S and g_T [4]

Configuration, i	$g_S = 0$, $g_T = 0$	$g_S = 0.1$, $g_T = 0$	$g_S = 0$, $g_T = 0.02$
1	44.14	46.44	43.40
2	55.24	53.32	55.82
3	0.622	0.238	0.780
4	0.0	0.0	0.0

where

$$\chi = \frac{1 + g_S}{\lambda - 2g_T} \text{ and } \lambda = \frac{g_A}{g_V} = -1.2695 \pm 0.0029$$

and where g_A, g_V, g_S, g_T are the axial, vector, scalar and tensor coupling constants respectively. Therefore, by measuring W_i, a combination of g_S and g_T may be obtained. Examples of the population probabilities assuming different values of g_S and g_T are shown in Table 2.

A more exciting possibility would be the observation of a non-zero population of the spin configurations not permitted by V-A theory. A non-zero value of W_4 would therefore be evidence of a possible small contribution of negative helicity to the $\bar{\nu}$ and hence mediation of the neutron decay by right-handed currents.[1] Using the left-right symmetric V+A model, the population of the fourth configuration would be [5]:

$$W_4 = \frac{(x + \lambda y)^2}{2(1 + 3\lambda^2 + x^2 + 3\lambda^2 y^2)} \tag{3}$$

with $x = \eta - \zeta$ and $y = \eta + \zeta$. Here $\eta \leq 0.036$ [6] is the mass-ratio squared of the two intermediate charged vector bosons and $\zeta \leq 0.03$ (C.L. 90%) [7] is the boson mass eigenstate mixing angle. In addition, this model expresses the neutrino helicity as:

$$H_{\bar{\nu}} = \frac{1 + 3\lambda^2 - x^2 - 3\lambda^2 y^2}{1 + 3\lambda^2 + x^2 - 3\lambda^2 y^2} \tag{4}$$

3 The proposed BOB experiment

Due to the small expected branching ratio for the bound beta-decay and the sensitivity of the experiment to neutron background, an intensive thermal neutron source with a through-going beam tube is required. Both the SR6 beam tube at the FRMII reactor in Garching with $\sim 1 \times 10^{14}$ ncm^{-2}s^{-1} maximum thermal-neutron flux [4] and the H6-H7 beam tube at the ILL reactor in Grenoble with $\sim 5 \times 10^{14}$ ncm^{-2}s^{-1} maximum thermal flux [9] would be suitable.

A schematic diagram of the proposed set up is shown in Fig. 1. The original hyperfine state and its spin projection on the axis is preserved by a small longitudinal magnetic field, $\mathbf{B}_1$. One of the four H(2s) hyperfine spin states can then be selected by a Lamb-Shift Spin Filter [8]. The spin filter first applies a transverse electric field, $\mathbf{E}_1$, to the atoms in order to reduce the lifetimes of the states. A longitudinal magnetic field, $\mathbf{B}_2$, splits the $2s_{1/2}$ and $2p_{1/2}$ states according to their values of $m_j = \pm 1/2$. If

[1] This hyperfine spin state could also be fed by decays of H(ns) states where n>3.

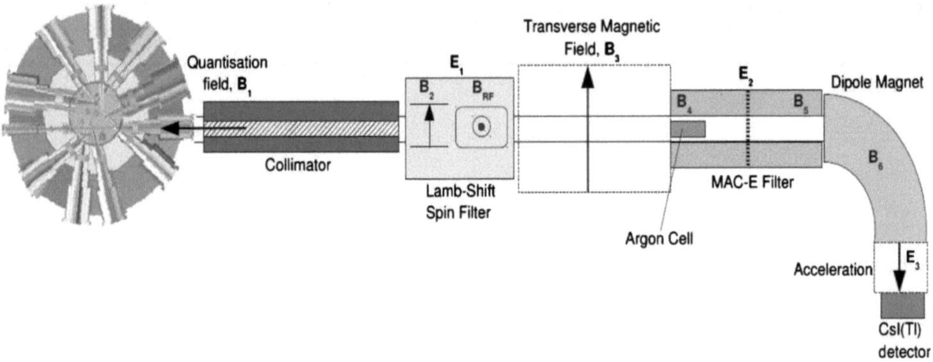

Fig. 1 Schematic diagram of the proposed set up for the Bound Beta-Decay Experiment on the experimental side. Not shown are a further transverse magnetic field followed by a beam stop located beyond the exit of the through-going beam tube to the left of the diagram. The symbols and roles of each component are explained in the text

$0.54\ \text{mT} \leq \mathbf{B}_2 \leq 0.61\ \text{mT}$, mixing of the $2s_{1/2}$ and $2p_{1/2}$ states occurs, resulting in the quenching of one pair of the hyperfine $2s_{1/2}$ spin states. An RF field, $\mathbf{B}_{RF}$, of ~1600 MHz is then applied, inducing the transition of the remaining hyperfine spin state to the ground state.

A transverse magnetic field, $\mathbf{B}_3$, then deflects protons and electrons from the three-body neutron beta-decay away from the beam axis before the hydrogen detection stage which begins at the argon cell. Here the metastable H(2s) atoms, are ionised to H^- using the charge exchange reaction in argon [10]:

$$H + Ar \rightarrow H^- + Ar^+ \tag{5}$$

The H^- ions which originate from metastable H(2s) atoms leave the argon cell with a kinetic energy of 320.1 eV, as opposed to H^- ions originating from H(1s) atoms which have a lower kinetic energy of 309.9 eV. This 10.2 eV energy difference has been observed experimentally by the BOB group. The cross-section of the charge exchange reaction is also expected to be approximately two orders of magnitude lower for the H(1s) atoms than for the H(2s) atoms [10]. A magnetic energy analyser system for the H^- is presently being developed and constructed at the TU München based on the MAC-E filter principle. This filter consists of two solenoids which produce longitudinal magnetic fields $\mathbf{B}_4$ and $\mathbf{B}_5$ which rotate the momentum vectors of the H^- ions until they are parallel with the beam axis. An electric counter field, $\mathbf{E}_2$, determines the minimum energy of the ions which may pass through the filter. In this way the H^- ions which originate from the H(2s) atoms may be descriminated from those originating from the H(1s) atoms. The ions are then deflected by a magnetic spectrometer to the detection stage of the experiment before being accelerated by the electric field, $\mathbf{E}_3$, to the detector.

Due to the very small branching ratio of this decay, a further essential element of the design of this experiment is the simulation of the background expected at the detector. Using both MCNP and GEANT4 simulation programs, the neutron, gamma and bound-beta decay H atom rates in the vacuum tube beyond the exit of the through-going beam tube are estimated. For example, MCNP simulations of the FRMII reactor have shown that the expected H(2s) event rate at the exit of the SR6 beam tube will be 3 s^{-1} [4]. Furthermore, simulation of different collimator geometries to block gamma background from the reactor core and various distances of the beam stop from the opposite exit of the beam tube allow these components to be designed in such a way as to minimise the neutron and gamma flux at the detector. In particular, it has recently been shown that the position of the beam stop has the greatest influence on neutron and gamma flux at the position of the detector, and that the flux of these particles may be reduced by a factor of 100 by moving the beam stop just 0.5 m away from the exit of the beam tube (Grünauer, F., Röhrmoser, A.: Private Communication).

4 Conclusion and outlook

An experiment to measure the bound beta-decay of the neutron has been outlined. Due to the complexity of this measurement, the experiment will be performed in two stages. The first stage of the experiment will observe and determine the branching ratio of this exotic decay mode of the free neutron for the first time by detection of the metastable H(2s) atoms produced in this decay. In the second stage of the experiment, the population probability of the hyperfine spin states of the H(2s) atoms will be measured. Population probability W_3 will allow the small g_S or g_T contribution to the weak force, including the sign, to be calculated. Measurement of a non-zero value of population probability W_4 would suggest the presence of physics beyound the standard model and would allow the measurement of η and ζ. Previous calculations [4] have shown that the upper limits of g_S or g_T and η or ζ could be reduced by a factor of ten within a few weeks of measuring time.

References

1. Nemenov, L.L.: Decay of the neutron into a hydrogen atom and an antineutrino. Sov. J. Nucl. Phys. **31**, 115–119 (1980)
2. Faber M., et al.: Continuum-state and bound-state β^- decay rates of the neutron. Phys. Rev. C **80**, 035503 (2009)
3. Nemenov, L.L., Ovchinnikova, A.A.: Effects of scalar and tensor interactions on the atomic decay of the neutron. $n \rightarrow H + \bar{\nu}$. Sov. J. Nucl. Phys. **31**, 115–119 (1980)
4. Schott, W., et al.: An experiment for the measurement of the bound-β-decay of the free neutron. Eur. Phys. J., A **30**, 603–611 (2006)
5. Byrne, J.: Two-body decay of the neutron: a possible test for the existence of right-handed weak currents. Eur. Phys. Lett. **56**, 633–638 (2001)
6. Gaponenko, A., et al.: Measurement of the muon decay parameter delta. Phys. Rev., D **71**, 071101 (2005)
7. Musser, J.R.: Measurement of the Michel Parameter ρ in Muon Decay. Phys. Rev. Lett. **94**, 101805 (2005)
8. McKibben, J.L., Lawrence, G.P., Olsen, G.G.: Nuclear spin filter. Phys. Rev. Lett. **20**, 1180 (1968)
9. Kessler, E.G., et al.: The GAMS4 flat crystal facility. Nucl. Instrum. Methods A **457**, 187–202 (2001)
10. Roussel, F., et al.: Electron capture, electron loss and deexcitation of fast H(^{2}2s) and H(^{1}2s) atoms in collision with molecular hydrogen. Phys. Rev., A **16**, 1854 (1977)

Hyperfine Interact (2012) 210:19–23
DOI 10.1007/s10751-011-0540-0

Muon decay in orbit spectra for $\mu-$e conversion experiments

Andrzej Czarnecki · Xavier Garcia i Tormo · William Joseph Marciano

Published online: 3 January 2012

Abstract We have determined in detail the electron spectrum in the decay of bound muons. These results are especially relevant for the upcoming $\mu - e$ conversion experiments.

Keywords Muon decay · Muonic atoms · Electron spectrum · Muon-electron conversion

1 Introduction

From the observation of neutrino oscillations, we now know that lepton flavors are not conserved. However, the mixing and small neutrino mass differences seen in oscillations have a negligible effect on charged-lepton flavor violating (CLFV) reactions. Thus, the CLFV reactions provide a discovery window for interactions beyond Standard Model expectations [1, 2].

Muons play a central role in searches for CLFV [1, 2], because they can be produced in large numbers and live relatively long. One reaction that can be probed with particularly high sensitivity is coherent muon-electron conversion in a muonic atom,

$$\mu^- + (A, Z) \to e^- + (A, Z), \tag{1}$$

where (A, Z) represents a nucleus of atomic number Z and mass number A. It has the advantage of producing just a single particle, a mono-energetic electron. It does

A. Czarnecki (✉) · X. Garcia i Tormo
Department of Physics, University of Alberta, Edmonton, AB, Canada T6G 2E1
e-mail: andrzejc@ualberta.ca

X. Garcia i Tormo
Institut für Theoretische Physik, Universität Bern, Sidlerstrasse 5, 3012 Bern, Switzerland

W. J. Marciano
Department of Physics, Brookhaven National Laboratory, Upton, NY 11973, USA

not have the problem of accidental background that plagues searches for the decay $\mu^+ \to e^+\gamma$, which can be mimicked by a positron from a normal muon decay and a photon coming from the radiative decay of a different muon, bremsstrahlung, or positron annihilation-in-flight. Various experiments have been performed over the years to search for the conversion [3]. The most stringent results come from the SINDRUM II Collaboration [4], which reports an upper limit of 7×10^{-13} for the branching ratio of the conversion process relative to muon capture in gold. Several new efforts are being planned. In the nearest future, the DeeMe Collaboration [5] has proposed to reach 10^{-14} sensitivity. Larger scale searches, Mu2e at Fermilab [6] and COMET at J-PARC [7], aim for sensitivities below 10^{-16}. In the long run, intensity upgrades at Fermilab and the proposal PRISM/PRIME at J-PARC may allow them to reach 10^{-18} sensitivity. A quite remarkable improvement of about four orders of magnitude, with respect to the current limit, is therefore envisaged.

The success of the conversion searches depends critically on control of the background events. The signal for the $\mu - e$ conversion process in (1) is a mono-energetic electron with energy $E_{\mu e}$, given by

$$E_{\mu e} = m_\mu - E_{\text{b}} - E_{\text{rec}}, \tag{2}$$

where m_μ is the muon mass, $E_{\text{b}} \simeq Z^2\alpha^2 m_\mu/2$ is the binding energy of the muonic atom, and $E_{\text{rec}} \simeq m_\mu^2/(2m_N)$ is the nuclear-recoil energy, with α the fine-structure constant and m_N the nucleus mass. The main physics background for this signal comes from the so-called muon decay in orbit (DIO), a process in which the muon decays in the normal way, $\mu^- \to e^-\overline{\nu}_e\nu_\mu$, while in the orbit of the atom. Whereas in a free muon decay, in order to conserve energy and three-momentum, the maximum electron energy is $m_\mu/2$, for DIO, the nucleus recoil can balance the electron's three-momentum taking basically no energy. This allows for the maximum electron energy to be $E_{\mu e}$, close to the full muon mass m_μ. Therefore, the high-energy electrons from the muon decay in orbit constitute a background for conversion searches.

2 Muon decay in orbit

Several theoretical studies of the muon decay in orbit have been published. Expressions describing the electron spectrum including relativistic effects in the muon wavefunction, the Coulomb interaction between the electron and the nucleus and a finite nuclear size have been available for some time [8–10]. However, the high-energy endpoint of the spectrum, which is the most important region for conversion-search experiments, was not studied in detail. Shanker [11, 12] did study the high-energy end of the electron spectrum, and presented approximate results which allow for a quick rough estimate of the muon decay in orbit contribution to the background in conversion experiments. We have performed a new evaluation of the DIO spectrum, considering in detail all the effects needed in the high-energy region [13]. Our results describe the background contribution for $\mu - e$ conversion searches, as well as a check on previous low- and high-energy partial calculations [10, 11] and an interpolation between them. It is worth emphasizing that not only the high-energy region is relevant for conversion experiments, but the full spectrum is necessary in order to study reconstruction errors in the detector.

Table 1 Values for muon energies E_μ, nuclear masses m_N, and endpoint energies $E_{\mu e}$

Nucleus	Z	E_μ (MeV)	m_N (MeV)	$E_{\mu e}$ (MeV)
C	6	105.557	11,188	105.06
Al	13	105.194	25,133	104.973
Si	14	105.121	26,162	104.91
Ti	22	104.394	44,588	104.272

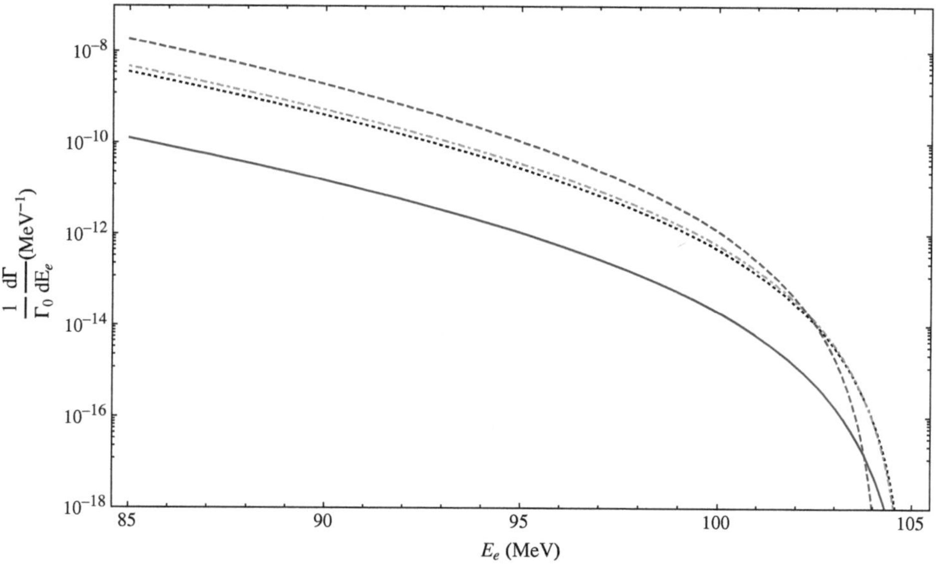

Fig. 1 Electron spectrum, normalized to the free-muon decay rate Γ_0. The *solid blue line* is for carbon, the *black dotted line* for aluminum, the *green dot-dashed line* for silicon and the *red dashed line* for titanium

To obtain the correct result for the high-energy tail of the spectrum it is crucial to include nuclear-recoil effects, since they modify the endpoint energy (see (2)). Also, to produce an on-shell electron with energy around m_μ, either the muon must be at the tail of the bound-state wavefunction or the produced electron must interact with the nucleus. This tells us that the full Dirac equation for the muon as well as the interaction of the outgoing electron with the field of the nucleus must be taken into account. Also, finite-nuclear-size effects will be most important in this region. Order α radiative corrections are not expected to significantly modify the results at the endpoint, and are not included in our results. Uncertainties in the modelling of finite nuclear-size effects induce errors in the spectrum that increase as we approach the endpoint, but those errors are never larger than a few percent.

3 Results and discussion

Here we present our results for the elements that are relevant for the upcoming conversion experiments. The DeeMe Collaboration plans to use a silicon-carbide target, whereas the Mu2e and COMET Collaborations are considering aluminum

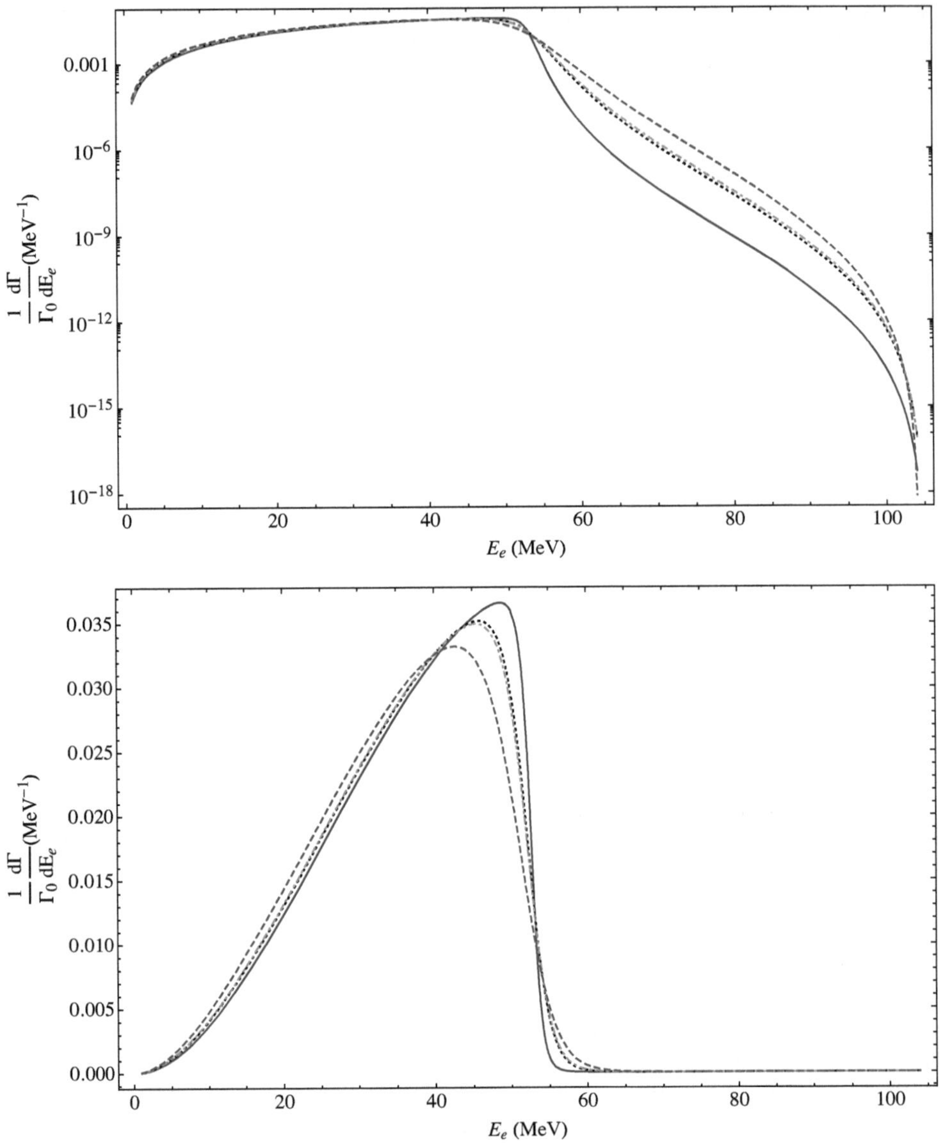

Fig. 2 Electron spectrum for the full range of E_e; see caption of Fig. 1

and titanium as targets. In Table 1 we give the values of the bound muon energy $E_\mu = m_\mu - E_\text{b}$ and the electron endpoint energy $E_{\mu e}$, for carbon, aluminum, silicon and titanium.

We present the results of the numerical evaluation of the spectra for those elements in Figs. 1 and 2. Regarding the nuclear distributions, a two-parameter Fermi distribution has been used for aluminum and titanium, a three-parameter Fermi distribution for silicon and a Fourier–Bessel expansion for carbon [13–15]. Those results are useful for assessing the DIO background events from carbon for the DeeMe experiment which will search primarily for conversion in silicon. They also illustrate the electron resolution requirements, as a function of stopping target, needed to reach future high sensitivity goals.

The high sensitivity that the upcoming conversion experiments will reach may also allow them to improve the present bounds on some exotic muon decays, like the decay of the muon into an electron and a majoron (a Goldstone boson that appears in models where lepton number is a spontaneously broken global symmetry) [16].

Acknowledgements This research was supported by Science and Engineering Research Canada (NSERC) and by the United States Department of Energy under Grant Contract DE-AC02-98CH10886.

References

1. Kuno, Y., Okada, Y.: Rev. Mod. Phys. **73**, 151 (2001). arXiv:hep-ph/9909265
2. Czarnecki, A., Marciano, W.J.: In: Roberts, B.L., Marciano, W.J. (eds.) Lepton Dipole Moments, p. 11. World Scientific, Singapore (2009); Adv. Ser. Dir. HEP, vol. 20; Kuno, Y.: ibid., p. 701; Okada, Y.: ibid., p. 683
3. Marciano, W.J., Mori, T., Roney, J.M.: Annu. Rev. Nucl. Part. Sci. **58**, 315 (2008)
4. Bertl, W.H., et al., SINDRUM II Collaboration: Eur. Phys. J. C **47**, 337 (2006)
5. Aoki, M.: PoS **ICHEP 2010**, 279 (2010)
6. Carey, R.M., et al., Mu2e Collaboration: Proposal to search for $\mu^- N \to e^- N$ with a single event sensitivity below 10^{-16}. Fermilab Proposal 0973 (2008)
7. Cui, Y.G., et al., COMET Collaboration: Conceptual design report for experimental search for lepton flavor violating $\mu^- - e^-$ conversion at sensitivity of 10^{-16} with a slow-extracted bunched proton beam (COMET). KEK Report 2009–2010 (2009)
8. Haenggi, P., Viollier, R.D., Raff, U., Alder, K.: Phys. Lett. **B51**, 119 (1974)
9. Watanabe, R., Fukui, M., Ohtsubo, H., et al.: Prog. Theor. Phys. **78**, 114 (1987)
10. Watanabe, R. et al.: At. Data Nucl. Data Tables **54**, 165 (1993)
11. Shanker, O.U.: Phys. Rev. **D25**, 1847 (1982)
12. Shanker, O.U., Roy, R.: Phys. Rev. **D55**, 7307 (1997)
13. Czarnecki, A., Garcia i Tormo, X., Marciano, W.J.: Phys. Rev. **D84**, 013006 (2011). arXiv: 1106.4756 (hep-ph)
14. De Vries, H., De Jager, C.W., De Vries, C.: At. Data Nucl. Data Tables **36**, 495–536 (1987)
15. Fricke, G., Bernhardt, C., Heilig, K., Schaller, L.A., Schellenberg, L., Shera, E.B., De Jager, C.W.: At. Data Nucl. Data Tables **60**, 177 (1995)
16. Garcia i Tormo, X., Bryman, D., Czarnecki, A., Dowling, M.: Phys. Rev. **D84**, 113010 (2011). arXiv:1110.2874 (hep-ph)

Hyperfine Interact (2012) 210:25–31
DOI 10.1007/s10751-012-0569-8

Static Stark effect in the molecular ion HD$^+$

Dimitar Bakalov · Stephan Schiller

Published online: 31 January 2012

Abstract We present preliminary results on the Stark effect of the hyperfine levels of the molecular ion HD$^+$.

1 Introduction

In previous papers [1] we have studied the Zeeman effect in precision spectroscopy of HD$^+$ ions and shown that it must be accounted for at the precision level of current rotational and vibrational spectroscopy experiments [2]. As an additional systematic effect we consider here the static Stark effect from slowly varying electric fields in ion traps.

We distinguish the dipole and quadrupole static Stark shifts. The present paper is restricted to discussions of the dominating dipole ones; the quadrupole corrections, which require more refined numerical calculations, will be discussed in details elsewhere. The focus is on the dipole Stark shift of the individual hyperfine levels, which is of importance for high-resolution spectroscopy and has not been considered in earlier papers on the subject, such as [3, 4], and the recent [5] (see also the references therein).

The work was partially supported by Grant 2-288 of the Bulgarian Scientific Fund. We also acknowledge support by the DFG within project Schi 431/11-1.

D. Bakalov (✉)
Institute for Nuclear Research and Nuclear Energy,
Bulgarian Academy of sciences, Sofia, Bulgaria
e-mail: dbakalov@inrne.bas.bg

S. Schiller
Institut für Experimentalphysik, Heinrich-Heine-Universität Düsseldorf,
40225 Düsseldorf, Germany

2 Theoretical approach

In the non-relativistic approximation the ro-vibrational states of HD^+ are labeled with the vibrational quantum number v, the total orbital momentum L and its projection on the z-axis L_z. Due to the spin interactions of the constituents of HD^+ each ro-vibrational state (vLL_z) is split into a multiplet of hyperfine states, labeled with the quantum numbers of the total angular momentum J, its z-axis projection J_z and an additional label n. The energy E^{vLnJ} of the hyperfine state $(vLnJJ_z)$ (which is is independent of J_z in absence of external fields) is the sum of the Coulomb energy and the hyperfine splitting: $E^{vLnJ} = E^{(\mathrm{NR})vL} + \Delta E^{(\mathrm{hfs})vLnJ}$. In the lowest order of perturbation theory L is conserved, and the state vector $|vLnJ_z\rangle$ is a linear combination of the vectors $|vLFSJJ_z\rangle$ with definite values of L and the angular momenta $\mathbf{F} = \mathbf{S}_p + \mathbf{S}_e$, $\mathbf{S} = \mathbf{F} + \mathbf{S}_d$ and $\mathbf{J} = \mathbf{S} + \mathbf{L}$, where $\mathbf{S}_x, x = p, d, e$ is the spin of particle x: $|vLnJJ_z\rangle = \sum_{FS} \beta_{FS}^{vLnJ} |vLFSJJ_z\rangle$. The amplitudes β_{FS}^{vLnJ} and the hyperfine energies $\Delta E^{(\mathrm{hfs})vLnJ}$ are calculated as eigenvectors and eigenvalues of the spin interaction Hamiltonian. Note that, although $\mathbf{F}^2$ and $\mathbf{S}^2$ are not strictly conserved, the quantum numbers F and S are approximately good, so that in what follows the label n will be identified with the pair (F, S).

In the dipole approximation, the interaction of HD^+ with the electric field $\mathbf{E}$ is given by $V = -\mathbf{E} \cdot \mathbf{d}$, where $\mathbf{d}$ is the dipole moment of HD^+:

$$\mathbf{d} = \sum_{x=p,d,e} eZ_x \left(\mathbf{R}_x - \mathbf{R}_{CM}\right), \ \mathbf{R}_{CM} = \sum_{x=p,d,e} \frac{M_x}{M_p + M_d + M_e} \mathbf{R}_x, \tag{1}$$

and $\mathbf{R}_x$, Z_x and M_x are the position vector, electric charge (in units e) and mass of the x-th particle. The interaction with the electric field removes the degeneration of E^{vLFSJ} with respect to $|J_z|$; the hyperfine energy level E^{vLFSJ} is shifted by the quantity

$$\Delta E^{vLFSJJ_z} = \sum_{q=0,\pm1} \left|E^q\right|^2 \sum_{v'L'} \sum_{F'S'J'J'_z} \frac{\left|\langle v'L'F'S'J'J'_z|d_q|vLFSJJ_z\rangle\right|^2}{E^{(\mathrm{NR})vL} - E^{(\mathrm{NR})v'L'}}, \tag{2}$$

where the contribution of the electronic excitations has been neglected. The matrix elements of the cyclic components $d_q, q = 0, \pm1$ of the vector $\mathbf{d}$ are

$$\begin{aligned} &\langle v'L'F'S'J'J'_z|d_q|vLFSJJ_z\rangle = \\ &d_{v'L',vL} \sum_{F_1S_1} (-1)^{J'+L'+S_1} \sqrt{2J+1}\, C_{JJ_z,1q}^{J'J'_z} \begin{Bmatrix} J' & 1 & J \\ L & S_1 & L' \end{Bmatrix} \beta_{F_1S_1}^{v'L'F'S'J'} \beta_{F_1S_1}^{vLFSJ}. \end{aligned} \tag{3}$$

Here $d_{v'L',vL}$ are the reduced matrix elements of the dipole moment in the non-relativistic approximation. Having in mind spectroscopy of HD^+, we evaluate the Stark shift of the laser-stimulated $(vLFSJJ_z) \to (v'L'F'S'J'J'_z)$ transition frequency through the difference of the Stark shifts of the final and initial state energy levels as given by (2). The cyclic components of the external field $\mathbf{E}$ satisfy $\left|E^{\pm1}\right|^2 = (|\mathbf{E}|^2 \sin^2\theta)/2$, $\left|E^0\right|^2 = |\mathbf{E}|^2\cos^2\theta$, where θ is the angle between the static electric field and the z-axis. Next (2) is transformed to display the static longitudinal and transversal polarizabilities:

$$\Delta E^{vLFSJJ_z} = -\frac{\mathbf{E}^2}{2}\left(\alpha^{(l)}_{vLFSJJ_z}\cos^2\theta + \alpha^{(t)}_{vLFSJJ_z}\sin^2\theta\right), \tag{4}$$

$$\alpha^{(l)}_{vLFSJJ_z} = -2\sum_{v'L'F'S'J'} \frac{\left|\langle v'L'F'S'J'J_z|d_0|vLFSJJ_z\rangle\right|^2}{E^{(\mathrm{NR})vL} - E^{(\mathrm{NR})v'L'}},$$

$$\alpha^{(t)}_{vLFSJJ_z} = -\sum_{v'L'F'S'J'}\sum_{q=\pm1} \frac{\left|\langle v'L'F'S'J', J_z{+}q|d_q|vLFSJJ_z\rangle\right|^2}{E^{(\mathrm{NR})vL} - E^{(\mathrm{NR})v'L'}}.$$

3 Numerical results

The computations presented are based on the numerical results about the hyperfine structure of HD$^+$ of [6]. The matrix elements of the dipole moment of HD$^+$ have been calculated in the Born–Oppenheimer approximation. The three-body non-relativistic wave function was taken in the form:

$$\Psi^{vLL_z}(\mathbf{R},\mathbf{r}) = \frac{1}{R}\chi^{vL}(R)\varphi_{1s\sigma}(\xi,\eta;R)\,Y_{LL_z}\left(\hat{R}\right), \tag{5}$$

where $Y_{LL_z}(\hat{R})$ are spherical harmonics, $\mathbf{R} = \mathbf{R}_p - \mathbf{R}_d$, $\mathbf{r} = \mathbf{R}_e - (M_d\mathbf{R}_d + M_p\mathbf{R}_p)/(M_d + M_p)$, while ξ and η are the prolate spheroidal coordinates of the electron [7]. Table 1 gives the numerical values of the reduced matrix elements $d_{v'L',vL} \equiv \langle v'L'||\mathbf{d}||vL\rangle$ for the lowest ro-vibrational states of HD$^+$, needed - together with the amplitudes β from [6] - for the evaluation of the Stark shift of these states. Comparison with the results of [1], obtained with the most accurate vibrational wave functions available to date, shows that the relative inaccuracy of our values does not exceed 1×10^{-3}; the difference in the signs of some matrix elements is due to the choice of irrelevant phase factors. In the rapidly converging sum over intermediate states in (2,4) the contribution of the states with $v' \neq v$ also does not exceed 10^{-3} the of the dominant contributions from the two states with $v' \neq v$. We note that for some hyperfine states the latter cancel each other and the resulting polarizabilities are strongly suppressed. Thus, although the approximation of (5) is less accurate that the approaches of [3, 4], it is sufficient for estimating the systematic shifts occurring in current ion trap experiments and for identifying transitions with particularly low Stark shift. Table 2 shows the longitudinal and transversal polarizabilities for states involved in some experimentally relevant laser-induced transitions [1]. The static polarizabilities of the hyperfine states vary significantly with the quantum numbers F, S, J, J_z and may lead to observable shifts of the spectral lines in high resolution experiments.

The numerical accuracy may be improved by an order of magnitude by using a wider basis set in the expansion of (5), which includes excited electronic states. Of

Table 1 Reduced matrix elements $\langle L' = L+1, v' \| \mathbf{d} \| Lv \rangle$ of the lowest ro-vibrational states of HD^+, in a.u. The notation $a(b)$ stands for $a \times 10^b$

vL	$v'=0$	$v'=1$	$v'=2$	$v'=3$	$v'=4$	$v'=5$	$v'=6$	$v'=7$	$v'=8$
00	0.3428	−0.3394(−1)	0.4364(−2)	0.9794(−3)	0.2953(−3)	−0.1080(−3)	−0.4520(−4)	−0.2087(−4)	−0.1039(−4)
10	−0.3720(−1)	0.3622	0.4842(−1)	0.7550(−2)	0.1932(−2)	−0.6425(−3)	−0.2538(−3)	−0.1137(−3)	−0.5615(−4)
20	−0.4461(−2)	−0.5321(−1)	−0.3824	0.5985(−1)	0.1068(−1)	−0.3023(−2)	−0.1089(−2)	−0.4594(−3)	−0.2178(−3)
30	−0.9713(−3)	−0.7737(−2)	0.6595(−1)	−0.4033	0.6978(−1)	−0.1382(−1)	−0.4242(−2)	−0.1633(−2)	−0.7295(−3)
40	−0.2876(−3)	−0.1921(−2)	0.1097(−1)	0.7710(−1)	−0.4251	−0.7881(−1)	−0.1700(−1)	−0.5585(−2)	−0.2277(−2)
50	−0.1038(−3)	−0.6273(−3)	0.3013(−2)	0.1424(−1)	0.8733(−1)	0.4478	−0.8725(−1)	−0.2024(−1)	−0.7055(−2)
60	−0.4305(−4)	−0.2447(−3)	0.1066(−2)	0.4239(−2)	0.1756(−1)	−0.9697(−1)	0.4717	−0.9530(−1)	−0.2356(−1)
70	−0.1971(−4)	−0.1086(−3)	0.4440(−3)	0.1603(−2)	0.5596(−2)	−0.2096(−1)	−0.1062	0.4970	−0.1031
80	−0.9736(−5)	−0.5321(−4)	0.2085(−3)	0.7069(−3)	0.2241(−2)	−0.7089(−2)	−0.2448(−1)	−0.1153	0.5239
01	0.4854	0.4573(−1)	−0.6087(−2)	−0.1387(−2)	−0.4219(−3)	−0.1552(−3)	−0.6531(−4)	−0.3029(−4)	−0.1513(−4)
11	−0.5496(−1)	−0.5130	−0.6515(−1)	−0.1052(−1)	−0.2732(−2)	−0.9169(−3)	−0.3645(−3)	−0.1641(−3)	−0.8133(−4)
21	0.6361(−2)	−0.7871(−1)	−0.5415	0.8041(−1)	0.1486(−1)	0.4269(−2)	0.1552(−2)	0.6589(−3)	0.3139(−3)
31	0.1364(−2)	−0.1105(−1)	0.9766(−1)	−0.5710	0.9361(−1)	0.1920(−1)	0.5983(−2)	0.2325(−2)	0.1045(−2)
41	0.4002(−3)	−0.2702(−2)	0.1569(−1)	0.1143	−0.6018	0.1056	0.2358(−1)	0.7868(−2)	0.3238(−2)
51	−0.1436(−3)	0.8741(−3)	−0.4242(−2)	−0.2038(−1)	−0.1296	0.6340	−0.1167	−0.2804(−1)	−0.9925(−2)
61	−0.5926(−4)	0.3388(−3)	−0.1487(−2)	−0.5977(−2)	−0.2517(−1)	−0.1442	0.6678	−0.1272	−0.3259(−1)
71	−0.2702(−4)	0.1496(−3)	−0.6156(−3)	−0.2240(−2)	−0.7901(−2)	−0.3009(−1)	−0.1582	0.7036	−0.1374
81	−0.1329(−4)	0.7305(−4)	−0.2878(−3)	−0.9816(−3)	−0.3136(−2)	−0.1002(−1)	−0.3519(−1)	−0.1719	0.7416
02	−0.5960	0.5328(−1)	−0.7338(−2)	0.1696(−2)	0.5208(−3)	−0.1928(−3)	−0.8156(−4)	−0.3799(−4)	−0.1905(−4)
12	−0.7021(−1)	0.6297	0.7579(−1)	−0.1266(−1)	−0.3339(−2)	0.1131(−2)	0.4523(−3)	0.2046(−3)	0.1018(−3)
22	0.7841(−2)	0.1007	0.6645	0.9339(−1)	0.1787(−1)	−0.5211(−2)	−0.1911(−2)	−0.8168(−3)	−0.3909(−3)
32	0.1656(−2)	0.1363(−1)	−0.1250	−0.7007	0.1086	−0.2305(−1)	−0.7294(−2)	−0.2861(−2)	−0.1294(−2)
42	0.4815(−3)	0.3284(−2)	−0.1939(−1)	0.1465	−0.7384	−0.1222	−0.2828(−1)	−0.9580(−2)	−0.3979(−2)
52	0.1717(−3)	0.1053(−2)	−0.5164(−2)	0.2522(−1)	0.1664	0.7779	−0.1348	−0.3357(−1)	−0.1207(−1)
62	0.7053(−4)	0.4056(−3)	−0.1794(−2)	0.7286(−2)	0.3118(−1)	−0.1852	0.8193	−0.1467	−0.3896(−1)
72	0.3204(−4)	0.1783(−3)	−0.7380(−3)	0.2706(−2)	0.9646(−2)	−0.3733(−1)	−0.2035	0.8631	−0.1581
82	0.1570(−4)	0.8672(−4)	−0.3434(−3)	0.1179(−2)	0.3794(−2)	−0.1225(−1)	−0.4372(−1)	−0.2215	0.9097
03	−0.6905	0.5842(−1)	0.8323(−2)	0.1953(−2)	0.6047(−3)	0.2253(−3)	0.9581(−4)	0.4484(−4)	0.2257(−4)
13	0.8445(−1)	−0.7294	0.8297(−1)	0.1434(−1)	0.3839(−2)	0.1312(−2)	0.5280(−3)	0.2399(−3)	0.1198(−3)
23	−0.9098(−2)	−0.1212	0.7696	−0.1021	−0.2021(−1)	−0.5984(−2)	−0.2215(−2)	−0.9526(−3)	−0.4581(−3)

Table 1 (continued)

vL	$v'=0$	$v'=1$	$v'=2$	$v'=3$	$v'=4$	$v'=5$	$v'=6$	$v'=7$	$v'=8$
33	0.1893(−2)	0.1584(−1)	0.1507	−0.8114	0.1184	0.2604(−1)	0.8367(−2)	0.3312(−2)	0.1508(−2)
43	0.5452(−3)	0.3758(−2)	0.2255(−1)	0.1768	−0.8549	0.1331	0.3190(−1)	0.1097(−1)	0.4602(−2)
53	−0.1932(−3)	−0.1194(−2)	−0.5918(−2)	−0.2937(−1)	−0.2010	0.9005	−0.1466	−0.3781(−1)	−0.1381(−1)
63	−0.7905(−4)	−0.4571(−3)	−0.2037(−2)	−0.8360(−2)	−0.3636(−1)	−0.2240	0.9483	−0.1592	−0.4381(−1)
73	−0.3579(−4)	−0.2000(−3)	−0.8330(−3)	−0.3078(−2)	−0.1108(−1)	−0.4360(−1)	−0.2464	0.9990	−0.1711
83	−0.1747(−4)	−0.9693(−4)	−0.3858(−3)	−0.1333(−2)	−0.4322(−2)	−0.1411(−1)	−0.5114(−1)	−0.2685	1.0528
04	0.7753	−0.6191(−1)	0.9122(−2)	−0.2172(−2)	−0.6784(−3)	−0.2544(−3)	−0.1087(−3)	−0.5114(−4)	−0.2585(−4)
14	−0.9822(−1)	0.8187	0.8777(−1)	−0.1570(−1)	−0.4265(−2)	−0.1470(−2)	−0.5955(−3)	−0.2719(−3)	−0.1363(−3)
24	−0.1020(−1)	−0.1411	−0.8637	−0.1078	−0.2209(−1)	−0.6641(−2)	−0.2480(−2)	−0.1073(−2)	−0.5186(−3)
34	−0.2091(−2)	−0.1779(−1)	0.1756	0.9104	−0.1248	−0.2843(−1)	−0.9274(−2)	−0.3705(−2)	−0.1698(−2)
44	−0.5969(−3)	−0.4158(−2)	0.2536(−1)	−0.2063	0.9591	−0.1400	−0.3477(−1)	−0.1215(−1)	−0.5142(−2)
54	−0.2103(−3)	−0.1309(−2)	0.6557(−2)	−0.3307(−1)	−0.2347	1.0100	−0.1538	−0.4115(−1)	−0.1527(−1)
64	−0.8570(−4)	−0.4982(−3)	0.2237(−2)	−0.9276(−2)	−0.4100(−1)	−0.2619	1.0636	−0.1667	−0.4760(−1)
74	−0.3868(−4)	−0.2170(−3)	0.9093(−3)	−0.3385(−2)	−0.1232(−1)	−0.4923(−1)	−0.2884	1.1202	−0.1788
84	−0.1881(−4)	−0.1048(−3)	0.4193(−3)	−0.1457(−2)	−0.4762(−2)	−0.1570(−1)	−0.5782(−1)	−0.3147	1.1805
05	−0.8537	0.6417(−1)	0.9777(−2)	0.2361(−2)	0.7440(−3)	0.2807(−3)	0.1206(−3)	0.5703(−4)	0.2896(−4)
15	0.1118	−0.9012	0.9080(−1)	0.1680(−1)	0.4633(−2)	0.1611(−2)	0.6566(−3)	0.3012(−3)	0.1516(−3)
25	−0.1120(−1)	−0.1607	0.9505	−0.1113	−0.2361(−1)	−0.7205(−2)	−0.2715(−2)	−0.1183(−2)	−0.5741(−3)
35	0.2261(−2)	0.1954(−1)	0.2003	−1.0016	0.1286	0.3034(−1)	0.1005(−1)	0.4052(−2)	0.1869(−2)
45	0.6396(−3)	0.4502(−2)	0.2790(−1)	0.2354	−1.0549	0.1439	0.3705(−1)	0.1315(−1)	0.5619(−2)
55	0.2240(−3)	0.1405(−2)	0.7109(−2)	0.3643(−1)	0.2681	−1.1108	0.1578	0.4378(−1)	0.1651(−1)
65	0.9095(−4)	0.5314(−3)	0.2404(−2)	0.1007(−1)	0.4523(−1)	0.2995	−1.1695	0.1706	0.5056(−1)
75	0.4096(−4)	0.2305(−3)	0.9716(−3)	0.3645(−2)	0.1340(−1)	0.5438(−1)	0.3302	−1.2317	0.1824
85	0.1986(−4)	0.1109(−3)	0.4461(−3)	0.1560(−2)	0.5136(−2)	0.1710(−1)	0.6397(−1)	0.3607	−1.2979

Table 2 Transversal and longitudinal static polarizabilities $\alpha^{(t,l)}_{vLFSJJ_z}$ (in a.u.) for the hyperfine sub-levels $\Delta E^{(\mathrm{hfs})vLFSJ}$ (in MHz) of selected ro-vibrational levels of HD^+

$\Delta E^{(\mathrm{hfs})vLFSJ}$	FSJ	$\lvert J_z \rvert$	$\alpha^{(t)}_{vLFSJ}$	$\alpha^{(l)}_{vLFSJ}$
Ro-vibrational level $(vL) = (42)$				
−650.324	013	0	42.7	−83.0
		1	32.2	−62.1
		2	0.8	0.8
		3	−51.6	105.5
−645.345	012	0	26.9	−51.6
		1	13.8	−25.4
		2	−25.4	53.1
−640.561	011	0	37.4	−72.5
		1	−17.6	37.4
69.582	102	0	48.9	−95.5
		1	24.8	−47.4
		2	−47.4	97.1
132.846	111	0	33.5	−64.8
		1	−15.6	33.5
151.073	112	0	29.9	−57.6
		1	15.3	−28.4
		2	−28.4	59.1
175.547	113	0	42.6	−83.0
		1	32.2	−62.1
		2	0.8	0.8
		3	−51.6	105.4
216.408	120	0	0.8	0.8
240.084	121	0	−32.0	66.2
		1	17.1	−32.0
267.985	122	0	−9.2	20.6
		1	−4.2	10.7
		2	10.7	−19.1
		0	11.3	−20.3
		1	8.7	−15.1
		2	0.8	0.8
290.217	123	3	−12.4	27.1
295.770	124	0	38.2	−74.1
		1	32.6	−62.9
		2	15.7	−29.2
		3	−12.3	27.0
		4	−51.6	105.6
Ro-vibrational level $(vL) = (01)$				
−707.913	012	0	59.5	−116.5
		1	30.2	−57.8
		2	−57.8	118.2
−704.061	011	0	−57.8	118.2
		1	30.2	−57.8
−699.728	010	0	0.9	0.9
80.016	101	0	98.8	−195.1
		1	−48.1	98.8
153.513	111	0	−33.8	70.2
		1	18.2	−33.8
165.298	110	0	0.9	0.9
183.702	112	0	59.6	−116.6
		1	30.2	−57.9
		2	−57.9	118.3

Table 2 (continued)

$\Delta E^{(\mathrm{hfs})vLFSJ}$	FSJ	$\lvert J_z \rvert$	$\alpha^{(t)}_{vLFSJ}$	$\alpha^{(l)}_{vLFSJ}$
269.293	121	0	8.0	−13.3
		1	−2.7	8.0
312.575	123	0	47.8	−93.1
		1	36.1	−69.6
		2	0.9	0.9
		3	−57.9	118.3
314.236	122	0	−57.8	118.2
		1	−28.5	59.5
		2	59.5	−116.5

interest is also the consideration of the spectra in the realistic case of superimposed external magnetic and electric fields.

References

1. Bakalov, D., Korobov, V.I., Schiller, S.: Magnetic field effects in the transitions of the HD$^+$ molecular ion and precision spectroscopy. J. Phys. B: At. Mol. Opt. Phys. **44**, 025003 (2011)
2. Bressel, U., Borodin, A., Shen, J., et al.: Addressing and manipulation of individual hyperfine states in cold trapped molecular ions and application to HD$^+$ frequency metrology. Phys. Rev. Lett., in review (2012)
3. Moss, R.E., Valenzano, L.: The dipole polarizability of the hydrogen molecular cation HD$^+$ and other isotopomers. Mol. Phys. **100**, 1527 (2002)
4. Karr, J.-Ph., Kilic, S., Hilico, L.: Energy levels and two-photon transition probabilities in the HD$^+$ ion. J. Phys. B: At. Mol. Opt. Phys. **38**, 853 (2005)
5. Koelemeij, J.C.J.: Infrared dynamic polarizability of HD$^+$ rovibrational states Phys. Chem. Chem. Phys. **13**, 18844 (2011)
6. Bakalov, D., Korobov, V.I., Schiller, S.: High-precision calculation of the hyperfine structure of the HD$^+$ ion. Phys. Rev. Lett. **97**, 243001 (2006)
7. Esry, B.D., Sadeghpour, H.R.: Adiabatic formulation of the heteronuclear hydrogen molecular ion. Phys. Rev. **A 60**, 3604 (1999)

Table 2 (continued)

Hyperfine Interact (2012) 210:33–37
DOI 10.1007/s10751-012-0566-y

Lorentz invariance on trial in the weak decay of polarized atoms

Stefan E. Müller

Published online: 31 January 2012

Abstract The invariance of the laws of physics under Lorentz transformations is one of the most fundamental principles underlying our current understanding of nature. In theories trying to unify the Standard Model with quantum gravity, this invariance may be broken, and dedicated high-precision experiments at low energy could be used to reveal such suppressed signals from the Planck scale. We will test Lorentz invariance searching for a dependence of the decay rate of spin-polarized nuclei on the daily, yearly or deliberate re-orientation of the spin. Observation of such a dependence would imply a breakdown of Lorentz invariance.

Keywords Lorentz invariance · Optical pumping · β^+-decay

1 Introduction

The invariance of physical laws under rotations in 3-dimensional space and velocity changes (boosts) of the reference frame of a physical system is commonly referred to as Lorentz invariance. Lorentz invariance is a basic building block of both the theory of Special Relativity, as well as the local quantum field theories that build the Standard Model of particle physics. In addition, Lorentz invariance is a local symmetry of General relativity, and is connected to CPT invariance via the CPT theorem which states that a Lorentz invariant local quantum field theory must also be invariant under CPT-transformations. It has been proven that any theory which violates CPT should also violate Lorentz invariance [1] (however, this argument does not necessarily hold the other way around). Certain theoretical models of quantum gravity contain terms which violate Lorentz invariance, and this could lead to manifestations of Lorentz invariance violation also in low-energy observables accessible in laboratory experiments. Kostelecký and coworkers have developed a

S. E. Müller (✉)
Kernfysisch Versneller Instituut, Zernikelaan 25, 9747AA Groningen, The Netherlands
e-mail: S.Mueller@kvi.nl

theoretical framework named "Standard Model Extension" (SME) that contains all the properties of the Standard Model and General Relativity, but additionally contains all terms violating Lorentz invariance and CPT symmetry via hidden background fields connected to the vacuum [2]. There are many experimental tests constraining the parameters of the SME [3], most of which are tests of QED or gravity and astrophysical observations. The weak sector is tested mostly in CPT tests of neutral meson or neutrino oscillations. The electroweak decay sector is up to now largely unexplored.

In a simplified phenomological description, the differential decay rate of a weakly decaying nucleus with nuclear spin **I** can be written as[1]

$$\frac{d\Gamma}{dEd\Omega} \sim \left(1 + A_0 \frac{\langle \mathbf{I} \rangle}{I} \cdot \frac{\mathbf{p}}{E}\right) + \xi_1 \left(1 + \xi_A \left(\hat{\mathbf{p}} \cdot \frac{\langle \mathbf{I} \rangle}{I}\right)\right) \hat{\mathbf{p}}\hat{\mathbf{n}} + \xi_2 \frac{\langle \mathbf{I} \rangle}{I} \hat{\mathbf{n}} + \xi_3 \hat{p}_i \left(\frac{\langle \mathbf{I} \rangle}{I}\right)_j \rho^{ij} \quad (1)$$

in which **p** and E are the momentum and energy of the decay electron, and the $\xi_{1,2,3,A}$ describe the coupling strengths to the hypothetical vector and rank 3 tensor fields $\hat{\mathbf{n}}$ and ρ^{ij} which violate Lorentz invariance by defining a "preferred" reference frame. The first term in (1) describes the Standard Model contribution to the decay rate. Any non-zero ξ would indicate a violation of Lorentz invariance. Precise measurements of the decay rate as a function of the experimental system's rotation against the preferred reference frame can therefore be used to probe a possible violation of Lorentz invariance by looking for changes in the decay rate when the experimental system rotates on the earth (daily variations), when the experimental system rotates with the earth around the sun (annual variations), or by deliberately reorienting the experimental system on shorter timescales (i.e. by periodical reorientation of the nuclear spin **I**).

2 Testing Lorentz symmetry at KVI

We are setting up a test of Lorentz invariance using short-lived alkali isotopes produced with the AGOR cyclotron at KVI. To reduce effects of contamination from other isotopes, it is desirable to stop the atoms in a buffer gas and transport them away from the production beamline. Nuclear polarization is obtained via optical pumping using a weak magnetic field and circularly polarized laser light. Changing the circular polarization of the laser light will reverse the nuclear polarization. Detecting a change in the decay lifetime when switching the nuclear polarization would indicate the presence of violation of Lorentz invariance. Figure 1 depicts a simulation of decay rate over time assuming a change in the decay lifetime at each reversal of the nuclear spin. A signal of this type can be looked for detecting the β-particle (or the corresponding annihilation photons at 511 keV for β^+ decays) or photons from decays of daughter particles. In addition, monitoring the fluorescence produced in the optical pumping process gives an indication of the number of atoms

[1] J. P. Noordmans, R. G. E. Timmermans (private communication)

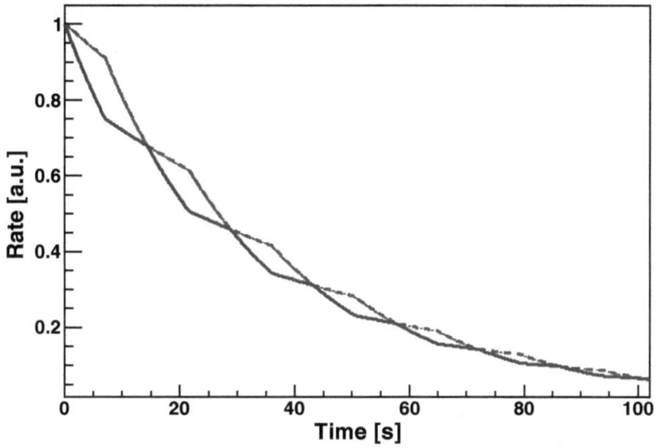

Fig. 1 Simulation of the decay rate of ^{80}Rb under periodic reversal of nuclear spin, assuming a change in decay lifetime of ±50%. The expected effect in lifetime change in the experiment is smaller by many magnitudes. *Solid* and *dashed lines* represent different sign (+ or −) of the change in lifetime for a given orientation of the nuclear spin

which have not decayed yet and may therefore also be used to reveal changes in the decay rate.

Efforts up to now are concentrated on the case of ^{80}Rb atoms ($\tau_{1/2} = 33.4$ s, see http://www.nndc.bnl.gov/chart/chartNuc.jsp). Other isotopes like ^{21}Na ($\tau_{1/2} = 22.5$ s) or ^{20}Na ($\tau_{1/2} = 0.448$ s) are also possible candidates to study Lorentz invariance in weak decays. One common requirement to all candidates is the pure Gamov-Teller decay. In this way, one can use the parity-violating asymmetry in the emission of the β-particle to monitor the nuclear polarization of the atoms.

2.1 Production and transport

^{80}Rb is produced in situ via a ^{82}Kr(p,3n)^{80}Rb reaction using a 10 bar krypton gas target. In this configuration, one can obtain up to 10^9 decays/s. The large pressure of the krypton gas reduces the diffusion of the ^{80}Rb atoms and thus prevents the adsorption at the inner wall of the production target tube. This makes it possible to transport the radioactive isotopes with the buffer gas out of the production target by means of a small circulation pump. During one circulation cycle, the gas flows past a stack of thin titanium foils, which are stacked at distances of few mm. The ^{80}Rb atoms diffuse to the surface of the foil, and are extracted from the buffer gas by adsorption at room temperature [4]. After pumping away the krypton gas, an electrical current through the titanium foils will heat them up to about 900 K, releasing the rubidium atoms to the optical cell in which they will subsequently be spin-polarized. The removal of the krypton gas is mandatory because the spin-polarization needs much lower buffer gas pressures (typically 10–100 mbar) than the 10 bar needed to produce the rubidium atoms.

2.2 Polarization via optical pumping and signal detection

The rubidium atoms are polarized via "optical pumping" (see e.g. [5]) using a weak magnetic field and circularly polarized laser light with a wavelength corresponding to the D1 or D2 transition of rubidium (795 or 780 nm). As shown in Fig. 2, the atoms will be forced into a "stretched" ground state with both the polarizations of the nucleus and the electron cloud aligned in the same direction along the magnetic

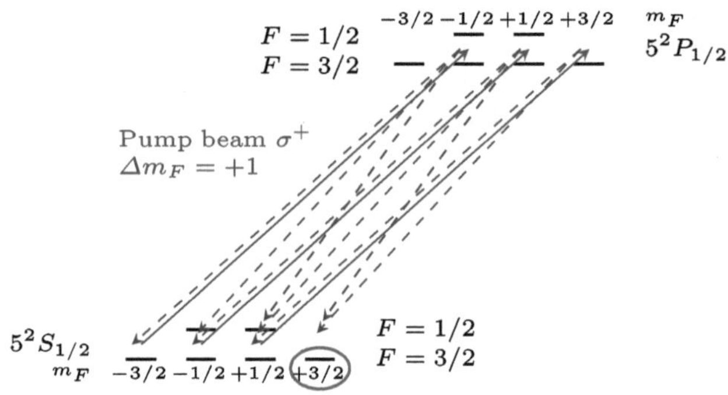

Fig. 2 Optical pumping of ^{80}Rb (I=1) using the $5^2S_{1/2} \rightarrow 5^2P_{1/2}$ (D1) transition. After several pumping cycles, atoms will end up in the "stretched" ground state with $F = 3/2, m_F = 3/2$. Nuclear and electronic polarization are both aligned along the magnetic field lines. Switching the helicity from σ^+ to σ^- will result in the atoms occupying the $F = 3/2, m_F = -3/2$ ground state, with nuclear and electronic polarizations both flipped by 180°

Fig. 3 Experimental setup to test Lorentz invariance at KVI (courtesy C. J. G. Onderwater)

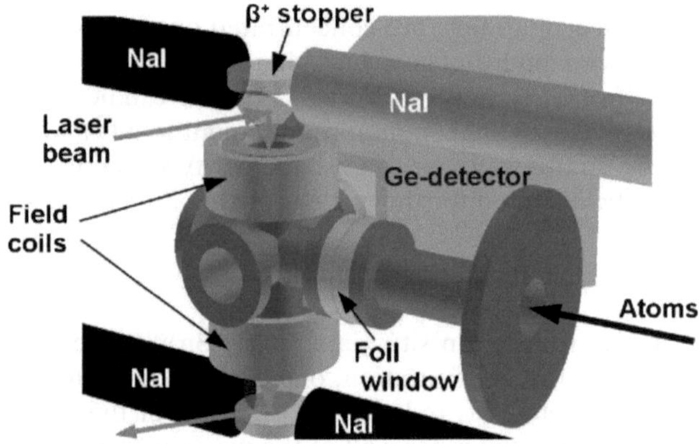

field lines. Changing the helicity of the laser light from σ^+ to σ^- will flip the nuclear and electronic polarization by 180°. Helicity switching is obtained by splitting a laser beam evenly into a σ^+ and a σ^- part and using a chopper wheel to rapidly switch between beams. Frequency stabilization of the laser is achieved by locking it to hyperfine transitions of stable rubidium in a vapor cell. The setup to detect the weak decay of the spin-polarized atoms is shown in Fig. 3. Two pairs of NaI-detectors (above and below the cell containing the radioactive atoms) are used to detect coincidences from 511keV-photons produced by β^+ particles emitted in the upward- or downward direction. The parity-violating asymmetry in the β^+-emission angle for spin-polarized atoms can be used to extract and monitor the amount of nuclear polarization created by the optical pumping process in the sample. A Ge-detector is looking from the side at the cell and is used to observe the decay rate by detecting β-particles or photons from the decay of daughter particles (e.g. in the 25% of the cases when a ^{80}Rb atom decays to an excited state of ^{80}Kr). The Ge-detector is thus providing the lifetime measurement used to test the Lorentz invariance.

3 Conclusions and outlook

Lorentz invariance is a keystone of modern physics, and the observation of violation of Lorentz invariance would highly affect our current understanding of nature. To date, no compelling evidence for Lorentz invariance violation has been found. At KVI, we are currently setting up an experiment which will provide a unique test of Lorentz invariance using the weak decay of spin-polarized atoms. Violation of Lorentz invariance would manifest itself by a change in the decay rate of the atoms while reverting the orientation of their nuclear spin. The corresponding theoretical development for interpreting the observables within the framework of the Standard Model Extension is underway. Tests for the production and transport of ^{80}Rb have been carried out. A laser setup to create nuclear polarization via optical pumping has been built, and the design of the detection system is ready. The next steps are to construct the detection system and test it together with the optical setup using a clean beam of ^{20}Na from the TRIμP isotope separator.

Acknowledgements This research was supported by the Stichting voor Fundamenteel Onderzoek der Materie (FOM) under Program 114 (TRIμP) and FOM projectruimte 08PR2636-1. The presented work has been done in collaboration with E. A. Dijck, J. P. Noordmans, C. J. G. Onderwater, R. G. E. Timmermans and H. W. Wilschut (all KVI Groningen).

References

1. Greenberg, O.W.: Phys. Rev. Lett. **89**, 231602 (2002)
2. Colladay, D., Kostelecky, V.A.: Phys. Rev. D **58**, 116002 (1998)
3. Kostelecky, V.A., Russell, N.: Rev. Mod. Phys. **83**, 11 (2011)
4. Roßbach, H., Eichler, B.: Adsorption von Metallen auf metallische Oberflächen und Möglichkeiten ihrer Nutzung in der Kernchemie. ZfK Report Rossendorf (1984)
5. Demtröder, W.: Laser spectroscopy: Basic Concepts and Instrumentation. Springer, Berlin (1998)

Hyperfine Interact (2012) 210:39–44
DOI 10.1007/s10751-011-0519-x

Strangeness physics at JLab

Satoshi N. Nakamura

Published online: 15 December 2011

Abstract In year 2000, spectroscopic study of hypernuclei with electron beam was started at the Jefferson Lab (JLab). There were many difficulties to be overcome to explore the full potential of this new experimental technique. A decade of efforts and progress for the hypernuclear study at JLab are reviewed.

Keywords Electro-production · Strangeness · Hypernuclei · Spectroscopy

1 Introduction

A hypernucleus is a nucleus which contains one or more hyperons as impurity. By introducing an impurity, 1) system can be studied using an impurity as a probe, and 2) system itself is modified by the existence of an impurity and it might open a door to totally new physics. The physics motivation of the hypernuclear physics is quite similar to it for exotic atoms though spatial size scale is much smaller (a few fm $\sim 10^{-15}$m) and the interaction to bind the system is the strong interaction rather than the electromagnetic force.

The lightest hyperon is Λ which consists of up, down and strange quarks. In order to produce Λ hypernuclei, the most straightforward way is introducing a strange quark from the outside of the system and convert a neutron (up, down, down quarks) to a Λ. The (K^-, π^-) reaction adopts this principle. A K^- meson has a strange quark which is exchanged with a down quark in a neutron to produce Λ and π^-. The other way is the pair production of strange and anti-strange quarks which (π^+, K^+) and (e,e'K^+) reactions use. Strangeness production in a nucleus can be tagged by detection of the emitted K^+ (anti-strangeness).

On behalf of the JLab hypernuclear collaboration.

S. N. Nakamura (✉)
Graduate School of Science, Tohoku University, Sendai, 980-8578, Japan
e-mail: nue@lambda.phys.tohoku.ac.jp

Spectroscopic study of Λ hypernuclei by the (e,e'K$^+$) reaction has unique features as; 1) The (e,e'K$^+$) reaction produced strangeness by the electromagnetic interaction while meson induced reactions do by the strong interaction. 2) The (e,e'K$^+$) reaction converts a proton to a Λ while (K$^-$, π^-), (π^+, K$^+$) reactions convert a neutron to a Λ. 3) Electromagnetic interaction can populate both spin-flip and spin-nonflip states while spin-zero particles such as π and K populate only spin-nonflip states in forward angles. 4) High quality and high intensity primary beam is available for electron and it contributes to high resolution in the missing mass spectroscopy.

The first (e,e'K$^+$) hypernuclear spectroscopic experiment was carried out successfully at Jefferson Lab's (JLab) Hall C in the spring of 2000 by the E89–009 (HNSS, HyperNuclear Spectrometer System) collaboration in order to prove the principle of the (e,e'K$^+$) hypernuclear spectroscopy.

2 Spectroscopic studies of hypernuclei at JLab

There are many interesting features for the (e,e'K$^+$) hypernuclear spectroscopy, but the experiment had been considered to be not realistic before the success of E89–009 at JLab. Difficulties of the experiment are summarized as follows. 1) Electron background from Bremsstrahlung and Møller scattering is very severe in the electron spectrometer. 2) Cross section of hypernuclei for the (e,e'K$^+$) reaction is much less than it for mesonic production. 3) Coincidence measurement is necessary for electro-produced K$^+$ and scattered electron.

Severe electron background deteriorates signal-to-noise ratio and detectors should be developed to operate under the background. Small cross section and coincidence measurement result in less statistics. Therefore, careful design of the experiment and high quality continuous electron beam are essentially important. Requirements for the electron beam are summarized as follows. 1) The electron energy should be higher than $\simeq$1.5 GeV to produce hypernuclei. 2) In order to obtain a hypernuclear mass as a missing mass, the electron beam energy, the momenta of scattered electron and K$^+$ should be measured precisely in coincidence. Therefore a continuous beam is necessary.

CEBAF at JLab provides a high energy (<6 GeV), high intensity (<100μA), small emittance (>10^{-9} m·rad) continuous electron beam. The electron beam energy is known with a precision of better than 10^{-4} which corresponds to a few 100 keV and thus it is not necessary to be measured by a beamline spectrometer. Before the upgrade of MAMI-C at Mainz, only CEBAF can provide such a high-quality electron beam which can be used for the (e,e'K$^+$) hypernuclear spectroscopy.

Taking advantages of CEBAF's electron beam, the JLab E89–009 experiment was performed in 2000. The existing spectrometer, Short Orbit Spectrometer (SOS), was used as a kaon spectrometer and an ENGE type split-pole spectrometer was installed in JLab Hall-C as a scattered electron spectrometer.

The E89–009 experiment demonstrated a great potential of the (e,e'K$^+$) reaction, obtaining a hypernuclear mass spectrum in the ^{12}C(e,e'K$^+$)$^{12}_{\Lambda}$B reaction [1, 2]. It achieved sub-MeV energy resolution for the first time as a hypernuclear reaction spectroscopy.

Though the E89–009 experiment was a great success as a pilot experiment to prove the principle of the (e,e'K$^+$) spectroscopy, it made clear that an improvement of the

Table 1 Kinematic conditions for the second generation hypernuclear experiments at JLab

Experiment (Hall)	E_e (GeV)	E_γ (GeV)	$P_{e'}$ (GeV/c)	P_K (GeV/c)	$\theta_{e'}$ (deg)	θ_K (deg)	W (GeV)	Q^2 (GeV^2/c^2)
E94–107(A)	4.0	2.2	1.8	2.0	6	6	2.2	0.07
E01–011(C)	1.8	1.5	0.3	1.2	4.7*	1–13	1.9	0.01

*Vertically tilted to the dispersion plane

hypernuclear spectrometer system is vitally needed to fully explore the potential of this new experimental technique. A new high-efficiency high-resolution kaon spectrometer and a method to reduce electron background need to be developed.

Two different experimental approaches were carried out as the second generation hypernuclear experiments at JLab. One is the E94–107 experiment in Hall A [3] and the other is the E01–011 experiment in Hall C [4].

The E94–107 experiment used two of high resolution spectrometers (HRS) which are the standard spectrometers in Hall A while the E01–011 experiment designed a new High-resolution Kaon Spectrometer (HKS), dedicated to this program. They based on basically the same experimental principle, but different choice of spectrometer system results in different kinematic conditions. HRS has a longer orbit than HKS and thus higher kaon momentum (2 GeV/c) is necessary to avoid serious K^+ decay loss. E94–107 adopted super-conducting septum magnets and E01–011 used a normal-conducting splitter magnet to place the kaon and electron spectrometers in forward angles to make the virtual photon yield as large as possible (Table 1).

Both experiments introduced new experimental techniques to improve the first generation experiment. E94–107 developed a Ring Image CHerenkov counter (RICH) [5] to select 2.0 GeV/c K^+s cleanly, a waterfall target as a proton target for the calibration and as the ^{16}O target for $^{16}_{\Lambda}N$ hypernucleus study.

E01–011 in Hall-C newly constructed a new kaon spectrometer, HKS with an intention to realize both large detection efficiency and high-resolution, simultaneously. It also adopted a new experimental configuration, 'tilt method' to avoid very forward electrons associated with Bremsstrahlung and Møller processes and thus a higher luminosity ($\propto$ target thickness $\times$ beam intensity) can be accepted.

In the pilot experiment, zero-degree electrons were measured to maximize the virtual photon yield. However, it caused serious electron background in the electron spectrometer. The counting rate of the e' spectrometer (ENGE) in E89–009 was over 200 MHz and it limited usable beam intensity and target thickness. In E01–011, the ENGE spectrometer was vertically tilted from the dispersion plane to avoid very forward Bremsstrahlung and Møller scattered electrons (tilt method). Virtual photon yield is reduced by this tilt method, however, background rate is highly suppressed so the virtual photon loss can be compensated by higher beam current and thicker target. The 200 MHz electron background in E89–009 was reduced to be only 1 MHz in E01–011 though luminosity was more than 200 times higher.

The second generation experiment in JLab Hall C, E01–011, took data for $^{12}_{\Lambda}B$, $^{7}_{\Lambda}He$ and $^{28}_{\Lambda}Al$ which is the first beyond p-shell hypernucleus studied by the (e,e'K^+) reaction in year 2005. It proved that HKS achieved the designed performance and the tilt method successfully reduced the electron background [6].

During the beam time of the second generation experiment E01–011 in Hall C, a new proposal was submitted to JLab PAC28 and accepted as E05–115 [7].

Table 2 Summary of hypernuclei studied at JLab

Target	Hypernuclei	Experiment (Hall)	Citation
p	Λ, Σ^0	E94–107(A), E01–011(C), E05–115 (C)	[8–13]
^{7}Li	${}^{7}_{\Lambda}$He	E01–011(C), E05–115(C)	[6](preliminary)
^{9}Be	${}^{9}_{\Lambda}$Li	E94–107(A), E05–115(C)	[15]
^{10}B	${}^{10}_{\Lambda}$Be	E05–115(C)	Under analysis
^{12}C	${}^{12}_{\Lambda}$B	E89–009(C), E94–107(A), E01–011(C), E05–115(C)	[1, 2, 8–13]
^{16}O	${}^{16}_{\Lambda}$N	E94–107(A)	[16]
^{28}Si	${}^{28}_{\Lambda}$Al	E01–011(C)	[10–13](preliminary)
^{52}Cr	${}^{52}_{\Lambda}$V	E05–115(C)	Under analysis

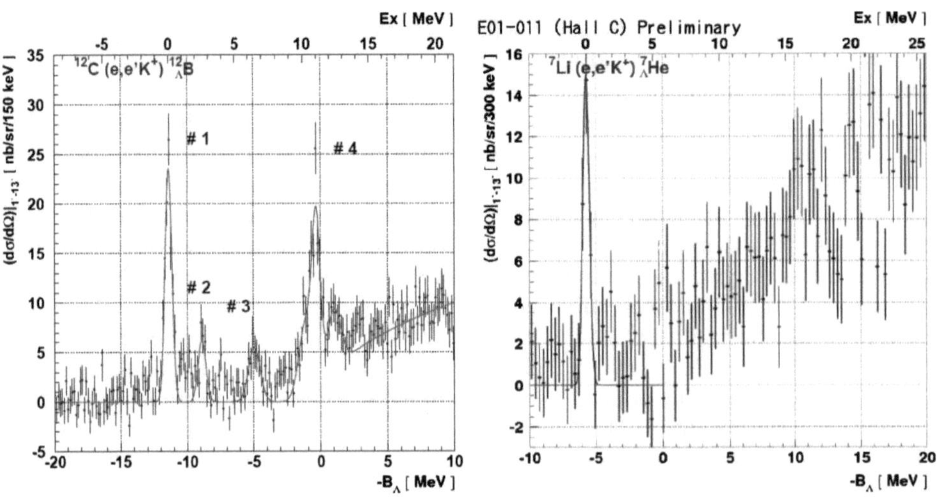

Fig. 1 Preliminary missing mass spectra of ${}^{12}_{\Lambda}$B and ${}^{7}_{\Lambda}$He by E01–011 [6, 10–13]

The third generation experiment, E05–115 adopted a new High-resolution Electron Spectrometer (HES) with a new SPLitter magnet (SPL) in addition to the established spectrometer HKS and the tilt method used in the second generation experiment, E01–011.

Introduction of new HES and SPL in addition to HKS allows us to measure medium heavy ${}^{52}_{\Lambda}$V hypernucleus of which electron background becomes > 10 times severer than the ${}^{12}_{\Lambda}$B measurement. The data taking of E05–115 with HKS and HES was successfully finished in year 2009.

Table 2 shows a summary of hypernuclei studied at JLab.

In this review, two results are picked up. Figure 1 left shows the binding energy spectra of ${}^{12}_{\Lambda}$B obtained by E01–011. Prominent peaks of Λ in s and p-orbits are observed as well as core excited states. The binding energy of the ${}^{12}_{\Lambda}$B ground state was observed as $11.40 \pm 0.01(stat) \pm 0.14(sys)$ MeV (E01–011 preliminary) and it is the consistent result with the emulsion result (11.45± 0.07 MeV) [14]. For (π^+,K$^+$) and (K$^-$,π^-) hypernuclear spectroscopy, the binding energy of ${}^{12}_{\Lambda}$C measured by emulsion is always used as the reference energy of the binding energies. It is important to measure the binding energy of its mirror hypernucleus, ${}^{12}_{\Lambda}$B by a totally independent method from emulsion to confirm that reaction spectroscopy and emulsion experiments give consistent result for $A = 12$ hypernuclear system.

The E01–011 experiment observed the ground state of ${}^{7}_{\Lambda}$He clearly for the first time (Fig. 1 right). It was known that charge symmetry breaking (CSB) terms in Λ-N potential are necessary to explain mass difference between ${}^{4}_{\Lambda}$H and ${}^{4}_{\Lambda}$He measured by emulsion. The comparison between a recent cluster calculation [17] and the ground state energies of the $A = 7$, $T = 1$ hypernuclear iso-triplet including ${}^{7}_{\Lambda}$He indicates that the CSB term in Λ-N potential is not necessary for the $A = 7$ system though it is necessary for the $A = 4$ system. It may be due to too naive treatment of the CSB term in Λ-N potential or problem of the old $A = 4$ hypernuclear data. Therefore it is important to re-examine the $A = 4$ system.

3 Future plans for hypernuclear study with electron beams

At JLab, Hall-A and Hall-C hypernuclear collaborations have started to form a new collaboration to perform more comprehensive program for various targets. A possibility of a new configuration of spectrometers such as HKS plus septum magnets is now under discussion.

Another interesting plan is the decay pion spectroscopy of electro-produced hypernuclei. So far, the (e,e'K$^+$) reaction has been used only for reaction spectroscopy. As a next step, decay products measurement from the hypernuclei was proposed [18, 19]. Some fraction of the hypernuclei produced by the (e,e'K$^+$) reaction directly or indirectory after nuclear fragmentation stopped in the target and decay in two-body to a normal nucleus by emitting a pion. Knowing the normal nucleus and pion masses, only pion momentum is necessary to obtain the hypernuclear mass. This method has a potential to measure masses of light hypernuclei very precisely. As already mentioned, $A = 4$ hypernuclear masses are important parameters to discuss the CSB term in Λ-N potential and this new method gives a change to measure ${}^{4}_{\Lambda}$H mass with a few 10 keV accuracy.

A feasibility study of the decay pion spectroscopy is already started at the recently upgraded MAMI-C accelerator by the A1 hypernuclear collaboration [20] at Mainz University.

In the next decade, the (e,e'K$^+$) spectroscopy will be applied to a wide variety of targets combined with a new experimental technique, the decay pion spectroscopy of electroproduced hypernuclei. Hypernuclear study with electron beams will be more important to provide complimentary information to the hypernuclei to be studied at J-PARC which is going to provide intense K beam soon.

Acknowledgements Daily collaborations with P. Achenbach, F. Garibaldi, O. Hashimoto, J.J. LeRose, A. Margaryan, P. Markowitz, J. Pochodzalla, J. Reinhold, L. Tang, S.A. Wood and other members of JLab hypernuclear are acknowledged. Supports from physics division of Thomas Jefferson National Laboratory (JLab) and supporing staff are acknowledged.

The hypernuclear programs at JLab are partly supported by Japan-MEXT Grant-in-aid for Scientific Research (16GS0201, 15684005, 12002001, 08239102, 09304028, 09554007, 11440070, 15204014), Japan-US collaborative research program, Core-to-core program (21002) and young researcher overseas visits program for vitalizing brain circulation (R2201) by JSPS, grant by INFN and Czech Rep. (202/08/0984), US-DOE contracts (DE-AC05-84ER40150, DE-FG02-99ER41065, DE-FG02-97ER41047, DE-AC02-06CH11357, DE-FG02-00ER41110, DE-AC02-98-CH10886) and US-NSF (013815).

References

1. Miyoshi, T., et al.: Phys. Rev. Lett. **90**, 232502 (2003)
2. Yuan, L., et al.: Phys. Rev. **C 73**, 044607 (2006)
3. Frullani, S., Garibaldi, F., Markowitz, P., Saito, T., et al.: JLab E94–107 Proposal (1994)
4. Hashimoto, O., Tang, L., Reinhold, J., Nakamura, S.N., et al.: JLab E01–011 Proposal (2001)
5. Garibaldi, F.: Nucl. Instrum. Methods **A 502**, 117 (2003)
6. Hashimoto, O., et al.: Nucl. Phys. **A 835**, 121 (2010)
7. Hashimoto, O., Tang, L., Reinhold, J., Nakamura, S.N., et al.: JLab E05–115 Proposal (2005)
8. Iodice, M., et al.: Phys. Rev. Lett. **99**, 052501 (2007)
9. Cusanno, F.: Ph.D. thesis, Universita' Roma Tre (2005)
10. Okayasu, Y.: Doctor thesis, Tohoku Univ. (2008)
11. Seva, T.: Ph.D. thesis, Zagreb Univ. (2009)
12. Matumura, A.: Doctor thesis, Tohoku Univ. (2010)
13. Baturin, P.: Ph.D. thesis, Florida International Univ. (2010)
14. Juric, M., et al.: Nucl. Phys. **B 52**, 1 (1973)
15. Cusanno, F., et al.: Nucl. Phys. **A 835**, 129 (2010)
16. Cusanno, F., et al.: Phys. Rev. Lett. **103**, 202501 (2009)
17. Hiyama, E., et al.: Phys. Rev. **C 80**, 054321 (2009)
18. Margaryan, A., Tang, L., Yuan, L., Nakamura, S.N., Reinhold, J., et al.: JLab E08–012 Proposal (2008)
19. Garibaldi, F., LeRose, J., Margaryan, A., Nakamura, S.N., Reinhold, J., Tang, L., Yuan, L., et al.: JLab E12–10–001 Proposal (2010)
20. Achenbach, P., et al.: Nucl. Phys. **A 835**, 313 (2010)

Hyperfine Interact (2012) 210:45–51
DOI 10.1007/s10751-012-0577-8

Determination of the $\Sigma(1385)^0/\Lambda(1405)$ ratio in p+p collisions at 3.5 GeV

E. Epple · L. Fabbietti for the HADES collaboration

Published online: 29 February 2012

Abstract The aim of the present analysis is to determine the relative production cross sections of the $\Lambda(1405)$ and $\Sigma(1385)^0$ resonances in p+p collisions at $E_{kin} = 3.5$ GeV measured with HADES. Upper and lower limits have been determined for the ratio $\sigma_{(\Sigma(1385)^0+p+K^+)}/\sigma_{(\Lambda(1405)+p+K^+)} = 0.76^{+0.54}_{-0.26}$. The knowledge of this ratio is an essential input for the analysis of the decay $\Lambda(1405) \rightarrow \Sigma^{\pm}\pi^{\mp}$, where an unambiguous separation of the $\Lambda(1405)$ and $\Sigma(1385)^0$ signals is not possible.

Keywords Strangeness · Resonances · Kaon-nucleon interaction

1 Introduction

Among the baryon resonances listed in the Particle Data Group tables [1] the $\Lambda(1405)$ exhibits exceptional characteristics. Indeed, effective field theories [2] describe it as a superposition of a $\Sigma\pi$ resonance and a $\bar{K}N$ bound state contributing to the resonance formation. It is assumed, moreover, that these pole contributions are populated with different strengths depending on the entrance reaction. Thus, it is rather important to access the spectral shape of the $\Lambda(1405)$ in different entrance channels.

The data discussed here have been collected with the **H**igh **A**cceptance **D**i-**E**lectron **S**pectrometer (HADES) [3] at GSI. In the measured p+p reaction ($E_{kin} = 3.5$ GeV, fixed target) the $\Lambda(1405)$ and the $\Sigma(1385)^0$ resonances are produced

E. Epple (✉) · L. Fabbietti
Excellence Cluster 'Origin and Structure of the Universe', 85748 Garching, Germany
e-mail: eliane.epple@ph.tum.de

L. Fabbietti
e-mail: laura.fabbietti@ph.tum.de

L. Fabbietti
Physik Department E12, Technische Universität München, 85748 Garching, Germany

together with a proton and a K^+ meson. The two charged decay channels of the $\Lambda(1405)$ resonance, $\Sigma^{\pm}\pi^{\mp}$, have been also analyzed for this data set and the results are reported in [4]. There, a separation of $\Lambda(1405)$ and $\Sigma(1385)^0$ signals is not possible. Hence, an external reference is needed to determine their relative yields. This reference is delivered by the analysis of the $\Lambda(1405)$ decay channel $\Sigma^0\pi^0$. Indeed, the two resonances can be partially disentangled in this decay channel, since the decay of the $\Sigma(1385)^0$ into $\Sigma^0\pi^0$ is forbidden, and limits for the ratio $\sigma_{(\Sigma(1385)^0+p+K^+)}/\sigma_{(\Lambda(1405)+p+K^+)}$ can be estimated.

2 Analysis procedure

The analysis presented here aims to reconstruct the reaction: $p + p \rightarrow p + K^+ + Y$, where $Y = \Lambda(1405)$ and $Y = \Sigma(1385)^0$ are of interest. When detecting the proton and the K^+ meson the spectral shape of the resonances can be reconstructed by the missing mass $MM_{(pK^+)}$. To further select the data sample and at least partially disentangle these two resonances, their charged decay products have to be detected as well. The analyzed decays are $\Lambda(1405) \rightarrow \Sigma^0\pi^0 \rightarrow (\Lambda\gamma)\pi^0$ and $\Sigma(1385)^0 \rightarrow \Lambda\pi^0$. The second step in the analysis consists in selecting a Λ hyperon in the final state. The decay $\Lambda \rightarrow p\pi^-$ (BR = 63.9 %) is considered, where the Λ candidates have been selected by applying the following cuts: (1) distance between the proton and pion track ($d_{p-\pi^-} < 18$ mm), (2) the distance of closest approach of the Λ to the primary vertex ($DCA_{\Lambda} < 23$ mm), (3) the constraint ($DCA_p < DCA_{\pi^-}$) and (4) a cut on the $p\pi^-$ invariant mass ($1106 < M_{(p,\pi^-)} < 1122\, MeV/c^2$). The obtained missing mass spectrum of a proton and a K^+ meson shown in Fig. 1 consists of two different sets of data. The left panel of Fig. 1 shows the 'HADES data set', where all charged particles were detected within HADES [3]. The right panel of Fig. 1 shows the 'WALL data set', where the proton from the Λ decay is detected in the Forward Wall[1] and the other three particles in HADES. These two data sets can be analyzed independently and provide complementary information as they cover different regions of the phase space (a resolution of $\sigma = 4\,\mathrm{MeV/c^2}$ and $2\,\mathrm{MeV/c^2}$ for the reconstructed Λ mass is obtained for the 'WALL' and 'HADES data set', respectively [5]). Both data sets show a similar resolution in the missing mass $MM_{(p,K^+)}$ ($\sigma = 24\,(18)\,\mathrm{MeV/c^2}$ and $\sigma = 21\,(15)\,\mathrm{MeV/c^2}$ at the Λ and Σ^0 pole positions, respectively). The possible reactions with a final state p+K^++X contributing to the missing mass $MM_{(p,K^+)}$ distribution are listed in Table 1. Besides several reactions involving strangeness production, the total measured yield also contains some misidentification background by p and π^+ identified as K^+ (cf. [6, Section IV]). It is however possible to model this background precisely by a side-band analysis of the reconstructed K^+ mass spectra. An exhaustive description of this data driven side-band method is available in [6]. To determine the exact yield of the misidentification background in the analyzed spectra the side-band data sample has to be compared to the measured sample. In this analysis we choose to compare the spectra by means of the $p - \pi^-$ invariant mass, as illustrated in Fig. 1 of [5],

[1] An external hodoscope placed 7 m downstream of the target with a time resolution of 500–700 ns.

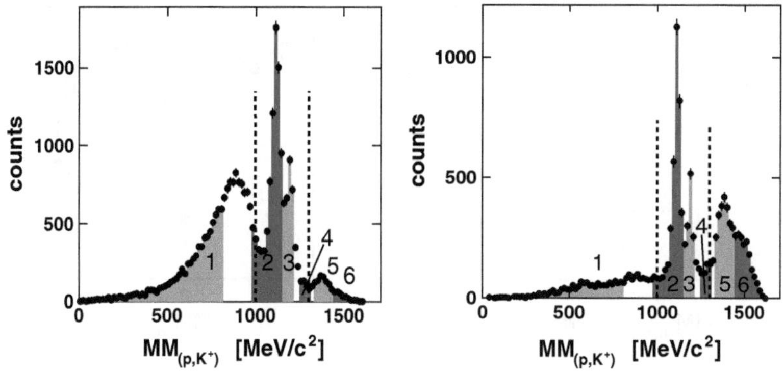

Fig. 1 The missing mass $MM_{(p,K^+)}$ for the 'HADES data set' (*left*) and the 'WALL data set' (*right*). The numbered areas show the six bins in the missing mass $MM_{(p,K^+,p,\pi^-)}$ distribution that have been simultaneously fitted. The *black dashed lines* separates the three intervals in which the validation of the misidentification background has been cross-checked independently

Table 1 Shown is a typical outcome of the fit

Number	Color-code	Final state X	Relative scaling factors
1)	Violet	$\Lambda(1405)K^+p$	0.266
2)	Light green	$\Sigma(1385)^0K^+p$	0.14
3)	Dark green	$\Lambda(1520)K^+p$	0.257
4)	Dark blue	ΛK^+p	1
5)	Red	$\Lambda K^+p\pi^0$	0.031
6)	Cyan	Σ^0K^+p	0.348
7)	Pink	$\Sigma^0K^+p\pi^0$	0.023
8)	Beige	$\Sigma^+K^+p\pi^-$	0.008

The relative scaling factors of the simulated channels p+p→X have been normalized to the ΛK^+p yield

where the $p-\pi^-$ invariant mass is displayed under the condition $MM_{(p,K^+)} >$ 0 MeV/c². The side-band data sample was additionally compared to two further $p-\pi^-$ invariant mass spectra, applying the following conditions: $MM_{(p,K^+)} > 1000$ MeV/c² and $MM_{(p,K^+)} > 1300$ MeV/c². These three intervals are defined between the vertical dashed lines, shown in Fig. 1. In this way, the yield of the side-band background could be cross-checked in one observable, independently for different data sub-samples selected by means of $MM_{(pK^+)}$. The precision of the side-band method is limited by the fact that the p- and π^+-misidentification samples are obtained by a side-band selection on the reconstructed kaon mass in two intervals rather distant from the nominal K^+ mass. Even by doing so, some real kaons might be in the two background samples. This contribution is responsible for the fact that the side-band background underneath the $M_{(p,\pi^-)}$ signal is not flat but shows a small $\Lambda(1116)$ signal [5]. The strength of physical signal in the side-band statistic can be evaluated by filtering simulations through the side-band analysis. In this way, one can determine the fraction of signal in the side-band that fulfills all selection criteria. In a later step of the analysis this simulated signal is subtracted from the misidentification background. The reactions given in Table 1 were each simulated with the same statistics of events. The final states $\Lambda/\Sigma^0K^+p\pi^-\pi^+$ and $\Sigma^+K^+p\pi^-\pi^0$ have not been included since studies based on simulations showed that their contribution to the

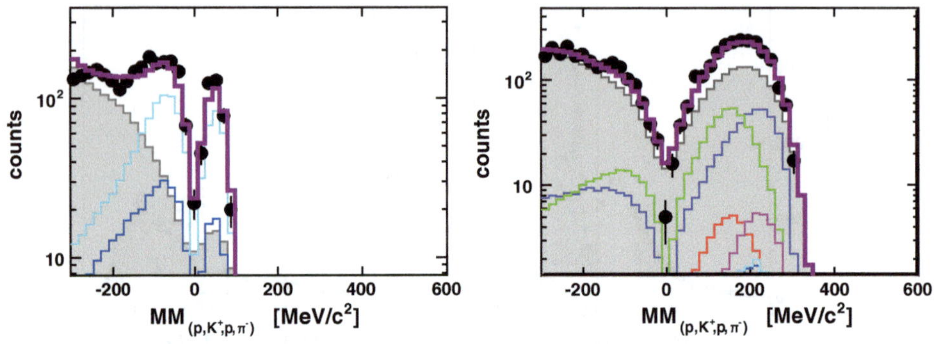

Fig. 2 An example of a fit result in two of the 14 fitted bins of the missing mass $MM_{(p,\pi^-,p,K^+)}$ for the 'WALL data set'. On the *left* the third interval displayed in Fig. 1 and on the *right* the fifth interval is displayed. The experimental points are shown in *black*, the sum of all fitted contributions is given in *purple*. The *gray* area displays the misidentification background. The *colored curves* correspond to the different simulated channels listed in Table 1

measured yield is not significant. Each channel has been simulated to determine the corresponding shape in the $MM_{(pK^+)}$ distribution. Their relative yield has been extracted by fitting the scaled sum of all channels to the experimental distribution. Six intervals have been defined in the missing mass $MM_{(p,K^+)}$ for the two data samples separately, as shown in Fig. 1 by the gray areas. For each of the 12 intervals an experimental distribution in $MM_{(p,\pi^-,p,K^+)}$ is obtained and later fitted with a function composed of the sum of all the simulated contributions listed in Table 1 each multiplied by a fit parameter and the misidentification background, of which the yield is quantitatively determined by the side-band method. During the fitting procedure the yield of $\Lambda(1520)$ has been by trend underestimated. Therefore the simulations could not describe the $MM_{(p,K^+)}$ spectrum in the mass region around 1520 MeV/c^2. An additional distribution corresponding to the interval of $1450 < MM_{(p,K^+)} < 1620$ MeV/c^2 for both data sets has been included in the fit to solve this problem. In total 14 experimental distributions were fitted simultaneously.

In Fig. 2 the experimental distributions of two out of the 14 bins are shown together with a result of the fit. The following constraints have been applied during the fit: $\sigma_{(\Lambda K^+ p\pi^0)}$ and $\sigma_{(\Sigma^0 K^+ p\pi^0)} < \sigma_{(\Lambda K^+ p)}$. A further constraint has been obtained from the analysis described in [4]. Here, the contribution from the final state $\Sigma^+ K^+ p\pi^-$ appears as a rather flat distribution under the $\Lambda(1405)$ resonance signal in the $MM_{(pK^+)}$ spectrum. The ratio of the resonance signal to the flat contribution has been determined. It results in the constraint $\sigma_{(\Sigma^+ K^+ p\pi^-)} < 0.3 \cdot (\sigma_{(\Lambda(1405)K^+ p)} + \sigma_{(\Sigma(1385)^0 K^+ p)})$. In order to describe both data sets simultaneously with the same simulations it has been necessary to include a non-isotropic angular distribution for the simulation of the channel $\Lambda K^+ p$. The reason for this is that the $\Lambda(1116)$ is produced with an angular anisotropy in p+p collisions, as discussed in [7]. The simulations were filtered with the Λ angular distribution in the p-p CM system extracted at a beam kinetic energy of 2.4 GeV [7]. The experimental data could have been described much better with this modified simulation. Recent data from the same experiment [6] have shown, that also the $\Sigma(1385)^+$ resonance is produced with an angular anisotropy. Thus in a second step the simulations for the final state $\Sigma(1385)^0 K^+ p$ were weighted with the angular distribution of the $\Sigma(1385)^+$

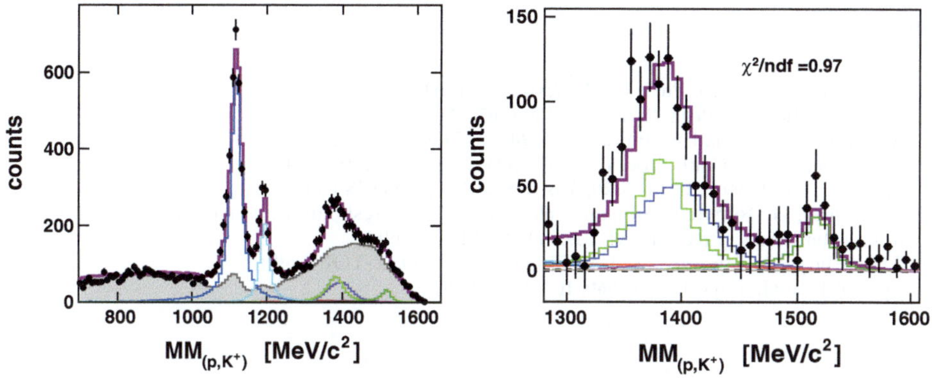

Fig. 3 Fitted distributions shown together with the experimental distribution of $MM_{(p,K^+)}$ (*left*) and for the resonance mass area (*right*), after subtracting the misidentification background. The experimental points are *black*, the sum of all fitted contributions is shown in *purple*. The *gray* area is the misidentification background. The *colored curves* correspond to the different simulated channels listed in Table 1

in the pp CM frame reported in [6] to study the impact on the fit parameters. A final result is shown in Fig. 3, where the $MM_{(p,K^+)}$ experimental distribution is shown together with the contributions of the misidentification background and the simulated channels for the whole range (left panel) and for the range, where the $\Lambda(1405)$ appears after subtraction of the misidentification background (right panel). The obtained χ^2 per degree of freedom of the simultaneous fit to the HADES and WALL data sets amounts to 2.1. This corresponds to the scaling factors for the different channels listed in Table 1. Systematic studies were carried out by varying the track cuts and the invariant mass interval for the $\Lambda(1116)$ (1), (2) and (4) by ± 10 and $\pm$ 20 % and varying the yield of the misidentification background by 4 to 6 %. Furthermore, the stability of the fit has been tested by assuming either a flat angular distribution for the $\Sigma(1385)^0$ production or the anisotropic one extracted from the $\Sigma(1385)^+$ production. Within these variations we have obtained a distribution of the $\sigma_{\Sigma(1385)^0 K^+ p}/\sigma_{\Lambda(1405) K^+ p}$ ratio of $0.76^{+0.54}_{-0.26}$ (syst.), where the systematic error is the σ value of the distribution obtained by the cut variations. These values can be compared with the result obtained by the ANKE collaboration at $E_{kin} = 2.8$ GeV [8], where 0.89 ± 0.46 (stat.)± 0.5 (syst.) has been determined as ratio. The extracted value for the ratio of $\Sigma(1385)^0/\Lambda(1405) = 1.1$ has been employed in the analysis described in [4].

Acknowledgements The author gratefully acknowledges support from the TUM Graduate School. The following funding are acknowledged: LIP Coimbra, Coimbra (Portugal): PTDC/FIS/113339/2009, SIP JUC Cracow, Cracow (Poland): NN202286038, NN202198639, HZ Dresden-Rossendorf, Dresden (Germany): BMBF 06DR9059D, TU Muenchen, Garching (Germany) MLL Muenchen DFG EClust: 153 VH-NG-330, BMBF 06MT9156 TP5 TP6, GSI TMKrue 1012, GSI TMFABI 1012, NPI AS CR, Rez (Czech Republic): MSMT LC07050, GAASCR IAA100480803, USC—S. de Compostela, Santiago de Compostela (Spain): CPAN:CSD2007-00042, Goethe Univ. Frankfurt (Germany): HA216/EMMI, HIC for FAIR (LOEWE), BMBF06FY9100I, GSI F&E01, CNRS/IN2P3 (France).

References

1. Nakamura, K., et al.: J. Phys. G **37**, 075021 (2010) (PDG)
2. Hyodo, T., Jido, D.: Prog. Part. Nucl. Phys. **67**, 55–98 (2012)
3. Agakishiev, G., et al.: Eur. Phys. J. A **4**, 243–277 (2009) (HADES Coll.)
4. Siebenson, J.: AIP Conf. Proc. **1322**, 389–393 (2010) (HADES Coll.)
5. Epple, E.: Int. J. Mod. Phys. A **26**, 616–618 (2011) (HADES Coll.)
6. Agakishiev, G.: Baryonic resonances close to the $\bar{K}$N threshold: the case of $\Sigma(1385)^+$ in pp collisions, submitted to PRC. arXiv:1109.6806 (HADES Coll.)
7. Abdel-Bary, M.: Eur. Phys. J. A **46**, 27–44 (2010) (COSY-TOF Coll.)
8. Zychor, I., et al.: Phys. Lett. B **660**, 167–171 (2008)

The HADES collaboration

G. Agakishiev[6], A. Balanda[3,†] D. Belver[16], A. V. Belyaev[6], A. Blanco[2], M. Böhmer[9], J. L. Boyard[14], P. Cabanelas[16], E. Castro[16], J. C. Berger-Chen[8], S. Chernenko[6], T. Christ[9], M. Destefanis[10], F. Dohrmann[5], A. Dybczak[3], E. Epple[8], L. Fabbietti[8,9], O. V. Fateev[6], P. Finocchiaro[1], P. Fonte[2,18], J. Friese[9], I. Fröhlich[7], T. Galatyuk[7,19], J. A. Garzón[16], R. Gernhäuser[9], C. Gilardi[10], M. Golubeva[11], D. González-Díaz[20], F. Guber[11], M. Gumberidze[14], T. Heinz[4], T. Hennino[14], R. Holzmann[4], A. Ierusalimov[6], I. Iori[22,24,†] A. Ivashkin[11], M. Jurkovic[9], B. Kämpfer[5,21], K. Kanaki[5], T. Karavicheva[11], I. Koenig[4], W. Koenig[4], B. W. Kolb[4], R. Kotte[5], A. Krása[15], F. Krizek[15], R. Krücken[9], H. Kuc[3,14], W. Kühn[10], A. Kugler[15], A. Kurepin[11], R. Lalik[8], S. Lang[4], J. S. Lange[10], K. Lapidus[8], T. Liu[14], L. Lopes[2], M. Lorenz[7], L. Maier[9], A. Mangiarotti[2], J. Markert[7], V. Metag[10], B. Michalska[3], J. Michel[7], E. Morinière[14], J. Mousa[12], C. Müntz[7], L. Naumann[5], J. Otwinowski[3], Y. C. Pachmayer[7], M. Palka[3], Y. Parpottas[12,13], V. Pechenov[4], O. Pechenova[7], J. Pietraszko[7], W. Przygoda[3], B. Ramstein[14], A. Reshetin[11], A. Rustamov[7], A. Sadovsky[11], P. Salabura[3], A. Schmah[8,17], E. Schwab[4], J. Siebenson[8], Yu. G. Sobolev[15], S. Spataro[23], B. Spruck[10], H. Ströbele[7], J. Stroth[4,7], C. Sturm[4], A. Tarantola[7], K. Teilab[7], P. Tlusty[15], M. Traxler[4], R. Trebacz[3], H. Tsertos[12], V. Wagner[15], M. Weber[9], C. Wendisch[5], J. Wüstenfeld[5], S. Yurevich[4], Y. V. Zanevsky[6]

[1]Istituto Nazionale di Fisica Nucleare - Laboratori Nazionali del Sud, 95125 Catania, Italy
[2]LIP-Laboratório de Instrumentação e Física Experimental de Partículas, 3004-516 Coimbra, Portugal
[3]Smoluchowski Institute of Physics, Jagiellonian University of Cracow, 30-059 Kraków, Poland
[4]GSI Helmholtzzentrum für Schwerionenforschung GmbH, 64291 Darmstadt, Germany
[5]Institut für Strahlenphysik, Forschungszentrum Dresden-Rossendorf, 01314 Dresden, Germany
[6]Joint Institute of Nuclear Research, 141980 Dubna, Russia
[7]Institut für Kernphysik, Goethe-Universität, 60438 Frankfurt, Germany
[8]Excellence Cluster 'Origin and Structure of the Universe', 85748 Garching, Germany

[9]Physik Department E12, Technische Universität München, 85748 Garching, Germany
[10]II.Physikalisches Institut, Justus Liebig Universität Giessen, 35392 Giessen, Germany
[11]Institute for Nuclear Research, Russian Academy of Science, 117312 Moscow, Russia
[12]Department of Physics, University of Cyprus, 1678 Nicosia, Cyprus
[13]Frederick University, 1036 Nicosia, Cyprus
[14]Institut de Physique Nucléaire (UMR 8608), CNRS/IN2P3 - Université Paris Sud, 91406 Orsay Cedex, France
[15]Nuclear Physics Institute, Academy of Sciences of Czech Republic, 25068 Rez, Czech Republic
[16]Departamento de Física de Partículas, Univ. de Santiago de Compostela, 15706 Santiago de Compostela, Spain
[17]Lawrence Berkeley National Laboratory, Berkeley, USA
[18]ISEC Coimbra, Coimbra, Portugal
[19]ExtreMe Matter Institute EMMI, 64291 Darmstadt, Germany
[20]Technische Univesität Darmstadt, Darmstadt, Germany
[21]Technische Universität Dresden, 01062 Dresden, Germany
[22]Dipartimento di Fisica, Università di Milano, 20133 Milano, Italy
[23]Dipartimento di Fisica Generale and INFN, Università di Torino, 10125 Torino, Italy
[24]Istituto Nazionale di Fisica Nucleare, Sezione di Milano, 20133 Milano, Italy
[†]Deceased

Hyperfine Interact (2012) 210:53–57
DOI 10.1007/s10751-011-0557-4

Recent results on K^- multinucleon absorption by FINUDA

Alessandra Filippi · Stefano Piano

Published online; 12 January 2012

Abstract An overview of the results achieved so far by the FINUDA experiment in the study of K^- absorption at rest by few nucleons is presented.

Keywords Kaon-nuclear bound systems · Kaon absorption reactions

1 Introduction

A negative kaon, when interacting with a nuclear system, usually gets absorbed in the nuclear periphery. K^-'s can therefore be used as hadronic probes to infer the nucleon densities in the tail regions. Conversely, information on nuclear densities and on composition in the nuclear peripheral region, as well as on the Fermi motion of the outmost nucleons, can be used as inputs to understand the K^- capture mechanism, and, consequently, the features of possible K^--nuclear attractive potentials.

K^- absorption processes could possibly be mediated by the formation of bound kaonic systems, few-body strange aggregates formed by nucleons and an antikaon, which could be kept together by rather strong potentials. Their existence was suggested almost a decade ago [1, 2] and in the last years a few experiments have been pursuing their search. However, conclusive observations are still awaited and the issue is still controversial [3–10]. In fact, according to some theoretical models [8], the possibility to observe such bound states, provided they really exist, is rather critical as their binding could likely be very shallow, and their widths so broad as to prevent their experimental observations.

A. Filippi (✉)
INFN Torino, via P. Giuria, 1 10125 Torino, Italy
e-mail: filippi@to.infn.it

S. Piano
INFN Trieste, via A. Valerio, 2 34127 Trieste, Italy
e-mail: stefano.piano@ts.infn.it

From the experimental point of view their search is based on the studies of reactions in which kaons are absorbed, preferably by light nuclei [6, 7], and in which nuclear-kaonic systems could be formed in intermediate steps. Such systems could undergo a non-mesonic decay in a pionless final state composed by a high momentum hyperon and a nucleon or a light nucleus (*i.e.* a deuteron or a triton), angularly correlated. The large momentum of the particles emitted in a multi-nucleon absorption (> 500 MeV/c) offers an experimental handle to clearly separate such reactions from those occurring on a single nucleon.

Results on many-nucleon absorptions are scarce. Apart from a milestone work on K^{-4}He reactions in bubble chamber [11], few other data exist, mainly with large errors. The oldest ones are related to the interaction of K^-'s on a hydrocarbon mixture, or on Ne (both in bubble chambers [12, 13]) or on nuclear emulsions, composed by mixtures of rather heavy nuclei [14, 15]. More recently, the E549 experiment added new valuable information on the K^4He reaction, studying also the three body absorption mechanism [16]. The contribution of mesonless kaonic absorption reactions is as sizeable as 20%, in medium-heavy nuclei. In ^{4}He [11], the non-mesonic production of Λ (and Σ^0) is about 12% (per K^-_{stop}), and for the non-mesonic production of Σ^+ and Σ^- the rates are 1% and 3.6%, respectively. The non-mesonic rates, as deduced for heavier targets [12, 13], depend almost linearly on A, which calls for the already mentioned peripheral nature of the absorption process.

The experimental signature of pionless multi-nucleon kaon absorption reactions, unfortunately, could not be fully exploited by old bubble chamber experiments, as they had not enough resolution to measure precisely high momentum particles. To provide thorough information on absorption processes full spectrometers are needed, with larger acceptance, resolution and P.ID. capabilities. Among them, the FINUDA experiment, installed at the DAΦNE ϕ-factory in LNF (Italy), had been thoroghly dedicated to the study of kaon induced interactions, and is presently delivering a complete set of new results.

In FINUDA negative kaons, produced in the $\phi(1020)$ decay, could be stopped in a set of eight targets (nuclei with $6 \leq A \leq 51$). In two data takings a total amount of about 1 fb^{-1} (e^+e^-) collisions were collected. The apparatus was a magnetic spectrometer with cylindrical symmetry able to track charged particles with a momentum resolution up to 0.6% FWHM, and a particle identification efficiency as large as 96% for pions and protons. It was completed by a time-of-flight system allowing the detection of neutrons with an efficiency of about 8% [17].

In the second FINUDA data taking (954 pb^{-1}) p-shell nuclei were used as targets: ^{6}Li, ^{7}Li, ^{9}Be, ^{13}C and ^{16}O. The heaviest nuclei are subject to largest Final State Interaction (FSI) effects among the emitted particles: as suggested by some Authors [9, 10], these rescatterings are likely to spoil a clean observation of the final state particles' kinematics and could imply severe distortions in the observed spectra. For this reason, after a first exploratory study performed on a mixture of light targets [18], the efforts were focussed on the investigation of kaon induced reactions on a single nuclear species, specifically starting from ^{6}Li, the lightest target available in FINUDA. In ^{6}Li, the FSI effects are considerably reduced.

In addition, the ^{6}Li nucleus features a di-cluster structure, as measured in pionic absorption experiments [19, 20], being formed by a rather loosely bound quasi-

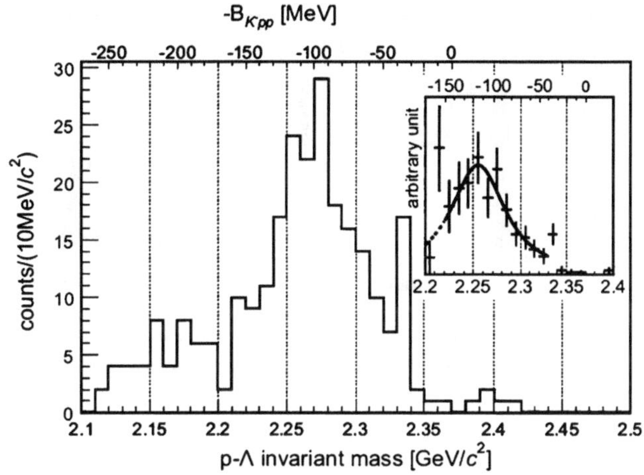

Fig. 1 Invariant mass of the (Λp) system, for K^- interactions in the ^{6}Li, ^{7}Li and ^{12}C targets of FINUDA [18]. In the inset the spectrum obtained after acceptance correction is shown, together with a fit with a gaussian convoluted Lorentzian function to extract the signal features

deuteron and quasi-α pair. Thus, studies of kaon absorption on single ^{6}Li subclusters, in particular the quasi-deuteron, can be performed [21].

In this paper, an overview of the results achieved by FINUDA in the semi-inclusive studies of final states composed by a hyperon (Λ, Σ^-) and one proton, produced by the K^- ^{6}Li interaction, is given. In the two cases, hints on the $K^-(pp)$ and $K^-(pn)$ absorption mechanism, respectively, can be gathered.

2 Study of the ^{6}Li$(K^-_{stop}, \Lambda p)X$ semi-exclusive reaction

Only a short summary is given here, as recent results have already been described in larger detail elsewhere [22, 23]. In 2005 a top-cited paper by FINUDA [18] pointed out the existence of a bump structure, visible in the (Λp) invariant mass spectrum at about 2255 MeV, 67 MeV wide, that could not be explained as due to the simple Quasi-Free (QF) two-nucleon absorption reaction $K^- pp \to \Lambda p$. Figure 1 reports the observed (Λp) invariant mass spectrum [18].

The events were characterized by a marked back-to-back correlation between the emitted Λ and p, which suggested as possible interpretation the existence of a $[K^- pp]$ state, 115 MeV bound, decaying in Λp. The analysis, performed over a sample of mixed light targets (^{6}Li, ^{7}Li and, mainly, ^{12}C), has been recently repeated using ^{6}Li data only, and leads basically to the same conclusions. Thanks to the larger available statistics a missing mass analysis can be performed, which helps disentangling the contributions of some QF two-nucleon induced reactions to the experimental sample. From this analysis, the $K^- pp \to \Sigma^0 p$ reaction, followed by the $\Sigma^0 \to \Lambda\gamma$ decay, appears not to be enough to explain completely the experimental spectra. The possible contribution of $\Sigma\Lambda$ conversion reactions is currently being evaluated; the contributions of some channels can be easily ruled out by studying the shape of several different simulated distributions: in some cases they cannot fit the experimental observations.

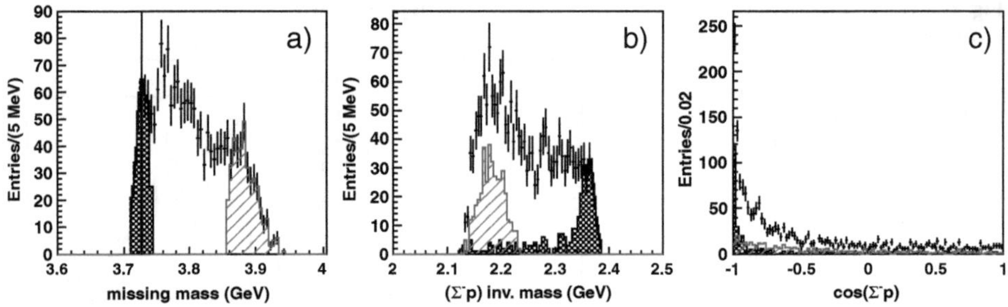

Fig. 2 $\Sigma^- p$ semi-inclusive sample: (**a**) missing mass spectrum of the ^{6}Li$(K^-_{stop}, \Sigma^- p)X'$ reaction; (**b**) invariant mass spectrum of the $(\Sigma^- p)$ system; (**c**) angular distribution of the angle between Σ^- and p

3 Study of the ^{6}Li$(K^-_{stop}, \Sigma^- p)X'$ semi-exclusive reaction

The starting point for the study of the $\Sigma^- p$ final state is the selection of a clean $np\pi^-$ sample. Details on the neutron selection criteria can be found in [24, 25]. Without kinematic cuts, but only the selection of high quality charged tracks, the signal from the Σ^- baryon, observed in its $\Sigma^- \to n\pi^-$ decay mode (B.R. = 99.85%), emerges with $S/B \simeq 1$. The background has therefore to be reduced trying to reject at best contributions from accidentals or competitory reactions which feed the $np\pi^-$ final state.

The main contaminations come from fake γ's wrongly identified as neutrons, or from neutrons detected after their rescattering in the apparatus materials. Also some physical channels can contaminate the selected sample, for instance one-nucleon absorptions followed by $\Sigma\Lambda$ conversions. After the application of suitable kinematic cuts, semi-inclusive $\Sigma^- p$ events ($S/B > 2$) can then be selected considering a narrow window (28 MeV) across the nominal Σ^- mass value. With the selected $\Sigma^- p$ semi-inclusive sample missing mass studies can be performed to single out the contribution of exclusive QF two-nucleon kaon absorptions. The missing mass plot of the selected events is shown in Fig. 2a, while the invariant mass distribution of the $(np\pi^-)$ events is displayed in Fig. 2b (all the distributions are not acceptance corrected). Centered at the ^{4}He mass (3.73 GeV), a peak appears in the missing mass distribution (filled area in the picture): these events belong to the exclusive $K^- {}^6\text{Li} \to \Sigma^- p {}^4\text{He}$ reaction. In the invariant mass distribution they are located on the highest mass side of the available phase space. Figure 2c, which reports the angular distribution between the proton and the $(n\pi^-)$ pair selected in the Σ^- mass region, shows that for these particles a marked back-to-back correlation exists.

The hatched area in Fig. 2a is populated, on the other hand, by events in which an additional pion is missing. These events fill the low mass region in the invariant mass plot, and they are distributed flatly over the angular correlation spectrum. More details on this analysis may be found in [26].

A sizeable part of the missing mass and invariant mass spectra cannot be explained just resorting to the two mentioned QF two-nucleon absorptions. Studies of the reactions filling the experimental spectra are currently underway. Possible contributions could be played by proton, pion and Σ^- rescattering reactions. However, their features hardly match those of all the available experimental distributions.

4 Summary

The amount of data collected by FINUDA on ^{6}Li targets allowed to perform a parallel analysis on the Λp and $\Sigma^- p$ final states, to study different aspects of the kaon absorption by two nucleons, namely $K^-(pp)$ and $K^-(pn)$. In both cases a sizeable part of the experimental spectra fail to be explained by simple QF two-nucleon absorption reactions. As well, in both cases events are characterized by a marked back-to-back correlation of the hyperon-proton pairs.

References

1. Akaishi, Y., Yamazaki, T.: Phys. Rev. **C65**, 044005 (2002)
2. Yamazaki, T., Akaishi, Y.: Nucl. Phys. **B535** 70 (2002)
3. Shevchenko, N.V., Gal, A., Mareš, J.: Phys. Rev. Lett. **98**, 082301 (2007)
4. Weise, W., Härtle, H.: Nucl. Phys. **A804** 173, (2008),
5. Hyodo, T., Weise, W.: Int. J. Mod. Phys. **E19**, 2612 (2010)
6. Mareš, J.: Nucl. Phys. **A804**, 296 (2008)
7. Mareš, J., Friedman, E., Gal, A.: Nucl. Phys. **A770**, 84 (2006)
8. Ramos, A., Oset, E.: Nucl. Phys. **B671**, 153 (2000)
9. Oset, E., Toki, H.: Phys. Rev. **C74**, 015207 (2006)
10. Magas, V.K., Oset, E., Ramos, A., Toki, H.: Phys. Rev. **C74**, 025206 (2006)
11. Katz, P.A., et al.: Phys. Rev. **D1**, 1267 (1970)
12. Vander Velde-Wilquet, C., et al.: Nuovo Cim. **39**, 538 (1977)
13. Moulder, J.W., et al.: Nucl. Phys. **35B**, 332 (1971)
14. Knight, W.L., et al.: Nuovo Cim. **32**, 598 (1964)
15. Davis, H., et al.: Nuovo Cim. **53A**, 313 (1968)
16. Suzuki, T., et al.: Phys. Rev. **C76**, 068202 (2007)
17. For a description of the FINUDA apparatus see, for instance, Zenoni, A.: In: Proceedings of the International School of Physics "E. Fermi" Course CLVIII (Hadron Physics)", SIF-IOS Press, 2005, p. 183
18. Agnello, M., et al.: Phys. Rev. Lett. **94**, 212303 (2005)
19. Favier, J., et al.: Nucl. Phys. **A169**, 540 (1971)
20. Cernigoi, C., et al.: Nucl. Phys. **A352**, 343 (1981); Nucl. Phys. **A456**, 599 (1986)
21. Akaishi, Y., Yamazaki, T.: Nucl. Phys. **A792**, 229 (2007)
22. Piano, S.: Nucl. Phys. **A835**, 35 (2010)
23. Filippi, A., Piano, S.: J. Phys.: Conf. Ser. **312**, 022002 (2011)
24. FINUDA Collaboration, Agnello, M., et al.: Phys. Lett. **B704**, 474 (2011)
25. Piano S.: In Proc. of the EXA11 Conference, Vienna, 5–9 September (2011, this Volume)
26. Agnello, M., et al.: In: Proc. of the STORI11 Conference, Lab. Nazionali di Frascati, http://pos.sissa.it/archive/conferences/150/049/STORI11_049.pdf 9–15 October (2011)

Hyperfine Interact (2012) 210:59–63
DOI 10.1007/s10751-011-0547-6

Momentum dependence of hadronic production of the ϕ-meson and its width in nuclear matter

Michael Hartmann · Burkhard Kämpfer · Yury T. Kiselev · Vacas K. Magas · Eduard Ya. Paryev · Andrey Polyanskiy · Luis Roca · Henry Schade · Colin Wilkin

Published online: 3 January 2012

Abstract Information on the properties of the ϕ meson in the nuclear environment has been derived from its production in proton collisions with C, Cu, Al, and Au nuclear targets. The experiment was carried out with 2.83 GeV protons at the Cooler Synchrotron COSY, with the ϕ being detected via its K^+K^- decay using

M. Hartmann (✉) · A. Polyanskiy
Institut für Kernphysik and Jülich Centre for Hadron Physics,
Forschungszentrum Jülich, 52425 Jülich, Germany
e-mail: m.hartmann@fz-juelich.de

B. Kämpfer · H. Schade
Institut für Strahlenphysik, Helmholtz-Zentrum Dresden-Rossendorf,
01314 Dresden, Germany

Yu. T. Kiselev · A. Polyanskiy
Institute for Theoretical and Experimental Physics, 117218 Moscow, Russia

V. K. Magas
Departament d'Estructura i Constituents de la Matèria and Institut de Ciències del Cosmos,
Universitat de Barcelona, 08028 Barcelona, Spain

E. Ya. Paryev
Institute for Nuclear Research, Russian Academy of Science,
117312 Moscow, Russia

L. Roca
Departamento de Fisica, Universidad de Murcia,
30071 Murcia, Spain

H. Schade
Institut für Theoretische Physik, TU Dresden,
01062 Dresden, Germany

C. Wilkin
Physics and Astronomy Department, UCL, London WC1E 6BT, UK

the ANKE magnetic spectrometer. The measured dependence of the production cross section on the nuclear mass number has been compared with calculations within three different nuclear models. These suggest a significant broadening of the width of the ϕ in medium relative to its vacuum value. The ANKE results obtained in the momentum range $0.6 < p_\phi < 1.6$ GeV/c are compared with data from photoproduction experiments at slightly higher momenta.

Keywords ϕ meson production · Nuclear medium effects

1 Introduction

The study of the effective masses and widths of light vector mesons in nuclear medium has received considerable attention in recent years [1, 2]. Nuclear collisions of elementary particles (γ, π, p), in so-called cold static matter, can show changes in the properties [3–6] of these mesons. Compared to ultra-relativistic heavy-ion collisions (hot and dense matter), the theoretical treatment of such experimental data is more transparent and more closely related to the elementary processes.

The narrow vacuum width of the $\phi(1020)$ ($\Gamma = 4.3$ MeV/c^2) allows one to investigate small changes in its properties. The main modification in nuclear matter is expected to be a broadening of the ϕ spectral function while the mass should hardly shift. Such a broadening should be directly testable by examining the invariant mass spectra. Owing to the small final-state interactions, the dileptonic decay $\phi \to e^+e^-$ has been proposed but such measurements are made difficult by the low e^+e^- branching ratio of about 10^{-4}. The KEK-PS-E325 collaboration has investigated the e^+e^- mass spectra in proton-induced reactions at 12 GeV [7]. From the study of the spectral shape of production on copper relative to that on carbon they deduced a small mass reduction of 3.4% and a width increase by a factor of 3.6 at normal nuclear density ρ_0. This corresponds to an in-medium ϕ width of about 11 MeV in the nuclear rest frame for their average ϕ momentum of about 1 GeV/c.

A much larger in-medium broadening was reported by the LEPS collaboration in ϕ photoproduction on Li, C, Al and Cu targets when detecting the K^+K^- decay [8]. The variation of the production cross section with atomic number A depends on the attenuation of the ϕ flux in the nucleus. This is governed in turn by the imaginary part of the ϕ in-medium self-energy or width. In the low-density approximation (LDA), this width can be related to an effective ϕN total cross section [9], though this is less obvious at higher densities where two-nucleon mechanisms are important. The big advantage of this approach is that one can exploit the large K^+K^- branching ratio ($\approx 50\%$). The ϕN total cross section of about 35 mb deduced at LEPS is significantly bigger than that in free space, viz. ≈ 10 mb [10–12], and would correspond to an in-medium width of about 97 MeV/c^2 in the nuclear rest for $p_\phi \approx 1.8$ GeV/c.

The CLAS collaboration at JLab also studied the variation of ϕ photoproduction on different targets ^{2}H, C, Ti, Fe, Pb [13]. From an analysis of the A-dependence, in the form of transparency ratios normalised to carbon (compare below), they extracted values of $\sigma_{\phi N}$ in the range of 16-70 mb for $p_\phi \approx 2$ GeV/c. The result is not inconsistent with the LEPS findings at slightly lower momenta, although the precision is limited due to the low statistics associated with the detected e^+e^- decay.

Fig. 1 Momentum dependence of the transparency ratios for the four nuclei studied

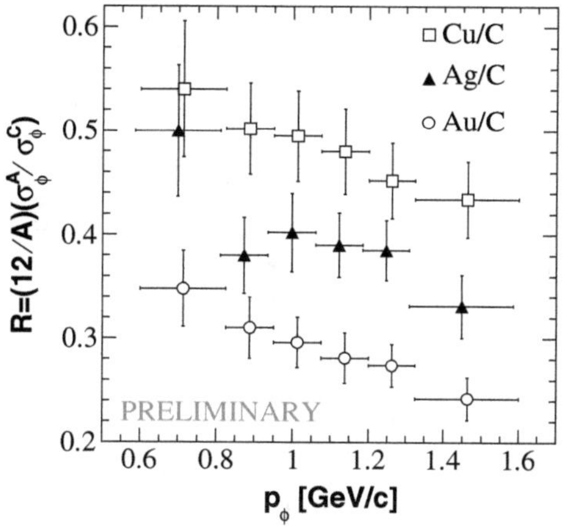

2 Experiment and results

The attenuation method was also used at COSY-ANKE but, in contrast to the photoproduction of LEPS and CLAS, the ϕ mesons were produced in proton-induced collisions at the Cooler Synchrotron of the Forschungszentrum Jülich. The beam energy of 2.83 GeV corresponds to an excess energy of about 76 MeV above the free NN threshold where few production channels are open. The magnetic spectrometer ANKE detected the ϕ-meson via its K^+K^- decay in the angular cone $\theta_\phi < 8°$, were secondary production processes are also expected to be less important.

In the initial stage we studied the nuclear transparency ratios normalised to carbon, $R = (12/A)(\sigma^A/\sigma^C)$, averaged over the ϕ momentum range 0.6–1.6 GeV/c [14, 15]. Here σ^A and σ^C are inclusive cross sections for ϕ production in pA ($A =$ Cu, Ag, Au) and pC collisions. The comparison of the ratios with model calculations [16, 17] yields an in-medium ϕ width of 33–50 MeV/c^2 in the nuclear rest frame at an average ϕ momentum of 1.1 GeV/c for normal nuclear density $\rho_0 = 0.16$ fm^{-3}.

The large number of reconstructed ϕ mesons for each target (7,000–10,000) allows the data to be put into six momentum bins of approximately equal statistics in order to carry out more detailed studies. In Fig. 1 the preliminary results on the momentum dependence of the measured transparency ratios are shown for different nuclei. However, to extract information on the in-medium ϕ width, a reaction model is essential and we consider here three approaches to the problem.

Model 1: The eikonal approximation of the Valencia group [16] uses the predicted ϕ self-energy [18, 19] for both one-step ($pN \to pN\phi$) and two-step production processes, with nucleon and Δ intermediate states.

Model 2: Paryev [17] developed the spectral function approach for ϕ production in both the primary proton-nucleon and secondary pion-nucleon channels.

Model 3: The Rossendorf BUU transport calculation [20] includes a variety of secondary ϕ production processes. In contrast to Models 1 and 2, where ϕ absorption is governed by its width, Γ_ϕ, Model 3 describes it in terms of an effective in-medium ϕN cross section $\sigma_{\phi N}$ that can be related to the ϕ width Γ_ϕ by using the standard low-density relation $\Gamma_0(p) = \rho_0\sigma_0^*(p)p/E_\phi$.

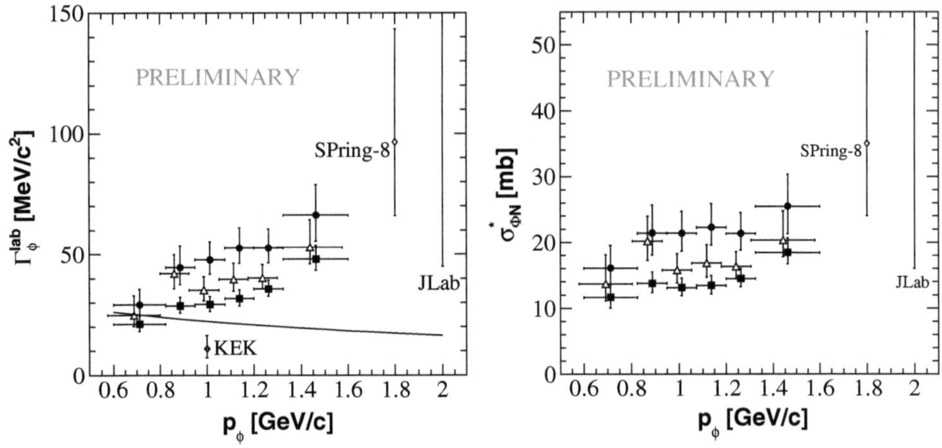

Fig. 2 *Left*: Momentum dependence of the in-medium ϕ width for normal nuclear density extracted using different models: Model 1 (*squares*), Model 2 (*circles*) and Model 3 (*triangles*). Experimental results from KEK-PS-E325 [7], Spring-8 [8] and JLab [13] are also plotted. The theoretical prediction of [18, 19] is shown by the *solid line*. *Right*: The effective ϕN total cross section deduced from the in-medium ϕ widths within the low-density approximation (see text)

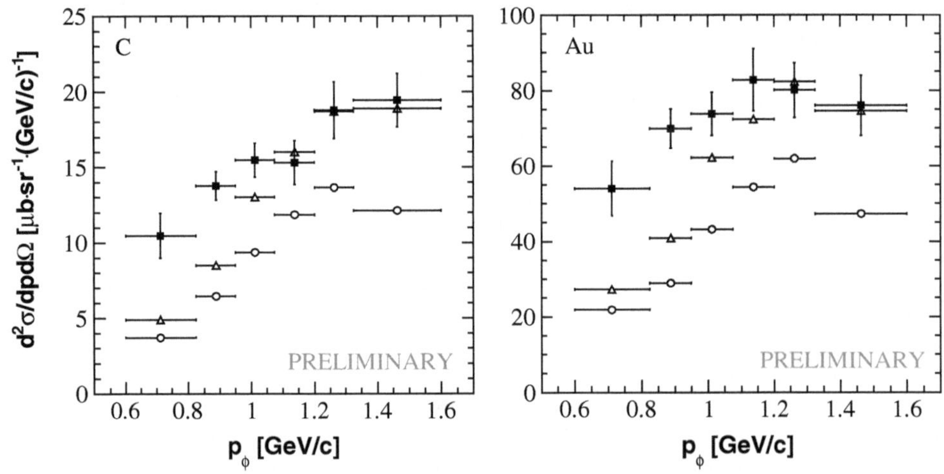

Fig. 3 Comparison of the measured double differential cross section for ϕ production at small angles (*full squares*) for carbon (*left*) and gold (*right*) nuclei with the predictions of Models 2 (*open circles*) and 3 (*open triangles*)

The in-medium ϕ width in the nuclear rest frame at normal nuclear density obtained in these models is presented in Fig. 2. Similar behaviour is seen for all three approaches and the differences come mainly from the choices in the descriptions of secondary production processes. The ϕ width extracted is not in disagreement with the Spring-8 [8] and JLab [13] results that have been determined at slightly higher momentum but it exceeds the Valencia prediction [18, 19].

In order to understand further the model calculations, the double differential cross sections for ϕ production have been evaluated within the ANKE acceptance window for different momentum bins (compare [15, 21]). The experimental results for carbon and gold nuclei are compared in Fig. 3 with the predictions of the Paryev and BUU calculations that used the extracted values of the in-medium ϕ width or $\sigma_{\phi N}$ cross section as input.

The BUU calculation describes rather well the high momenta, where direct ϕ production dominates. Both models strongly underestimate ϕ production at low momenta. This suggests that some process, whose contribution to the ϕ production cross sections increases for low ϕ momenta and with the size of the nucleus, is not included in the models. It is not obvious how such an extra process would effect the transparency ratios and therefore further theoretical studies of secondary production reactions are clearly needed to extract the maximum information from these experiments.

Acknowledgements We wish to thank the COSY machine crew and other members of the ANKE collaboration for their help during the preparation and running of the experiments reported here. The work was supported in part by the BMBF, DFG, COSY-FFE and RFBR.

References

1. Hayano, R.S., Hatsuda, T.: Rev. Mod. Phys. **82**, 2949 (2010)
2. Leupold, S., Metag, V., Mosel, U.: Int. J. Mod. Phys. E **19**, 147 (2010)
3. Oset, E., Ramos, A.: Nucl. Phys. A **679**, 616 (2001)
4. Cabrera, D., Vicente Vacas, M.J.: Phys. Rev. C **67**, 045203 (2003)
5. Klingl, F., Waas, T., Weise, W.: Phys. Lett. B **431**, 254 (1998)
6. Mosel, U.: In: Menze, D.W., Metsch, B.Ch. (eds.) Proc. Baryons '98, p. 629. World Scientific, Singapore (1999); arXiv:nucl-th/9811065 (1998)
7. Muto, R., et al.: Phys. Rev. Lett. **98**, 042501 (2007)
8. Ishikawa, T., et al.: Phys. Lett. B **608**, 215 (2005)
9. Dover, C.B., Hüfner, J., Lemmer, R.H.: Ann. Phys. (N.Y.) **66**, 248 (1971)
10. Sibirtsev, A., Hammer, H.W., Meißner, U.-G., Thomas, A.W.: Eur. Phys. J. A **29**, 209 (2006)
11. Behrend, H.-J., et al.: Nucl. Phys. B **144**, 22 (1978)
12. Lipkin, H.J.: Phys. Rev. Lett. **16**, 1015 (1966)
13. Wood, M.H., et al.: Phys. Rev. Lett. **105**, 112301 (2010)
14. Polyanskiy, A., et al.: Phys. Lett. B **695**, 74 (2011)
15. Hartmann, M., et al.: AIP Conf. Proc. **1322**, 349 (2010)
16. Magas, V.K., Roca, L., Oset, E.: Phys. Rev. C **71**, 065202 (2005)
17. Paryev, E.Ya.: J. Phys. G **36**, 015103 (2009)
18. Cabrera, D., Vicente Vacas, M.J.: Phys. Rev. C **67**, 045203 (2003)
19. Cabrera, D., et al.: Nucl. Phys. A **733**, 130 (2004)
20. Schade, H.: University of Dresden PhD thesis (2010)
21. Polyanskiy, A., et al.: Proceedings of the XIV International Conference on Hadron Spectroscopy, Munich (2011) (to be published in eConf C)

Hyperfine Interact (2012) 210:65–69
DOI 10.1007/s10751-012-0600-0

Strangeness production in AA collisions at SIS18

Norbert Herrmann · Yvonne Leifels for the FOPI collaboration

Published online: 30 March 2012

Abstract Strange particles allow to probe the properties of hot and dense baryonic matter. At incident energies close to the production threshold medium effects on the strange hadrons are magnified. New data on strangeness production obtained with the recently upgraded FOPI setup are presented.

Keywords Heavy ion reactions · Strangeness production · Hypernuclei

1 Introduction

Modifications of hadron properties in dense baryonic matter are a current subject of intensive research in hadron physics. Various theoretical approaches [1–3] agree qualitatively in predicting, for example, modifications of masses and coupling constants for kaons and anti-kaons. Due to the density dependence of the KN/($\overline{\text{K}}$N) potential, the K^- effective mass is expected to drop, whereas the mass of K^+ mesons is predicted to rise with increasing density of nuclear matter. However, the depth of the potential especially of the anti-kaons is currently a matter of controversy (for a recent review see [4]).

From general arguments one can conclude how the medium-modifications of the KN($\overline{\text{K}}$N) potentials influence particle production yields and phase space distributions in heavy ion collisions. However, access to the depth of the in-medium potentials is only possible via microscopic transport codes modeling the dynamical evolution of the reaction. Comparing simultaneously several observables to model calculations

N. Herrmann
Physikalisches Institut der Universität Heidelberg, Philosophenweg 12, 69120 Heidelberg, Germany
e-mail: herrmann@physi.uni-heidelberg.de

Y. Leifels (✉)
GSI Darmstadt, Planckstr. 1, 64291 Darmstadt, Germany
e-mail: y.leifels@gsi.de

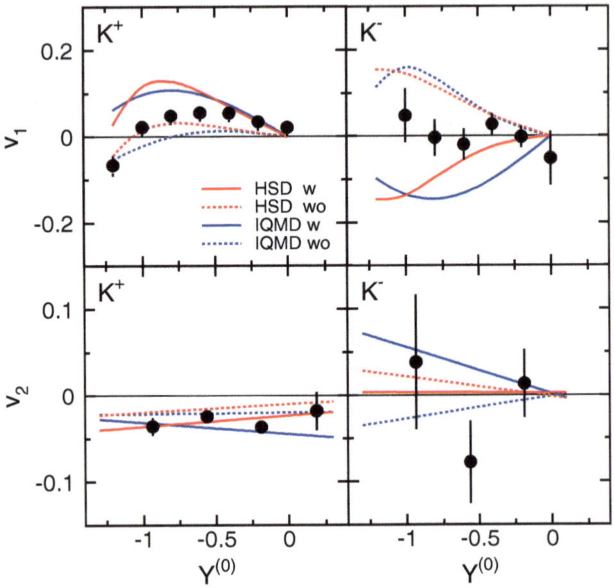

Fig. 1 Directed v_1 and elliptic v_2 flow of charged kaons as function of normalized rapidity $Y^{(0)} = y/y_{cm} - 1$ for NiNi collisions at 1.9 AGeV. The *left/right* panels show v_1 (*upper*) and v_2 (*lower*) of positive/negative kaons. *Black dots* represent the measured data points. *Lines* are results of model predictions. *Dotted lines* are predictions from model calculations assuming no modification of the (anti)kaon-nucleon interaction in medium, *full lines* are model predictions with in medium modifications

is mandatory to resolve ambiguities in the interpretation [4]. In the following new data obtained with the FOPI detector at the SIS18 accelerator of GSI are shown for the reaction Ni+Ni at an incident energy of 1.91 AGeV adding new information towards the goal of extracting the fundamental properties of the interaction of strange particles in dense hadronic matter.

2 Kaon flow

The measurement of the phase space distribution of kaons with respect to the reaction plane is referred to as 'flow' and provides constraints to the in-medium potentials independent of the production cross section [5]. The excellent kaon identification capabilities of the new time-of-flight barrel based on Multi-gap–Resistive–Plate–Chambers (MRPC) [6] surrounding the central drift chamber allows to continue the FOPI efforts [7] to determine the in-medium potential for kaons and anti-kaons by the measurement of kaon and anti-kaon flow in heavy ion collisions.

The direction of the reaction plane is determined by the transverse momentum method [8] making use of particles detected in the forward wall of FOPI only in order to avoid autocorrelations. The data are corrected for the finite reaction plane resolution employing the Ollitrault formalism [9].

Model calculations have to take into account the acceptance of the detector when compared to the data. The data shown in Fig. 1 are limited to the acceptance of the central drift chamber CDC covering a polar angle range of $26° < \Theta_{lab} < 120°$. For kaons additional constraints have to be put on the momenta necessary to obtain sufficient particle identification in the data. In the polar angular range from $30.5° < \Theta_{lab} < 52°$ (MRPC acceptance) kaons are included up to a momentum of 1.0 GeV/c, in the range $53° < \Theta_{lab} < 120°$ (plastic barrel acceptance) the laboratory momentum is limited to 0.5 GeV/c.

Older experimental data for positive kaons and protons for central collisions [7] are reproduced by the new measurements. The new results compared to HSD model predictions confirm the old findings for central collisions: The HSD transport code is only able to reproduce the observations when a repulsive KN potential is considered. A potential of $U_{\text{pot,KN}}(\rho_0) = 20$ MeV at normal nuclear matter density and linearly rising with baryon density as predicted by various theoretical models e.g. [10] is describing the data. The height of the repulsive potential is consistent with the result from the analysis of pion induced reactions measured by the FOPI collaboration [11]. In addition, the HSD model describes the proton flow data measured in the same acceptance range reasonably well.

The sensitivity of the flow variables to medium properties is largest at intermediate impact parameters for which sizeable kaon and anti-kaon flows have been predicted [12]. In the new dataset the measured impact parameter range is much wider than previously corresponding to a sharp-cutoff impact parameter equivalent of $b_{\text{geo}} = 5\,fm$. In Fig. 1 the new data on kaon and anti-kaon flows are compared to the predictions of microscopic transport models. Two different models have been applied, IQMD [13] and HSD [14]. One difference between both models is the depth of the K and $\overline{\text{K}}$ potentials. In contrast to HSD a value of $U_{\text{pot,KN}} = 40$ MeV at normal nuclear matter densities is applied in IQMD. For the $\overline{\text{K}}$N interaction in matter HSD uses a G-Matrix approach [15] resulting in $U_{\overline{\text{K}}\text{N}}(\rho_0, p = 0) \approx 50$ MeV, IQMD uses a different Ansatz for the density dependent $\overline{\text{K}}$N potential with $U_{\overline{\text{K}}\text{N}}(\rho_0, p = 0) \approx$ 90 MeV [10].

The measured directed flow values of kaons are described consistently by both models without a KN potential, which is in contrast to the findings for central events [7]. The measured $v_2(y)$ values are described by both models. However, the elliptic flow of kaons is only weakly sensitive to the depth of the KN-potential at these energies. The directed sideward flow of anti-kaons measured in the FOPI acceptance is shown in the upper right panel of Fig. 1. Both models predict a strong rapidity dependence that is at variance with the experimental data. Since the slope of the theoretical predictions changes from negative when no potential is present to positive when the in-medium potential is activated it is tempting to conclude that the measured vanishing side flow of the anti-kaons is caused by a weaker potential as currently implemented in the models. The data for the elliptic flow $v_2(y)$ of Anti-kaons are shown for completeness only. The experimental error bars are still too large to draw definite conclusions.

3 Hyper-triton search

Currently the identification of quasi-bound and bound states carrying strangeness is a very active field with controversial interpretations (see e.g. [4]). Production of light hyper-nuclei in heavy-ion reactions is a topic of interest on its own [16], but from an experimental perspective known hyper-nuclei provide a way to gauge the experimental reconstruction procedures targeted at more exotic states.

The hyper-triton, which is lightest known hyper-nucleus, decays weakly with a half life of 200–300 ps. Its two particle decay channel ${}_{\Lambda}\text{t} \rightarrow {}^3\text{He} + \pi^-$ is accessible

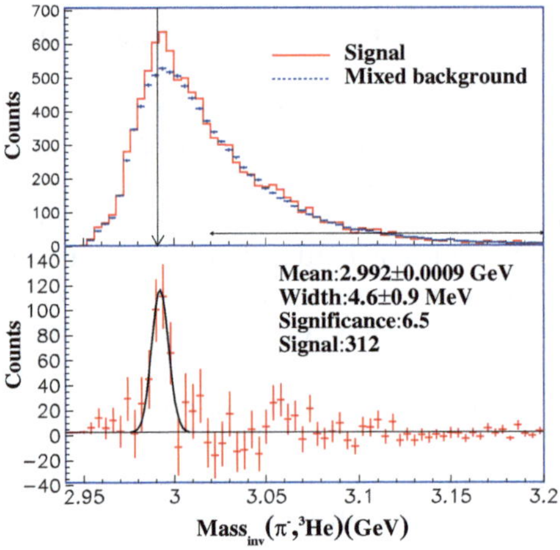

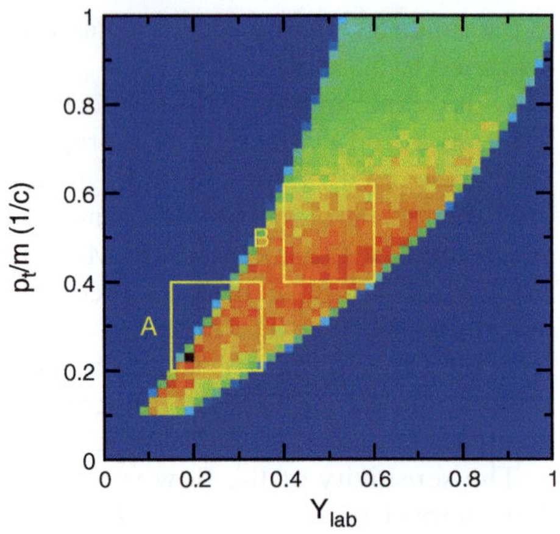

Fig. 2 *Left:* Reconstructed invariant mass of ^{3}He $-$ π^- pairs. In the upper panel the solid histograms and crosses denote the data (*red*) and the scaled mixed-events background (*blue*). The background distribution is normalized to the data over the range indicated by the horizontal arrow. The subtracted spectrum is shown in the lower panel. *Right:* Phase space of identified ^{3}He. The yellow boxes represent the phase space regions named A and B in the text

with the FOPI set-up. The combination of time-of-flight detectors and momentum-measurement in the central drift chamber of the FOPI setup allows to distinguish Z=1 and Z=2 particles in the phase space region depicted in the right panel of Fig. 2. Selecting off-vertex π^-—mesons, calculating the invariant mass of π^- and ^{3}He pairs, requesting the proper decay topology and constraining the data to the kinematical limits of region A gives rise to the the solid, red histogram in the upper panel of Fig. 2 (left). The dashed, blue line in Fig. 2 corresponds to the combinatorial background, which was reconstructed by the mixed-event method and normalized to the signal spectrum in the region depicted by the horizontal line (3.03–3.6 GeV/c^2). The red histogram in the lower panel of Fig. 2 shows the distribution obtained after subtracting the normalized background from the signal spectrum. The error bars correspond to statistical fluctuations. A total of about 300 hyper-triton candidates is reconstructed in 56 million semi-central collisions in region A while no evidence for hypertritons is found in region B. The lifetime of the hyper-triton can be estimated from the radial distance of the decay vertex to the primary vertex r_s and $\beta\gamma$ of the hyper-triton, ie. $t = r_s/(\beta\gamma)$. When taking into account detection efficiencies an average life time $\tau = 273 \pm 40$ ps is deduced, which is in agreement with other experimental data.

The characteristics of the mesonic two particle decay of the hyper-triton is dominated by the heavy ^{3}He nucleus. Hence, the phase space where ^{3}He are identified shown in the right panel of Fig. 2 is nearly equivalent to the phase space region of the reconstructed hyper-tritons. Surprisingly, hyper-tritons are not distributed in phase space like other light clusters (e.g. t, ^{3}He). This finding is in variance with a thermal model Ansatz that would predict the same ratio in regions A and B. To which extend the observation can be explained by the interaction of strange particles produced in the fireball with the cold matter in the spectator parts of the nuclei is a matter of ongoing research [17].

4 Conclusion

We presented data on strangeness production in heavy ion Ni+Ni collisions at 1.9 AGeV measured with the upgraded FOPI setup. The high statistics and high quality data allows to determine the directed flow of charged kaons and the investigation of strongly and weakly decaying states including strangeness. Hyper-tritons have been reconstructed by their two-particle mesonic decay and their phase space population is being investigated.

Acknowledgements This work was supported by the German BMBF under Contract No. 06HD190I, by the Korea Science and Engineering Foundation (KOSEF) under Grant No. F01-2006-000-10035-0, by the mutual agreement between GSI and IN2P3/CEA, by the Hungarian OTKA under Grant No. 47168, within the Framework of the WTZ program (Project RUS 02/021), by DAAD (PPP D/03/44611), by DFG (Projekt 446-KOR-113/76/04) and by the EU, 6th Framework Program, Integrated Infrastructure: Strongly Interacting Matter (Hadron Physics), Contract No. RII3-CT-2004-506078.

References

1. Brown, G.E., Lee, C.H., Rho, M., Thorsson, V.: Nucl. Phys. A **567**, 937 (1994)
2. Weise, W.: Nucl. Phys. A **610**, 35C (1996)
3. Schaffner-Bielich, J., Koch, V., Effenberger, M.: Nucl. Phys. A **669**, 153 (2000)
4. Hartnack, C., Oeschler, H., Leifels, Y., Bratkovskaya, E.L., Aichelin, J.: Phys. Rept. **510**, 119 (2012)
5. Li, G.-Q., Ko, C.M., Li, B.-A.: Phys. Rev. Lett. **74**, 235–238 (1995)
6. Kis, M., et al.: [FOPI Collaboration]. Nucl. Instrum. Meth. A **646**, 27 (2011)
7. Crochet, P., et al.: [FOPI Collaboration]. Phys. Lett. B **486**, 6 (2000)
8. Danielewicz, P., Odyniec, G.: Phys. Lett. **B157**, 146–150 (1985)
9. Ollitrault, J.-Y.: [nucl-ex/9711003]
10. Schaffner-Bielich, J.: J. Phys. G **27**, 337 (2001)
11. Benabderrahmane, M.L., et al.: [FOPI Collab.]. Phys. Rev. Lett. **102**, 182501 (2009)
12. Mishra, A., Bratkovskaya, E.L., Schaffner-Bielich, J., Schramm, S., Stöcker, H.: Phys. Rev. **C70**, 044904 (2004). [arXiv:nucl-th/0402062]
13. Hartnack, C., Puri, R.K., Aichelin, J., Konopka, J., Bass, S.A., Stoecker, H., Greiner, W.: Eur. Phys. J. **A1**, 151 (1998). [arXiv:nucl-th/9811015]
14. Cassing, W., Bratkovskaya, E.L.: Phys. Rept. **308**, 65 (1999)
15. Tolos, L., Ramos, A., Polls, A.: Phys. Rev. **C65**, 054907 (2002)
16. Abelev, B.I., et al.: [STAR collaloration]. Science **328**, 58 (2010)
17. Gaitanos, T., Lenske, H., Mosel, U.: Phys. Lett. **B675**, 297–304 (2009)

4 Conclusion

Hyperfine Interact (2012) 210:71–75
DOI 10.1007/s10751-012-0602-y

Population of the X(2265) resonance in the $p + p \rightarrow X + K^+$ reaction at $T_p = 2.5$ GeV

Ken Suzuki · Paul Kienle · Marco Maggiora · Toshimitsu Yamazaki

Published online: 20 March 2012

Abstract DISTO data of $p + p \rightarrow p + \Lambda + K^+$ reactions at several beam energies were studied. We found in the missing-mass $\Delta M(K^+)$ spectrum and the Λp invariant-mass spectrum at 2.85 GeV a broad resonance with $M = 2267$ MeV/c^2 and $\Gamma = 118$ MeV, associated with a mono energetic kaon spectrum, indicating that a dibaryonic state (X) with strangeness -1 is produced in the $p + p \rightarrow X + K^+$ reaction with X decaying into $p + \Lambda$. The energy dependence of the X production rate and the preliminary result of the recently analyzed $p + p \rightarrow p + \Lambda + K^+$ data at $T_p = 2.5$ GeV is presented.

Keywords $\bar{K}$ nuclei · $\bar{K}N$ interaction · Strange dibaryon

This work is partly supported by Austrian Fonds zur Fördering der wissenschaftlichen Forschung (FWF) (contract number: P21457), the DFG cluster of excellence Origin and Structure of the Universe of Technische Universität München and by Grant-in-Aid for Scientific Research of Monbu- Kagakusho of Japan.

K. Suzuki (✉)
Stefan Meyer Institute for Subatomic Physics of the Austrian Academy of Sciences, Vienna, Austria
e-mail: ken.suzuki@oeaw.ac.at

P. Kienle
Excellence Cluster Universe, Technische Universität München, Garching, Germany

M. Maggiora
Dipartimento di Fisica Generale "A. Avogadro" and INFN, Torino, Italy

T. Yamazaki
Department of Physics, University of Tokyo, Tokyo, Japan

T. Yamazaki
RIKEN Nishina Center, Saitama, Japan

1 Introduction

The $\bar{K}N$ interaction is fundamental for low energy hadron physics with strangeness. Precise measurements of the strong interaction energy level shift and width using kaonic hydrogen x-rays [1] in 1997 gave strong constraints for the $\bar{K}N$ interaction at threshold. This lead Akaishi and Yamazaki to study the possible existence of kaonic nuclear bound states [2, 3] and to predict the dibaryon system $(\bar{K}NN)_{S=0,I=1/2}$ often called K^-pp to be the simplest antikaonic nuclear bound system with a mass $M = 2322$ MeV/c^2, an antikaon binding energy $B_K = 48$ MeV and a width $\Gamma_{\Sigma\pi p} =$ 61 MeV.

The phenomenological $\bar{K}N$ interaction constructed by Akaishi and Yamazaki using kaonic hydrogen x-ray spectra, low energy $\bar{K}N$ scattering data, as well as the mass and width of the $\Lambda(1405)$ resonance as a $\bar{K}N$ quasi bound state shows a deep potential. The strongly bound nature of the K^-pp system was supported by three-body Faddeev calculations [4–6]. On the other hand chiral dynamics leads to a much shallower binding [7, 8].

2 DISTO data at 2.85 GeV

In order to search for the K^-pp, we studied the $p + p \to p + \Lambda + K^+$ final state process, which consists of the "ordinary" three body process to $p\Lambda K^+$ and possibly the K^-pp production process in an "exotic" two-body process via the following processes (1) and (2),

$$p + p \to \underline{p + \Lambda^*} + K^+, \tag{1}$$

$$\hookrightarrow X = K^-pp \to p + \Lambda. \tag{2}$$

The probability that the $\Lambda^*(1405)$ and p merge and form a kaonic nuclear state is despite the large momentum transfer in the pp reaction proposed to be large [9] due to the high density nature of the deeply bound K^-pp system. In the DISTO data at $T_p = 2.85$ GeV 177k exclusive $p\Lambda K^+$ events are available [10–12]. The formation of X was examined simultaneously by a missing-mass spectroscopy of $\Delta M(K^+)$ and by a invariant-mass spectroscopy of $M(\Lambda p)$ of the Λp decay channel of the K^-pp.

In order to obtain an efficiency-compensated presentation of the experimental data, the experimental spectral distribution is bin-by-bin divided by the Monte-Carlo data which simulated the ordinary $p\Lambda K^+$ process with a uniform phase-space distribution, folded with the DISTO acceptance, digitized and then fed to the same reconstruction and analysis chain.

We introduce then two essential cuts to enhance the relative contribution of the exotic process over the ordinary process. For the ordinary process the maximum momentum of p is 0.75 GeV/c and the dominating proton group at large backward angles $[\cos\theta_{cm}(p) < -0.9]$ has a small transverse momentum $P_T < 0.3$ GeV/c, on the contrary the exotic process involving the decay of X has a transverse momentum of around 0.4 GeV/c. Therefore we first introduced a cut on the proton-angle, namely "large-angle proton" (LAP) cut: $|\cos\theta_{cm}(P)| < 0.6$ and "small-angle proton" (SAP) cut: $|\cos\theta_{cm}(P)| > 0.6$. The spectra with SAP showed basically a structure-less flat distribution. On the contrary the LAP spectra showed a broad resonance structure.

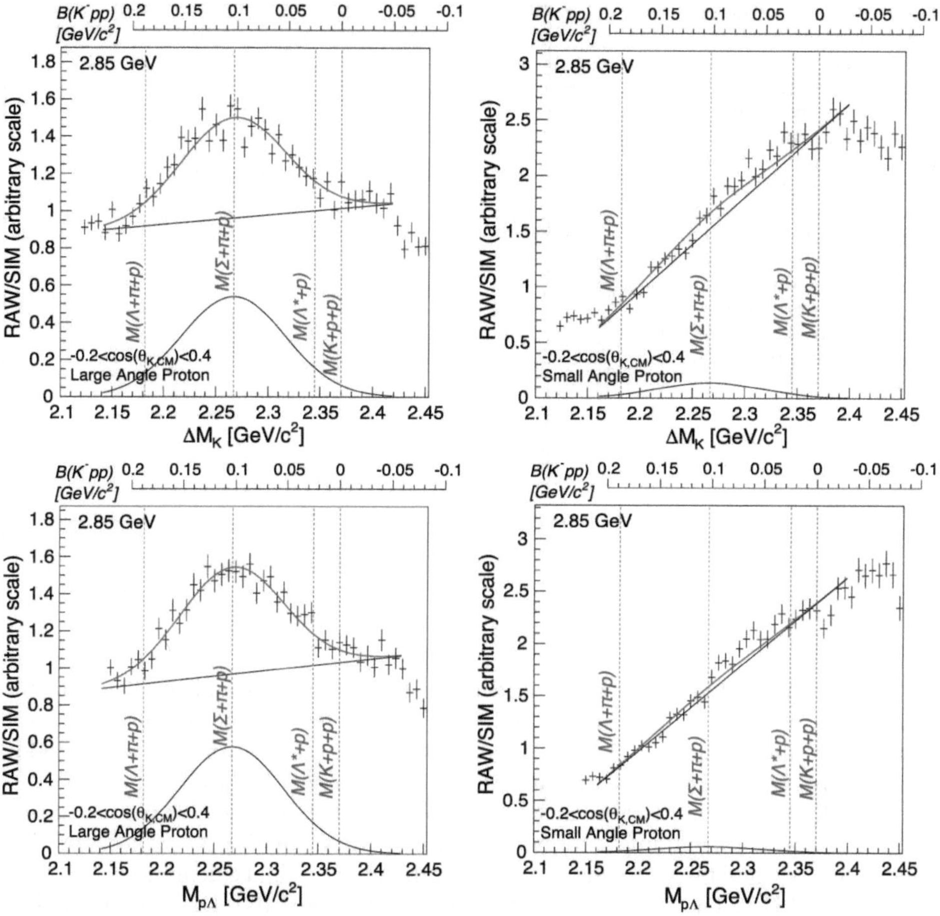

Fig. 1 Missing-mass spectra $\Delta M(K^+)$ (*top*) and invariant-mass spectra $M(p\Lambda)$ (*bottom*) for the "large-angle-proton" cut (*left*) and "small-angle-proton" cut (*right*), respectively, from the exclusive data sample of $pp \to p\Lambda K^+$ at 2.85 GeV where the kaon angle cut: $-0.2 < \cos\theta_{CM}(K^+) < 0.4$ is applied. The ordinate is in arbitrary units. The broad resonance structure was observed with LAP whereas the SAP spectra showed no structure. The thresholds of some relevant decay channels are shown by vertical broken lines

We additionally observed in the kaon phase space distribution a mono-energetic component of $P_{CM}(K^+) \sim 0.4$ GeV/c and applied a cut on the kaon angle, namely $-0.2 < \cos\theta_{cm}(K) < 0.4$ in order to select this mono-energetic component as the second cut. Applying this cut, while general features of the spectra stayed the same, the broad structure was further pronounced, indicating that the observed broad structure comes indeed from the "exotic" two-body process: $p + p \to K^+ + X$ and the X has a baryon number 2 and a strangeness -1. Figure 1 shows missing-mass spectra $\Delta M(K^+)$ (top) and invariant-mass spectra $M(p\Lambda)$ (bottom) for the "large-angle-proton" cut (left) and "small-angle-proton" cut (right), with the kaon angle cut. The nice agreement of the spectral shapes of the $\Delta M(K^+)$ and the $M(p\Lambda)$ proves that the selection of $p\Lambda K^+$ data sample is sufficiently pure and "background-free".

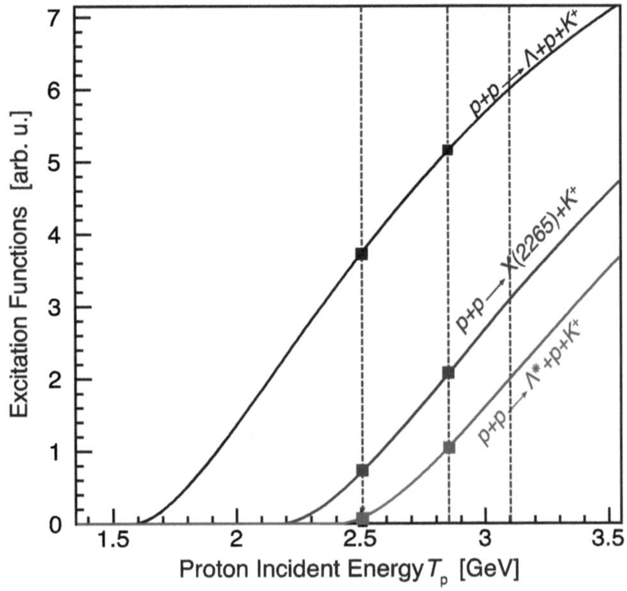

Fig. 2 Relative excitation functions of the reactions $pp \to p\Lambda K^+, \to X(2265)K^+$ and $\to p\Lambda^*(1405)K^+$. The *curves* are drawn by using a phase-space compatible universal formula. The *three vertical lines* indicate the beam energies of relevant experiments: the DISTO data at 2.85 GeV and 2.5 GeV and the FOPI experiment at 3.1 GeV. The production mechanism of the $X(2265)$ can be examined by the 2.5 GeV data

The spectra of Fig. 1 left were used to evaluate the mass and the width of the resonance. We used a gaussian plus a linear function to describe the resonance and the background (ordinary $p\Lambda K^+$ process), respectively. The resonance was found to have a mass $M_x = 2267$ MeV/c^2 and a width $\Gamma_x = 118$ MeV/c^2 [10–12].

We call the observed resonance $X(2265)$ hereafter. The $X(2265)$ mass corresponds to antikaon binding energy, B_K, of 103 MeV for the K^-pp system. The binding energy and the width are much larger than the original calculation of Akaishi and Yamazaki [2]. The peak-like structure reported by the FINUDA experiment [13] has within the errors the same mass $M = 2255 \pm 7$ MeV/c^2, $B_K = 115 \pm 7$ MeV, however, the width $\Gamma = 67 \pm 14$ MeV/c^2 is smaller than our observation using the DISTO data.

3 Energy dependence

For a further study of the nature of $X(2265)$ we investigate the energy dependence of the production rate of the $X(2265)$ in the $pp \to K^+X(2265)$ two-body reaction [14]. We take a semi-empirical universal function for hadron production cross section as a function of the center-of-mass energy ($\sqrt{s}$) [15] as expressed by $\sigma = \sigma_0(1 - s_0/s)^{1.8}(s_0/s)^{1.5}$ [15], where $\sqrt{s_0}$ is the production threshold and σ_0 is a scaling factor. We draw this excitation function for the reactions $pp \to p\Lambda K^+, \to X(2265)K^+$ and $\to p\Lambda^*(1405)K^+$ in Fig. 2. Since the scaling factor σ_0 is not known for the latter two cases, the σ_0 is set equal for all cases. Only the relative comparison of scales has a meaning.

The DISTO experiment has collected the $pp \to p\Lambda K^+$ data also at $T_p = 2.5$ GeV and 2.145 GeV. From the relation in Fig. 2 one expects, if $X(2265)$ is produced similarly as other hyperons, the production threshold of the $X(2265)$ is low enough and the $X(2265)$ could be populated as much as 33% at $T_p = 2.5$ GeV compared with the population at $T_p = 2.85$ GeV. However, if the K^-pp production in pp collision goes along with the Λ^* doorway scenario (1, 2) [9], namely the $\Lambda^*(1405)$ participates in high momentum collisions with the proton to form the K^-pp object at small $\Lambda(1405)$-p distances with a large $\Lambda^*p \to K^-pp$ sticking probability, and therefore the presence of the $\Lambda^*(1405)$ is essential, one expects $X(2265)$ is hardly populated at 2.5 GeV as the bombarding energy of 2.50 GeV is too close to the threshold of the $p + p \to \Lambda^*(1405) + p + K^+$ reaction. In summary, one can examine these possible scenarios of the production mechanism of the $X(2265)$ by analyzing the 2.5 GeV data and comparing the relative production rates with the 2.85 GeV data. Our preliminary analysis of the 2.5 GeV data shows basically no $X(2265)$ resonance, that fits to the latter picture [14].

4 Outlook

We measured at the end of 2009 the same reaction: $p + p \to p + \Lambda + K^+$ using FOPI apparatus at GSI, Darmstadt, Germany at $T_p = 3.1$ GeV [16, 17]. The beam energy was chosen to optimize the signal to background ratio and also to see the background shape separately. Yet we are still struggling to evaluate this data, though the analysis status was reported at several conferences [18]. The FOPI data is very important for cross-checking the DISTO observation. According to the present scenario, FOPI might see $X(2265)$ at 3.1 GeV with double the cross section as that at 2.85 GeV (Fig. 2). Also the DISTO data at 2.85 GeV will be further studied with respect to the influence of $K^+\Lambda$ final state interactions. They make a vital contribution to the long standing quest of the kaonic nuclear states.

References

1. Iwasaki, M., et al.: Phys. Rev. Lett. **78**, 3067 (1997)
2. Akaishi, Y., Yamazaki, T.: Phys. Rev. C **65**, 044005 (2002)
3. Yamazaki, T., Akaishi, Y.: Phys. Lett. B **535**, 70 (2002)
4. Shevchenko, N.V., Gal, A., Mareš, J., Révai, J.: Phys. Rev. C **76**, 044004 (2007)
5. Shevchenko, N.V., Gal, A., Mares, J.: Phys. Rev. Lett. **98**, 082301 (2007)
6. Ikeda, Y., Sato, T.: Phys. Rev. C **76**, 035203 (2007)
7. Hyodo, T., Weise, W.: Phys. Rev. C **77**, 035204 (2008)
8. Doté, A., Hyodo, T., Weise, W.: Nucl. Phys. A **804**, 197 (2008)
9. Yamazaki, T., Akaishi, Y.: Phys. Rev. C **76**, 045201 (2007)
10. Yamazaki, T., et al.: Hyperfine Interact. **193**, 181 (2009)
11. Maggiora, M., et al.: Nucl. Phys. A **835**, 43 (2010)
12. Yamazaki, T., et al.: Phys. Rev. Lett. **104**, 132502 (2010)
13. Agnello, M., et al.: Phys. Rev. Lett. **94**, 212303 (2005)
14. Kienle, P., et al.: arXiv:1102.0482v1 [nucl-ex]
15. Sibirtsev, A.: Phys. Lett. B **359**, 29 (1995)
16. FOPI Collaboration: Experimental proposal to GSI (2007)
17. Suzuki, K., et al.: Nucl. Phys. A **827**, 312c (2009)
18. Suzuki, K., Berger, M., et al.: Prog. Theor. Phys. Suppl. **186**, 351 (2010)

Hyperfine Interact (2012) 210:77–82
DOI 10.1007/s10751-011-0542-y

Consistency of $\Lambda\Lambda$ hypernuclear events

A. Gal · D. J. Millener

Published online: 30 December 2011

Abstract Highlights of $\Lambda\Lambda$ emulsion events are briefly reviewed. Given three accepted events, shell-model predictions based on p-shell Λ hypernuclear spectroscopic studies are shown to reproduce $B_{\Lambda\Lambda}(^{10}_{\Lambda\Lambda}\text{Be})$ and $B_{\Lambda\Lambda}(^{13}_{\Lambda\Lambda}\text{B})$ in terms of $B_{\Lambda\Lambda}(^{6}_{\Lambda\Lambda}\text{He})$. Predictions for other species offer judgement on several alternative assignments of the $^{13}_{\Lambda\Lambda}\text{B}$ KEK-E176 event, and on the assignments $^{11}_{\Lambda\Lambda}\text{Be}$ and $^{12}_{\Lambda\Lambda}\text{Be}$ suggested recently for the KEK-E373 HIDA event. The predictions of the shell model, spanning a wide range of A values, are compared with those of cluster models, where the latter are available.

Keywords Hypernuclei · Shell model · Cluster models

1 Introduction

$\Lambda\Lambda$ hypernuclei provide valuable information on the $\Lambda\Lambda$ interaction and how it fits into our understanding of the baryon-baryon interaction. Although the existence of $\Lambda\Lambda$ hypernuclei nearly rules out a stable H dibaryon, a ΞN dominated H resonance might affect the systematics of $\Lambda\Lambda$ binding energies. Only three emulsion events presented serious candidates for $\Lambda\Lambda$ hypernuclei before 2001: $^{10}_{\Lambda\Lambda}\text{Be}$ [1, 2], $^{6}_{\Lambda\Lambda}\text{He}$ [3] and $^{13}_{\Lambda\Lambda}\text{B}$ [4, 5]. The $\Lambda\Lambda$ binding energies $B_{\Lambda\Lambda}$ deduced from these events indicated that the $^1S_0$ interaction $V_{\Lambda\Lambda}$ was strongly attractive, with a $\Lambda\Lambda$ excess binding energy $\Delta B_{\Lambda\Lambda} \sim 4.5$ MeV, although it had been realized that the binding energies of $^{10}_{\Lambda\Lambda}\text{Be}$

A. Gal (✉)
Racah Institute of Physics, The Hebrew University, Jerusalem 91904, Israel
e-mail: avragal@savion.huji.ac.il

D. J. Millener
Physics Department, Brookhaven National Laboratory, Upton, NY 11973, USA

and ${}^{6}_{\Lambda\Lambda}$He were inconsistent with each other [6]. Here, the $\Lambda\Lambda$ excess binding energy is defined as

$$\Delta B_{\Lambda\Lambda}\left({}^{A}_{\Lambda\Lambda}Z\right) = B_{\Lambda\Lambda}\left({}^{A}_{\Lambda\Lambda}Z\right) - 2\bar{B}_{\Lambda}\left({}^{(A-1)}_{\Lambda}Z\right), \tag{1}$$

where $\bar{B}_{\Lambda}$ is the $(2J+1)$-average of B_{Λ} values for the ${}^{(A-1)}_{\Lambda}Z$ hypernuclear core levels. For comparison, $\Delta B_{\Lambda N}({}^{5}_{\Lambda}\mathrm{He}) = 1.73 \pm 0.13$ MeV, implying the unnatural ordering $\Delta B_{\Lambda\Lambda} > \Delta B_{\Lambda N}$. This perception changed in 2001 when a uniquely assigned ${}^{6}_{\Lambda\Lambda}$He hybrid-emulsion event [7], with updated values [8]

$$B_{\Lambda\Lambda}\left({}^{6}_{\Lambda\Lambda}\mathrm{He}\right) = 6.91 \pm 0.16\ \mathrm{MeV}, \quad \Delta B_{\Lambda\Lambda}({}^{6}_{\Lambda\Lambda}\mathrm{He}) = 0.67 \pm 0.17\ \mathrm{MeV}, \tag{2}$$

ruled out the high value of $\Delta B_{\Lambda\Lambda}$ from the dubious earlier ${}^{6}_{\Lambda\Lambda}$He event [3], restoring thus the expected hierarchy $\Delta B_{\Lambda\Lambda} < \Delta B_{\Lambda N}$. Both capture at rest formation $\Xi^- + {}^{12}\mathrm{C} \to {}^{6}_{\Lambda\Lambda}\mathrm{He} + t + \alpha$ and weak decay ${}^{6}_{\Lambda\Lambda}\mathrm{He} \to {}^{5}_{\Lambda}\mathrm{He} + p + \pi^-$, in this so called NAGARA event, yield consistently with each other the values listed in (2). Neither ${}^{6}_{\Lambda\Lambda}$He nor ${}^{5}_{\Lambda}$He have excited states that could bias the determination of $B_{\Lambda\Lambda}({}^{6}_{\Lambda\Lambda}\mathrm{He})$.

Accepting the NAGARA event calibration of $V_{\Lambda\Lambda}$, we review and discuss (i) particle stability for lighter $\Lambda\Lambda$ hypernuclei; (ii) reinterpretation of the events assigned ${}^{10}_{\Lambda\Lambda}$Be and ${}^{13}_{\Lambda\Lambda}$B; (iii) several alternative assignments for the ${}^{13}_{\Lambda\Lambda}$B event; and (iv) plausibility of the assignments ${}^{11}_{\Lambda\Lambda}$Be or ${}^{12}_{\Lambda\Lambda}$Be proposed for the recently reported HIDA event [8]. In the course of doing so, we compare $B_{\Lambda\Lambda}$ values derived from emulsion events with shell-model predictions [9] and with selected few-body cluster calculations [6, 10, 11] where the latter exist.

2 Onset of $\Lambda\Lambda$ hypernuclear stability

From the very beginning it was recognized that $\Lambda\Lambda$ and $\Lambda\Lambda N$ were unbound [12, 13]; if $\Lambda\Lambda N$ were bound, the existence of a $nn\Lambda$ bound state would follow. The existence of a ${}^{4}_{\Lambda\Lambda}$H bound state was claimed by AGS-E906 [14], from correlated weak-decay pions emitted sequentially by $\Lambda\Lambda$ hypernuclei produced in a (K^-, K^+) reaction on ^{9}Be. However, the ${}^{4}_{\Lambda\Lambda}$H interpretation is controversial [15, 16]. Several post-2001 calculations exist for ${}^{4}_{\Lambda\Lambda}$H. A Faddeev–Yakubovsky 4-body calculation finds no bound state [17], whereas a stochastic-variational (SV) 4-body calculation finds it to be bound by as much as 0.4 MeV [18]. The more comprehensive s-shell Λ- and $\Lambda\Lambda$-hypernuclear SV calculation of [19] finds ${}^{4}_{\Lambda\Lambda}$H to be particle stable by as little as a few keV, which would be insufficient to maintain particle stability once $V_{\Lambda\Lambda}$ is renormalized to reproduce the recently updated (smaller) value of (2) for $\Delta B_{\Lambda\Lambda}({}^{6}_{\Lambda\Lambda}\mathrm{He})$.

Regardless of whether ${}^{4}_{\Lambda\Lambda}$H is particle-stable or not, there is a general consensus that the mirror $\Lambda\Lambda$ hypernuclei ${}^{5}_{\Lambda\Lambda}$H–${}^{5}_{\Lambda\Lambda}$He are particle-stable, with $\Delta B_{\Lambda\Lambda} \sim 0.5 - 1$ MeV [20, 21], or larger owing to the $\Lambda\Lambda - \Xi N$ coupling which is particularly effective here [19, 22, 23]. In addition, substantial charge symmetry breaking effects are expected in these systems, resulting in a higher binding energy of ${}^{5}_{\Lambda\Lambda}$He by up to 0.5 MeV with respect to ${}^{5}_{\Lambda\Lambda}$H [23, 24]. Figure 1 demonstrates how $\Delta B_{\Lambda\Lambda}$ values for the $A = 5, 6$ systems, calculated over a broad range of $V_{\Lambda\Lambda}$ strengths, are nearly linearly correlated with only a small offset. Thus, the stability of ${}^{6}_{\Lambda\Lambda}$He ensures stability for ${}^{5}_{\Lambda\Lambda}$H.

Fig. 1 Faddeev calculations of $\Delta B_{\Lambda\Lambda}({}^{6}_{\Lambda\Lambda}\mathrm{He})$ *vs.* $\Delta B_{\Lambda\Lambda}({}^{5}_{\Lambda\Lambda}\mathrm{H}, {}^{5}_{\Lambda\Lambda}\mathrm{He})$ [20], see text

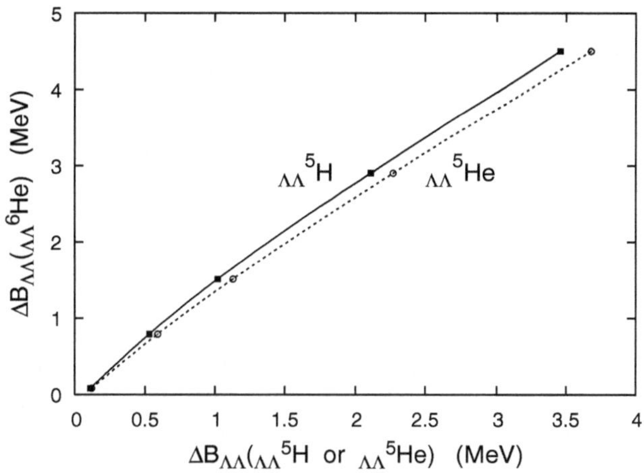

Table 1 Doublet splittings ΔE^{th} and ΔE^{exp} (in keV) from [25, 26], where $\Delta E^{\mathrm{th}}_{\mathrm{alt}}$ uses ESC04a–inspired $\Lambda - \Sigma$ coupling

	J^{π}_{up}	J^{π}_{low}	ΔE^{th}	$\Delta E^{\mathrm{th}}_{\mathrm{alt}}$	ΔE^{exp}
${}^{9}_{\Lambda}\mathrm{Be}^*$	$3/2^+$	$5/2^+$	44	49	43 ± 5
${}^{10}_{\Lambda}\mathrm{B}_{\mathrm{g.s.}}$	2^-	1^-	120	34	≤ 100
${}^{11}_{\Lambda}\mathrm{B}_{\mathrm{g.s.}}$	$7/2^+$	$5/2^+$	267	243	262.9 ± 0.2
${}^{12}_{\Lambda}\mathrm{C}_{\mathrm{g.s.}}$	2^-	1^-	153	167	161.4 ± 0.7
${}^{13}_{\Lambda}\mathrm{C}^*$	$5/2^+$	$3/2^+$	31	47	–

Note the sensitivity of $\Delta E^{\mathrm{th}}({}^{10}_{\Lambda}\mathrm{B}_{\mathrm{g.s.}})$ to the model used for $\Lambda - \Sigma$ mixing. The ${}^{9}_{\Lambda}\mathrm{Be}^*$ and ${}^{13}_{\Lambda}\mathrm{C}^*$ excited doublets are discussed in Section 4

3 Ingredients of hypernuclear shell model

Shell-model predictions for ΛΛ hypernuclei have been given recently [9] using (1) in which $\Delta B_{\Lambda\Lambda}({}^{A}_{\Lambda\Lambda}Z)$ is replaced by a constant $V_{\Lambda\Lambda}$ matrix element, identified with $\Delta B_{\Lambda\Lambda}({}^{6}_{\Lambda\Lambda}\mathrm{He})$ of (2).[1] Calculations of $B_{\Lambda\Lambda}({}^{A}_{\Lambda\Lambda}Z)$ require then the knowledge of $\bar{B}_{\Lambda}({}^{(A-1)}_{\Lambda}Z)$ involving single-Λ hypernuclear ground-state (g.s.) binding energies plus g.s. doublet splittings $\Delta E_{\mathrm{g.s.}}$ for $J_{\mathrm{core}} \neq 0$. Table 1 lists $\Delta E_{\mathrm{g.s.}}$ values relevant for the calculations reviewed here, exhibiting remarkable agreement between theory and experiment.

4 Interpretation of ${}^{10}_{\Lambda\Lambda}$Be and ${}^{13}_{\Lambda\Lambda}$B emulsion events

The $B_{\Lambda\Lambda}$ values of both ${}^{10}_{\Lambda\Lambda}\mathrm{Be}$ (17.5 ± 0.4 MeV) [2] and ${}^{13}_{\Lambda\Lambda}\mathrm{B}$ (28.2 ± 0.7 MeV) [27] were extracted assuming that their π^- weak decay proceeds to the g.s. of the respective daughter Λ hypernuclei. This led to $\Delta B_{\Lambda\Lambda} \sim 4 - 5$ MeV, substantially

[1] A straightforward modification for ${}^{10}_{\Lambda\Lambda}\mathrm{Be}_{\mathrm{g.s.}}(0^+)$, with a nuclear core ${}^{8}\mathrm{Be}$ unstable to α emission, is discussed in [9].

Table 2 Reinterpreted $B_{\Lambda\Lambda}^{\rm exp}$ values (in MeV) and predictions based on the NAGARA event for ${}_{\Lambda\Lambda}^{6}$He

	$B_{\Lambda\Lambda}^{\rm exp}$		$B_{\Lambda\Lambda}^{\rm CM}$		$B_{\Lambda\Lambda}^{\rm SM}$
	Eqs. (3) and (4)	Eq. (5)	Bodmer et al. [6]	Hiyama et al. [11]	Gal and Millener [9]
${}_{\Lambda\Lambda}^{10}$Be	14.5 ± 0.4	14.94 ± 0.13	14.35 ± 0.19	14.74 ± 0.19	14.97 ± 0.22
${}_{\Lambda\Lambda}^{13}$B	23.3 ± 0.7		–	–	23.21 ± 0.21

The error on $B_{\Lambda\Lambda}^{\rm exp}({}_{\Lambda\Lambda}^{6}\rm He)$ is incorporated into the predicted values

higher than for ${}_{\Lambda\Lambda}^{6}$He (NAGARA). However, as realized by Danysz et al. [1], the decay could proceed to excited states of the daughter Λ hypernucleus which deexcites then rapidly to the g.s. emitting unobserved γ radiation. This reduces the apparent $B_{\Lambda\Lambda}$ and $\Delta B_{\Lambda\Lambda}$ values by the Λ hypernuclear excitation energy involved in the π^- weak decay. Consistency with ${}_{\Lambda\Lambda}^{6}$He is restored upon accepting the following weak decays:

$$ {}_{\Lambda\Lambda}^{10}\mathrm{Be} \to {}_{\Lambda}^{9}\mathrm{Be}^{*}(3/2^+, 5/2^+; 3.04\ \mathrm{MeV}) + p + \pi^-, \tag{3} $$

$$ {}_{\Lambda\Lambda}^{13}\mathrm{B} \to {}_{\Lambda}^{13}\mathrm{C}^{*}(3/2^+, 5/2^+; 4.9\ \mathrm{MeV}) + \pi^-, \tag{4} $$

with rates comparable to those for decays to ${}_{\Lambda}^{9}\mathrm{Be}_{\rm g.s.}(1/2^+)$ and ${}_{\Lambda}^{13}\mathrm{C}_{\rm g.s.}(1/2^+)$, respectively. The doublet splittings of ${}_{\Lambda}^{9}$Be* and ${}_{\Lambda}^{13}$C* are listed in Table 1.

${}_{\Lambda\Lambda}^{10}$Be also fits the Demachi-Yanagi event observed in KEK-E373 [28], with $B_{\Lambda\Lambda} = 11.90 \pm 0.13$ MeV [8] determined from the assumed formation reaction kinematics. The approximately 6 MeV difference between this and the Danysz et al. [1, 2] value for $B_{\Lambda\Lambda}({}_{\Lambda\Lambda}^{10}\mathrm{Be})$ is reconciled by assuming that the Demachi-Yanagi event corresponds to formation of the first excited state ${}_{\Lambda\Lambda}^{10}$Be*,

$$ \Xi^- + {}^{12}\mathrm{C} \to {}_{\Lambda\Lambda}^{10}\mathrm{Be}^*(2^+; \approx 3\ \mathrm{MeV}) + t, \tag{5} $$

which decays to ${}_{\Lambda\Lambda}^{10}\mathrm{Be}_{\rm g.s.}$ by emitting unseen γ ray, the energy of which has to be added to the apparent $B_{\Lambda\Lambda}$ value deduced by assuming a g.s. formation. It is not clear why the formation of ${}_{\Lambda\Lambda}^{10}$Be* should be comparable or enhanced with respect to that of ${}_{\Lambda\Lambda}^{10}\mathrm{Be}_{\rm g.s.}$.

The $B_{\Lambda\Lambda}^{\rm exp}$ values corresponding to (3)–(5) are listed in Table 2 together with predictions made in cluster model (CM) and shell model (SM) calculations, all of which use $\Lambda\Lambda$ interactions normalized to $B_{\Lambda\Lambda}({}_{\Lambda\Lambda}^{6}\mathrm{He}) = 6.91 \pm 0.16$ MeV [8]. For ${}_{\Lambda\Lambda}^{13}$B, assuming charge symmetry, the ${}_{\Lambda}^{12}\mathrm{B}_{\rm g.s.}$ doublet splitting input was identified with that of ${}_{\Lambda}^{12}\mathrm{C}_{\rm g.s.}$ from Table 1. It is seen that both CM and SM calculations

Table 3 Reassignments of the ${}_{\Lambda\Lambda}^{13}$B KEK-E176 event

	$B_{\Lambda\Lambda}^{\rm exp}$ [27]	$B_{\Lambda\Lambda}^{\rm CM}$ [11]	$B_{\Lambda\Lambda}^{\rm SM}$ [9]
${}_{\Lambda\Lambda}^{11}$Be	17.53 ± 0.71	18.23 ± 0.19	18.40 ± 0.28
${}_{\Lambda\Lambda}^{12}$B	20.60 ± 0.74	–	20.85 ± 0.20

$B_{\Lambda\Lambda}$ values are in MeV

Table 4 Assignments suggested for the KEK-E373 HIDA event

	$B^{\mathrm{exp}}_{\Lambda\Lambda}$ [8]	$B^{\mathrm{CM}}_{\Lambda\Lambda}$ [11]	$B^{\mathrm{SM}}_{\Lambda\Lambda}$ [9]
${}^{11}_{\Lambda\Lambda}$Be	20.83 ± 1.27	18.23 ± 0.19	18.40 ± 0.28
${}^{12}_{\Lambda\Lambda}$Be	22.48 ± 1.21	–	20.72 ± 0.20

$B_{\Lambda\Lambda}$ values are in MeV

reproduce the reinterpreted $B_{\Lambda\Lambda}$ values of ${}^{10}_{\Lambda\Lambda}$Be and ${}^{13}_{\Lambda\Lambda}$B. The SM agrees well with the Hiyama et al. CM calculation [10, 11], and the SM calculation has no match for ${}^{13}_{\Lambda\Lambda}$B.

5 Alternative interpretations of the ${}^{13}_{\Lambda\Lambda}$B event

The emulsion event assigned to ${}^{13}_{\Lambda\Lambda}$B [4, 5] has been carefully scrutinized by the KEK-E176 Collaboration [27]. Several alternative assignments were pointed out, two of which that do not require Λ hypernuclear excitation in the π^- weak decay of the $\Lambda\Lambda$ hypernuclear g.s. are listed in Table 3. Comparison with model calculations suggests that such reassignments cannot be ruled out, although a ${}^{13}_{\Lambda\Lambda}$B assignment shows a higher degree of consistency between the $B_{\Lambda\Lambda}$ values derived from formation and from decay. In particular, the accepted formation reaction $\Xi^- + {}^{14}\mathrm{N} \rightarrow {}^{13}_{\Lambda\Lambda}\mathrm{B} + p + n$ was shown to occur naturally in Ξ^- capture at rest in light nuclei emulsion [5].

6 Interpretation of the KEK-E373 HIDA event

The KEK-E373 Collaboration has recently presented evidence from the HIDA event for another $\Lambda\Lambda$ hypernucleus, tentatively assigned to either ${}^{11}_{\Lambda\Lambda}$Be or to ${}^{12}_{\Lambda\Lambda}$Be [8]. The associated $B^{\mathrm{exp}}_{\Lambda\Lambda}$ values, together with model predictions, are listed in Table 4. We note that since no experimental data exist on ${}^{11}_{\Lambda}$Be, the required input for evaluating $B^{\mathrm{SM}}_{\Lambda\Lambda}({}^{12}_{\Lambda\Lambda}\mathrm{Be})$ was derived within the SM approach [9]. It is clear from the table that neither of the proposed assignments is favorable, although the relatively large experimental uncertainties do not completely rule out either of these.

7 Conclusion

It was shown how the three acceptable $\Lambda\Lambda$ emulsion events, corresponding to ${}^{6}_{\Lambda\Lambda}$He, ${}^{10}_{\Lambda\Lambda}$Be and ${}^{13}_{\Lambda\Lambda}$B, can be made consistent with each other, in good agreement with CM and with SM calculations of $B_{\Lambda\Lambda}$. Other possible assignments for the KEK-E176 ${}^{13}_{\Lambda\Lambda}$B event were discussed, and the assignments proposed for the recently reported HIDA event were found unfavorable. It was pointed out that simple shell-model estimates, making use of Λ-hypernuclear spectroscopic data and analysis, are sufficient for discussing the world data of $\Lambda\Lambda$ hypernuclear events. A relatively weak $\Lambda\Lambda$ interaction, with $(1s_\Lambda)^2$ matrix element of magnitude $\Delta B_{\Lambda\Lambda}({}^{6}_{\Lambda\Lambda}\mathrm{He}) = 0.67 \pm 0.17$ MeV, describes well the data in the observationally accessible mass range $6 \leq A \leq 13$. Comparably weak $\Lambda\Lambda$ interactions are obtained also in recent theoretical models, in Nijmegen extended soft-core (ESC) models [24, 29] and in lowest order χEFT [30]. Less well determined is the $\Lambda\Lambda$ coupling to the slightly higher ΞN channel, with appreciable model dependence in ESC models [24, 29]. The observation of $A = 5$ $\Lambda\Lambda$ hypernuclei would add valuable new information on this issue.

Acknowledgement Useful discussions with Emiko Hiyama on hypernuclear cluster-model calculations are gratefully acknowledged.

References

1. Danysz, M., et al., European K^- Collaboration: Nucl. Phys. **49**, 121 (1963)
2. Dalitz, R.H., Davis, D.H., Fowler, P.H., Montwill, A., Pniewski, J., Zakrzewski, J.A.: Proc. R. Soc. Lond. A **426**, 1 (1989)
3. Prowse, D.J.: Phys. Rev. Lett. **17**, 782 (1966)
4. Aoki, S., et al., KEK-E176 Collaboration: Prog. Theor. Phys. **85**, 1287 (1991)
5. Dover, C.B., Millener, D.J., Gal, A., Davis, D.H.: Phys. Rev. C **44**, 1905 (1991)
6. Bodmer, A.R., Usmani, Q.N., Carlson, J.: Nucl. Phys. A **422**, 510 (1984)
7. Takahashi, H., et al., KEK-E373 Collaboration: Phys. Rev. Lett. **87**, 212502 (2001)
8. Nakazawa, K.: Nucl. Phys. A **835**, 207 (2010)
9. Gal, A., Millener, D.J.: Phys. Lett. B **701**, 342 (2011)
10. Hiyama, E., Kamimura, M., Motoba, T., Yamada, T., Yamamoto, Y.: Phys. Rev. C **66**, 024007 (2002) and earlier work cited therein
11. Hiyama, E., Kamimura, M., Yamamoto, Y., Motoba, T.: Phys. Rev. Lett. **104**, 212502 (2010)
12. Dalitz, R.H.: Phys. Lett. **5**, 53 (1963)
13. Tang, Y.C., Herndon, R.C.: Phys. Rev. Lett. **14**, 991 (1965)
14. Ahn, J.K., et al., BNL-E906 Collaboration: Phys. Rev. Lett. **87**, 132504 (2001)
15. Kumagai-Fuse, I., Okabe, S.: Phys. Rev. C **66**, 014003 (2002)
16. Randeniya, S.D., Hungerford, E.V.: Phys. Rev. C **76**, 064308 (2007)
17. Filikhin, I.N., Gal, A.: Phys. Rev. Lett. **89**, 172502 (2002)
18. Nemura, H., Akaishi, Y., Myint, K.S.: Phys. Rev. C **67**, 051001(R) (2003)
19. Nemura, H., Shinmura, S., Akaishi, Y., Myint, K.S.: Phys. Rev. Lett. **94**, 202502 (2005)
20. Filikhin, I.N., Gal, A.: Nucl. Phys. A **707**, 491 (2002)
21. Filikhin, I.N., Gal, A., Suslov, V.M.: Phys. Rev. C **68**, 024002 (2003)
22. Myint, K.S., Shinmura, S., Akaishi, Y.: Eur. Phys. J. A **16**, 21 (2003)
23. Lanskoy, D.E., Yamamoto, Y.: Phys. Rev. C **69**, 014303 (2004)
24. Yamamoto, Y., Rijken, Th.A.: Nucl. Phys. A **804**, 139 (2008)
25. Millener, D.J.: Nucl. Phys. A **835**, 11 (2010)
26. Tamura, H., et al.: Nucl. Phys. A **754**, 58c (2005); **835**, 3 (2010)
27. Aoki, S., et al., KEK-E176 Collaboration: Nucl. Phys. A **828**, 191 (2009)
28. Ahn, J.K., et al.: AIP Conf. Proc. **594**, 180 (2001)
29. Rijken, Th.A., Nagels, M.M., Yamamoto, Y.: Nucl. Phys. A **835**, 160 (2010)
30. Polinder, H., Haidenbauer, J., Meißner, U.-G.: Phys. Lett. B **653**, 29 (2007)

Hyperfine Interact (2012) 210:83–87
DOI 10.1007/s10751-012-0594-7

A double-kaonic nuclear system, K^-K^-pp, to be formed in pp collisions

Maryam Hassanvand · Yoshinori Akaishi · Toshimitsu Yamazaki

Published online: 3 March 2012

Abstract Based on the idea that the simplest double-$\bar{K}$ nuclear cluster, K^-K^-pp, can be produced in the $p + p \to K^+ + K^+ + \Lambda^* + \Lambda^* \to K^+ + K^+ + K^-K^-pp$ reaction, where $\Lambda^* = \Lambda(1405)$ is a quasi-bound state K^-p, we have calculated the differential cross sections for this process and found out that a peak of K^-K^-pp dominates in the mass spectrum when the cluster is a deeply bound and dense system, helped by a very large momentum transfer $Q \sim 1.8$ MeV/c. The appropriate incident proton energy for this process is around 7 GeV. We found that, the deeper and denser the bound system is, the larger its population becomes. The K^--K^- repulsion inside K^-K^-pp is shown to give only a small change on the bound-state structure and the cross section.

Keywords Exotic nuclei · Kaon nuclear cluster · Double-kaonic nuclear system

1 Introduction

Exotic few-body nuclear systems involving $\bar{K}$ (K^- or $\bar{K}^0$) as a constituent have been investigated theoretically [1] based on phenomenologically constructed $\bar{K}N$ interactions, which reproduce low-energy $\bar{K}N$ scattering data [2], kaonic hydrogen atom data [3] and the binding energy and decay width of $\Lambda(1405)$ ($\equiv \Lambda^*$). They are characterized by a strongly attractive $I = 0$ part, which arises essentially from the

M. Hassanvand (✉)
Department of Physics, Isfahan University of Technology, Isfahan, Iran
e-mail: hassanvand@riken.jp

Y. Akaishi
College of Science and Technology, Nihon University, Funabashi, Chiba 274-8501, Japan

T. Yamazaki
Department of Physics, University of Tokyo, Bunkyo-ku, Tokyo 113-0033, Japan

deep bound state of $\Lambda(1405)$, but is fully reconciled with the "repulsive-like" low-energy scattering lengths and kaonic hydrogen level shift. The strongly attractive K^--p interaction was predicted to cause not only deep binding of K^- in proton-rich systems, but also enormous shrinkage of K^--bound nuclei [1, 4].

The simplest kaonic nuclear system, K^-pp, was predicted to exist with a large binding energy $E_K = 48$ MeV and a moderate $\Sigma\pi$ decay width, $\Gamma_K = 61$ MeV [5]. Later, Faddeev calculations on K^-pp [6, 7] were carried out, yielding similar values, whereas an extremely shallow binding was predicted by chiral theories [8]. More recently, the structure of K^-pp was studied comprehensively by a variational method [9], where a molecular nature of K^-pp and a super-strong nuclear force, caused by covalency of a migrating $\bar{K}$, was revealed [10]. In the same papers, a new reaction mechanism of enormous sticking of Λ^* with a participating proton into a dense $\Lambda^*p \equiv K^-pp$, when a Λ^* is produced in a $p + p \to p + \Lambda^* + K^+$ reaction with a short collision length, was pointed out and formulated. In fact, quite recently, a large formation of $X(2265)$ in $p + p \to K^+ + X$ with $X \to p + \Lambda$ was reported as an indication of K^-pp [11]. If $X = K^-pp$, M_X gives a binding energy of 2265 ± 7 MeV and a width of 118 ± 15 MeV. The binding energy is even larger than the original prediction, and is close to the one claimed in a FINUDA experiment on stopped K^- [12].

This finding of DISTO triggered us to study possible production of a double-$\bar{K}$ nuclear cluster, K^-K^-pp, in a similar reaction process, $p + p \to \Lambda^* + \Lambda^* + K^+ + K^+ \to K^-K^-pp + K^+ + K^+$ [13, 14]. The predicted binding energy and partial decay width for K^-K^-pp are 117 MeV and 35 MeV, respectively, and the $p - p$ rms distance is very much reduced to 1.3 fm [15]. Thus, the addition of a $\bar{K}$ increases the binding energy and the nucleon density. Since the observed binding energy of K^-pp indicates the $\bar{K}N$ interaction strength to be 25 % increased than the AY value, we study the reaction spectra by varying the K^-K^-pp energy in a wide range, because our main purpose is to provide a tool to find a K^-K^-pp by calculating the spectral shape and cross section for any possible value of the binding energy .

2 Formulation

We apply the Green's function method to obtain the cross section of the production of two Λ^*'s in pp collision. A detailed description is given in [13, 14].

$$\frac{d^6\sigma}{d\Omega_1 d\Omega_2 dE dE_1} = \beta(k)\frac{d^4\sigma^{\text{elem}}}{d\Omega_1 d\Omega_2} S(E) \tag{1}$$

where $\beta(k)$ is a kinematical factor and $S(E)$ is the spectral function:

$$S(E) = -\frac{1}{\pi}\text{Im}\left[\int\int d\mathbf{r}' d\mathbf{r}\, f^*(\mathbf{r}')\left\langle \mathbf{r}'|\frac{1}{E - H_{\Lambda^*\Lambda^*} + i\epsilon}|\mathbf{r}\right\rangle f(\mathbf{r})\right], \tag{2}$$

and

$$f(\mathbf{r}) = \frac{\beta}{r}\exp\left(-\frac{r}{\beta} + i\mathbf{Q}\mathbf{r}\right), \tag{3}$$

Fig. 1 Shape of the potentials, taken from [14]. The *solid black curve* is for the case without K^--K^- repulsion. The *blue solid curve* is the K^--K^- potential according to [16] and the *red solid and dotted curves* are the K^--K^- and total potentials with finite-size correction

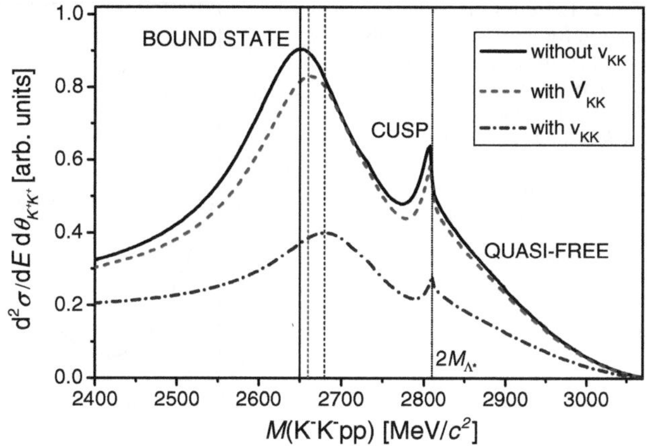

The interaction between two Λ^*'s to represent the binding energy and width of K^-K^-pp is taken as the following form:

$$H_{\Lambda^*\Lambda^*} = -\frac{\hbar^2}{2\mu_{\Lambda^*\Lambda^*}}\nabla^2 + U_{\Lambda^*\Lambda^*}, \tag{4}$$

where

$$U_{\Lambda^*\Lambda^*}(r) = (V_0 + iW_0)\left(\frac{r}{b}\right)^2 \exp\left(-\frac{\mathrm{r}}{\mathrm{b}}\right). \tag{5}$$

The differential cross section at typical incident proton energies and for various binding energies are shown in [13, 14]. By increasing the incident energies we obtained a larger population for K^-K^-pp, and we found that a more compact system with a larger binding energy is more favorably populated.

In the previous work [15], the K^--K^- interaction was neglected because of the lack of information. Now we take into account the K^--K^- repulsion effect according to [16] as:

$$v_{\bar{K}\bar{K}}(r_{KK}) = v_0 \exp\left[-\left(\frac{r_{KK}}{b_{KK}}\right)^2\right] \mathrm{MeV}, \tag{6}$$

where $v_0 = 313$ MeV and $b_{KK} = 0.47$ fm. Adding this repulsive interaction to $H_{\Lambda^*-\Lambda^*}$, it would reduce the binding energy of K^-K^-pp by 34 MeV and so the cross section values. This interaction should be modified and the finite size of the Λ^*'s should be included. We use a double folding potential model (DF) to obtain the K^--K^- repulsive interaction as follows:

$$U_{DF}(\mathbf{r}) = \int\int \rho_K(\mathbf{r_K})\rho_K(\mathbf{r'_K})v(\mathbf{r'})d\mathbf{r_K}d\mathbf{r'_K}, \tag{7}$$

where $\rho_K(\mathbf{r_K})$ is the K^- distribution inside Λ^* and has a Gaussian form. $v(\mathbf{r'})$ represents a K^--K^- repulsion, $\mathbf{r}$ is for $\Lambda^* - \Lambda^*$, $\mathbf{r_K}$ for the $K^- - p$ in each Λ^* and

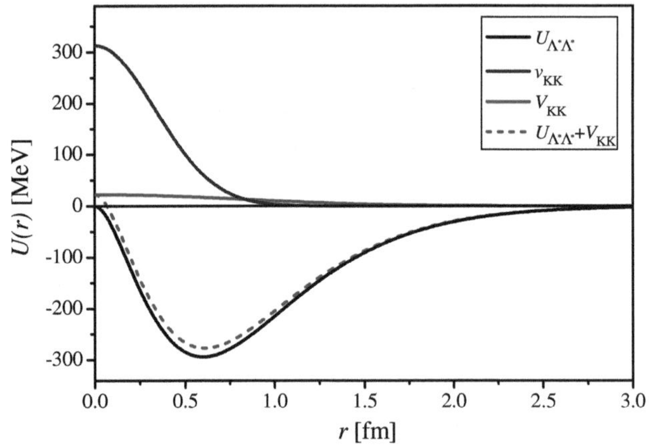

Fig. 2 Differential cross section for various potentials, taken from [14]. The *solid curve* is for the case without K^--K^- repulsion, the *dash-dotted curve* is for the "$U_{\Lambda^*\Lambda^*} + v_{KK}$" potential and the dashed curve for "$U_{\Lambda^*\Lambda^*} + V_{KK}$" case

$\mathbf{r}'_{\mathbf{K}}$ the K^--K^- distance. Knowing the R_{rms} distance of Λ^* [9], we finally obtain the K^--K^- repulsion:

$$V_{KK}(r) = \frac{v_0}{F^3} \exp\left[-\left(\frac{r}{Fb_{KK}}\right)^2\right], \tag{8}$$

$$F = \sqrt{1 + 2\left(\frac{a}{b_{KK}}\frac{M_p}{M_p + m_K}\right)} \sim 2.4 \tag{9}$$

Effectively, the inclusion of the finite size of Λ^* enlarges the range parameters, b_{KK}, by a factor of $F = 2.4$, and weakens the repulsive potential strength by a factor of $F^{-3} = 0.072$. The repulsion interaction in [16] and that with finite size correction are shown in Fig. 1. Clearly, the finite-size correction reduces the original potential only slightly, and so, it doesn't change the cross section much.

Finally, Fig. 2 shows the differential cross sections for these various potentials. The black solid curve is for the case of $E = -150$-i75 MeV (as a standard case which is deeper than the original prediction [15] and consistent with DISTO data [11]), $U_{\Lambda^*-\Lambda^*}$, when we consider no repulsion interaction. The blue dash-dotted curve is for the case without finite-size correction, $v_{\bar{K}\bar{K}}(r_{KK})$, and the red dashed curve is for the case with the above finite-size correction, $V_{KK}(r)$. It causes a reduction of binding energy of K^-K^-pp by 13 MeV. Thus, the K^--K^- interaction has no significant role in the structure of K^-K^-pp.

3 Conclusion

In this work, we present the possibility of the formation of the simplest anti-kaon system, K^-K^-pp. We predict that a bound state peak is clearly been populated. The compactness of this system is vitally important. For a more compact K^-K^-pp a larger population is expected. Deeper binding is more favorable. Adding K^--K^- repulsion causes only a slight reduction in the cross section.

References

1. Akaishi, Y., Yamazaki, T.: Phys. Rev. **C 65**, 044005 (2002)
2. Martin, A.D.: Nucl. Phys. **B 179**, 33 (1981)
3. Iwasaki, M., et al.: Phys. Rev. Lett. **78**, 3067 (1997)
4. Doté, A., Horiuchi, H., Akaishi, Y., Yamazaki, T.: Phys. Rev. **C 70**, 044313 (2004)
5. Yamazaki, T., Akaishi, Y.: Phys. Lett. **B 535**, 70 (2002)
6. Shevchenko, N.V., et al.: Phys. Rev. Lett. **98**, 082301 (2007)
7. Ikeda, Y., Sato, T.: Phy. Rev. **C 76**, 035203 (2007)
8. Doté, A., Hyodo, T., Weise, W.: Phys. Rev. **C 79**, 014003 (2009)
9. Yamazaki, T., Akaishi, Y.: Phys. Rev. **C 76**, 045201 (2007)
10. Yamazaki, T., Akaishi, Y.: Proc. Jpn. Acad. Ser. **B 83**, 144 (2007)
11. Yamazaki, T., et al.: Phys. Rev. Lett. **104**, 132502 (2010)
12. Agnello, M., et al.: Phys. Rev. Lett. **94**, 212303 (2005)
13. Yamazaki, T., Akaishi, Y., Hassanvand, M.: Proc. Jpn. Acad. **B 87**, 362 (2011)
14. Hassanvand, M., Akaishi, Y., Yamazaki, T.: Phys. Rev. **C 84**, 15207 (2011)
15. Yamazaki, T., Doté, A., Akaishi, Y.: Phys. Lett. **B 587**, 167 (2004)
16. Kanada-En'yo, Y., Jido, D.: Phys. Rev. **C 78**, 025212 (2008)

Hyperfine Interact (2012) 210:89–92
DOI 10.1007/s10751-011-0531-1

Energy dependence of K^--"pp" effective potential derived from coupled-channel Green's function

Takahisa Koike · Toru Harada

Published online: 23 December 2011

Abstract We investigate the energy dependence of a single-channel effective potential between the K^- and the "pp"-core nucleus, which can be obtained as an K^--"pp" equivalent local potential from a coupled-channel model for $\bar{K}(NN)$-$\pi(\Sigma N)$ systems. It turns out that the imaginary part of the resultant potential near the $\pi\Sigma N$ decay threshold can well approximate the phase space suppression factor of $K^-pp \to \pi\Sigma N$ decay modes. The effects on the pole position of the $\pi(\Sigma N)$ state in the $\pi\Sigma N$ channel are also discussed.

Keywords Kaonic nuclei · Equivalent local potential

1 Introduction

In order to elucidate the nature of antikaon-nucleon ($\bar{K}N$) interaction in high density nuclear matter, it is an important subject to clarify whether the "deeply-bound kaonic nucleus" exists or not. In particular, the $[\bar{K} \otimes \{NN\}_{I=1}]_{I=1/2}$, $J^\pi = 0^-$ bound state, which is called "K^-pp" here, is suggested to be the lightest and most fundamental kaonic nucleus [1]. A new experimental search of K^-pp via the ^{3}He(in-flight K^-, n) reaction has been planned at J-PARC as E15 experiment [2]. We theoretically have discussed the expected inclusive and semi-exclusive spectra for the ^{3}He(in-flight K^-, n) reaction within the framework of the distorted-wave

T. Koike (✉)
Strangeness Nuclear Physics Laboratory, RIKEN Nishina Center, Wako-shi, Saitama 351-0198, Japan
e-mail: tkoike@riken.jp

T. Harada
Research Center for Physics and Mathematics, Osaka Electro-Communication University, Neyagawa, Osaka 572-8530, Japan

impulse approximation (DWIA) . We have employed a phenomenological single-channel K^--"pp" (complex) effective potential between the K^- and the "pp"-core nucleus [3, 4], which has the form of

$$U^{\mathrm{eff}}(E;r) = (V_0 + i\,W_0\,f(E))\exp[-(r/b)^2], \tag{1}$$

where V_0 and W_0 are strength parameters in real and imaginary parts, respectively, and b is the range parameter. $f(E)$ is the phase space suppression factor of the $K^-pp \to \pi\Sigma N$ decay modes [5] where E is the energy measured from the K^-pp threshold.

In the single-channel framework, we have shown that behavior of the ^{3}He(in-flight K^-, n) spectrum can be understood in the "moving pole" picture [4]; a pole of the K^-pp bound state moves in the complex energy plane as a function of the energy E on the real axis because the K^--"pp" effective potential is considerably energy-dependent. A trajectory of its moving pole governs the shape of the spectrum. Thus, the validity of our calculations relies in part on whether the energy dependence of (1) is appropriate near the $\pi\Sigma N$ decay threshold or not. To examine this subject, we have extended the previous K^-pp single channel description to the $\bar{K}(NN)$-$\pi(\Sigma N)$ coupled-channel (CC) description because the energy dependence of (1) should originate from eliminating the $\pi\Sigma N$ channel in such a CC scheme.

In this article, we investigate a single-channel K^--"pp" equivalent local potential which is derived from CC Green's functions for $K^-(pp)$-$\pi(\Sigma N)$ systems, and evaluate the energy dependence of this effective potential to be compared with a phenomenological one [4] which is determined within (1).

2 Coupled-channel model

Let us consider the following CC Green's function G_{ij} for $K^-(pp)$-$\pi(\Sigma N)$ systems;

$$\begin{bmatrix} E_1 - T_1^{(l)} - U_{11}(r) & -U_{12}(r) \\ -U_{21}(r) & E_2 - T_2^{(l)} - U_{22}(r) \end{bmatrix} \begin{bmatrix} G_{11}^{(l)}(r,r') & G_{12}^{(l)}(r,r') \\ G_{21}^{(l)}(r,r') & G_{22}^{(l)}(r,r') \end{bmatrix} = \delta(r'-r)\mathbb{1}, \tag{2}$$

where the channel 1 (2) refers to the K^- and pp-core system (the π and ΣN-core system): $T_{1,2}^{(l)}$ denote the kinetic energies for 1, 2, and the energy $E_2 = E_1 - E_{\mathrm{th}}(\pi\Sigma N)$ where $E_{\mathrm{th}}(\pi\Sigma N) \simeq -100$ MeV below the K^-pp threshold. U_{11}, U_{22} and $U_{12}(= U_{21})$ denote the diagonal and coupling potentials, respectively. For simplicity, we assume an energy-independent Gaussian form as

$$\begin{bmatrix} U_{11}(r) & U_{12}(r) \\ U_{21}(r) & U_{22}(r) \end{bmatrix} = \begin{bmatrix} V_1 + i\,W_1 & V_c \\ V_c & V_2 + i\,W_2 \end{bmatrix} \exp[-(r/b)^2], \tag{3}$$

where $V_{1,2}$ and V_c denote the strength parameters of the diagonal and coupling potentials, respectively, and $W_{1,2}$ describe effects of the other decay modes. These parameters in (3) should be determined so as to reproduce the proper values of the binding energy and width which are obtained by single-channel calculations with U^{eff} in (1). If we use CC Green's functions in (2), we expect to confirm the previous results [3, 4] for the ^{3}He(in-flight K^-, n) spectrum within the DWIA calculation.

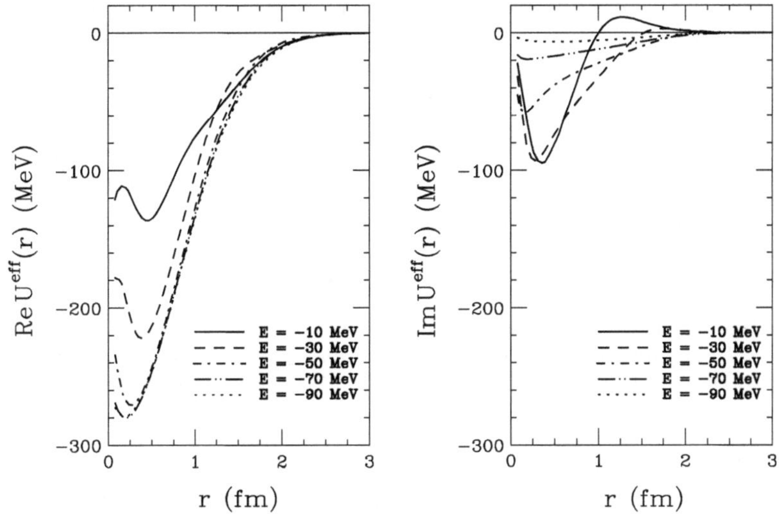

Fig. 1 Real (*left*) and imaginary (*right*) parts of the equivalent local potential $\tilde{U}_{11}^{\rm eff}$ for the K^--"pp" channel. Potential parameters of (3) are used as $V_1 = -300$ MeV, $V_2 = -150$ MeV, $V_c = -100$ MeV, $W_1 = W_2 = 0$ MeV and $b = 1.09$ fm. The energy E, which is measured from the K^-pp threshold, is varied form −10 MeV to −90 MeV

Now we define the K^--"pp" equivalent local potential for channel 1, $\tilde{U}_{11}^{\rm eff}$ as

$$\{E_1 - T_1^{(l)} - \tilde{U}_{11}^{\rm eff}(E; r)\}\, G_{11}^{(l)}(E; r, r') = \delta(r' - r), \tag{4}$$

where $G_{11}^{(l)}$ is the (1,1) component of a solution of (2). Then, we get

$$\tilde{U}_{11}^{\rm eff}(E; r)\, G_{11}^{(l)}(E; r, r') = U_{11}(r)\, G_{11}^{(l)}(E; r, r') + U_{12}(r)\, G_{21}^{(l)}(E; r, r'). \tag{5}$$

By multiplying the initial wave function $\phi_1(r')$ as a bound state in channel 1, and by integrating over r', we obtain the expression of the equivalent local potential as

$$\tilde{U}_{11}^{\rm eff}(E; r) = U_{11}(r) + U_{12}(r)\, \frac{\displaystyle\int_0^\infty G_{21}^{(l)}(E; r, r')\, \phi_1(r')\, dr'}{\displaystyle\int_0^\infty G_{11}^{(l)}(E; r, r')\, \phi_1(r')\, dr'}, \tag{6}$$

which is defined under the boundary condition at *every* E on the physical axis, because Green's functions can be calculated numerically at every points in the complex energy plane. This is the advantage that we used Green's functions rather than wave functions.

3 Results and discussion

Figure 1 shows the calculated equivalent local potential $\tilde{U}_{11}^{\rm eff}$ with $V_1 = -300$ MeV, $V_2 = -150$ MeV, $V_c = -100$ MeV, $W_1 = W_2 = 0$ MeV and $b = 1.09$ fm, in which the potential parameters are determined so that the binding energy and width of K^-pp bound state are consistent with those calculated by Yamazaki and Akaishi [1].

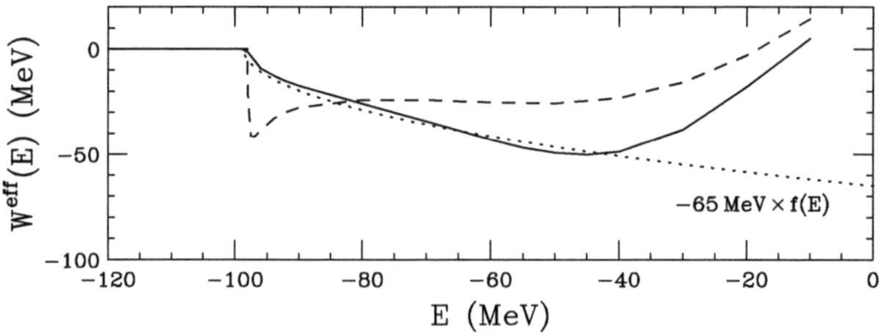

Fig. 2 Energy dependence of the imaginary part $W^{eff}(E)$ of the K^--(pp) equivalent local potential $\tilde{U}^{eff}_{11}$. The *solid* and *dashed lines* denote the cases with $V_2 = -150$ MeV and -300 MeV in (3), respectively. The *dotted line* denotes values obtained by $W_0 \times f(E)$ with $W_0 = -65$ MeV

We find that the strength of the imaginary part of $\tilde{U}^{eff}_{11}$ becomes shallower as E goes to $E_{th}(\pi\Sigma N)$. This behavior is approximately equivalent to that of the energy dependence obtained from the phase space factor in (1), whereas the shape of $\tilde{U}^{eff}_{11}$ is not a Gaussian function exactly. Therefore, we estimate the potential strength of $(V^{eff}(E), W^{eff}(E))$ which is obtained with the help of the volume integrals;

$$V^{eff}(E) + i\,W^{eff}(E) = \int_0^\infty \tilde{U}^{eff}_{11}(E; r)\, r^2 dr \Big/ \int_0^\infty \exp[-(r/b)^2]\, r^2 dr. \qquad (7)$$

Figure 2 shows values of $W^{eff}(E)$ as a function of the energy E, together with the fitted curve of $W_0 \times f(E)$ in (1). The results are as follows:

- If no bound state exists in the $\pi(\Sigma N)$ channel ($V_2 = -150$ MeV), Im $\tilde{U}^{eff}_{11}$ approximates to the phase space factor $f(E)$ that is used in (1) near the $K^- pp \to \pi\Sigma N$ decay threshold.
- If a bound state exists in the $\pi(\Sigma N)$ channel ($V_2 = -300$ MeV), the energy dependence of Im $\tilde{U}^{eff}_{11}$ considerably differs from that of (1) due to modification of the phase volume via a pole which is located near the $\pi\Sigma N$ threshold.

In summary, we have investigated the energy dependence of the single-channel K^--"pp" equivalent local potential $\tilde{U}^{eff}_{11}$ derived from the $\bar{K}(NN)$-$\pi(\Sigma N)$ model. It has turned out that the imaginary part of $\tilde{U}^{eff}_{11}$ near the $\pi\Sigma N$ decay threshold can well approximate the phase space suppression factor of $K^- pp \to \pi\Sigma N$ decay modes. If potential parameters in (3) are replaced by the energy-dependent ones, as obtained from chiral $\bar{K}N$-πY dynamics [6], their energy dependence would additionally contribute to the effective potential. The further detailed investigation is now in progress.

References

1. Yamazaki, T., Akaishi, Y.: Phys. Lett. **B535**, 70 (2002)
2. Iwasaki, M., et al.: J-PARC proposal. http://j-parc.jp/NuclPart/pac_0606/pdf/p15-Iwasaki.pdf
3. Koike, T., Harada, T.: Phys. Lett. **B652**, 262 (2007)
4. Koike, T., Harada, T.: Phys. Rev. **C80**, 055208 (2009)
5. Mareš, J., Friedman, E., Gal, A.: Phys. Lett. **B606**, 295 (2005)
6. Jido, D., et al.: Nucl. Phys. **A725**, 181 (2003)

Hyperfine Interact (2012) 210:93–96
DOI 10.1007/s10751-011-0545-8

K^- nuclear quasi-bound states in a chirally motivated coupled-channel approach

Jiri Mareš

Published online: 5 January 2012

Abstract K^- nuclear optical potentials are constructed from in-medium $\bar{K}N$ scattering amplitudes within a chirally motivated coupled-channel model. The strong energy and density dependence of the scattering amplitudes at and below threshold leads to K^- potential depths $-\mathrm{Re}V_{K^-}(\rho_0) \approx 80 - 100$ MeV. Self consistent calculations of K^- nuclear quasi-bound states are discussed.

Keywords K^- nuclear states · Kaon-nucleon interaction

1 Introduction

Meson-baryon interactions at low energies can be systematically evaluated within an SU(3) chiral approach combined with coupled channel T-matrix resummation techniques [1–3]. A typical scale for the nuclear-matter depth of chirally motivated K^- nuclear potentials is 100 MeV [4], although considerably shallower potentials have also been derived [1, 5].

In our present work [6, 7], we construct the K^- meson self energy operator (SE) from in-medium subthreshold K^-N scattering amplitudes within the chirally motivated model [3]. We demonstrate how to incorporate the strong energy and density dependence of the K^-N scattering amplitude at and near threshold into self consistent evaluation of the K^- nuclear potential V_{K^-}.

This work was supported by the GACR Grant No. 202/09/1441, as well as by the EU initiative FP7, HadronPhysics2, under Project No. 227431.

J. Mareš (✉)
Nuclear Physics Institute, 250 68 Řež, Czech Republic
e-mail: mares@ujf.cas.cz

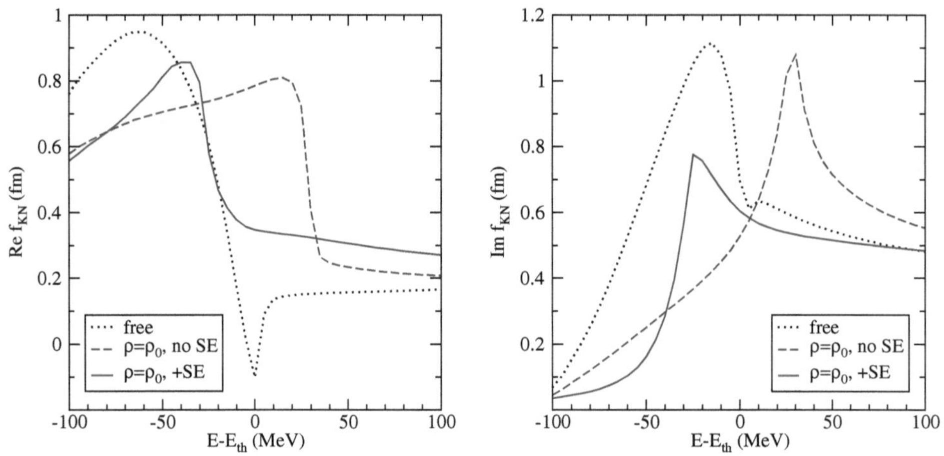

Fig. 1 Energy dependence of the c.m. reduced amplitude f_{K^-N} in model TW1 below and above threshold (*left*: real part, *right*: imaginary part). *Dotted lines*: free-space amplitude; *dashed lines*: Pauli blocked amplitude (no SE) for $\rho_0 = 0.16\ \mathrm{fm}^{-3}$; *solid lines*: including also hadron self energies (+SE) at ρ_0

2 In-medium $\bar{K}N$ amplitudes

The interaction of a K^- meson with a nucleus is described by the Klein Gordon (KG) equation of the form

$$\left[\nabla^2 - 2\mu\,(\mathcal{B}_K + V_C) + (V_C + \mathcal{B}_K)^2 + 4\pi\left(1 + \frac{m_K - B_K}{m_N}\right) F_{K^-N}\left(\sqrt{s}, \rho\right)\rho\right]\psi = 0\,, \tag{1}$$

where μ is the K^- – nucleus reduced mass, V_C is the Coulomb potential, s is the Mandelstam variable equal to the square of the total K^-N energy in the 2-body c.m. frame, $\mathcal{B}_K = B_K + \mathrm{i}\Gamma_K/2$ is a complex binding energy, and the K^- energy $\omega_K = m_K - \mathcal{B}_K - V_C$.

The K^-N scattering amplitude $F_{K^-N}(\sqrt{s}, \rho)$ is calculated in a separable form: $F_{K^-N} = g(p) f_{K^-N}(\sqrt{s}) g(p')$, within chirally motivated meson-baryon coupled-channel model [3]. When the elementary K^-N system is embedded in the nuclear medium one has to consider Pauli blocking in the intermediate states (marked 'no SE'). The other version (marked '+SE') adds self consistently hadron self energies [8]. Details of the model are given in [7].

It is to be noted here that the K^- nuclear potential V_{K^-} appears as an argument in the expressions for $\sqrt{s}$ and p^2, which in turn serve as arguments in the expression for the self energy, and thus for V_{K^-}. This suggests to calculate V_{K^-} self consistently.

Figure 1 illustrates a typical resonance-shape energy dependence of the reduced scattering amplitude $f_{K^-N}(\sqrt{s}, \rho) = 1/2[f_{K^-p}(\sqrt{s}, \rho) + f_{K^-n}(\sqrt{s}, \rho)]$, corresponding to the interaction of the K^- meson with symmetric nuclear matter. At threshold, the real part of the '+SE' amplitude is about half of that 'no SE', corresponding to a depth $-\mathrm{Re}V_{K^-}(\rho_0) \approx 40$ MeV, in agreement with Refs. [1, 5].

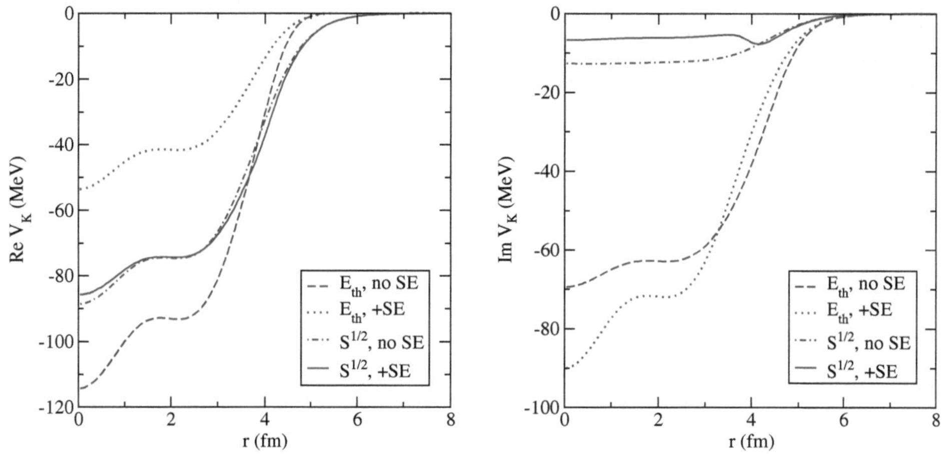

Fig. 2 K^--nuclear potentials in Ca (*left*: real part, *right*: imaginary part), calculated with static RMF nuclear densities and TW1 chiral amplitudes at threshold (E_{th}) and with $\sqrt{s}$, in both in-medium versions: 'only Pauli blocking' (no SE) and including also hadron self energies (+SE)

3 Kaonic nuclei

Within the above formulation, we calculated quasi bound K^- states in selected nuclei across the periodic table. We solved KG equation in a static approximation, as well as fully dynamically upon taking into account the polarization of the nuclear core by the strongly bound antikaon. We included also phenomenological 2N absorption terms to simulate $K^-NN \to YN$ absorption modes. We found that a p-wave interaction generated by the $\Sigma(1385)$ subthreshold resonance plays a secondary role in K^- nuclear states. The detailed discussion of the results is given in [7]. Here, we present selected results of the static calculations within LO model TW1 [7] fitted to the new SIDDHARTA data of the shift and width in the K^- hydrogen atom [9].

Figure 2 shows the K^--nuclear potentials in Ca at threshold (E_{th}) and for $\sqrt{s} = E_{th} - B_N - \xi_N B_K - 15.1(\rho/\rho_0)^{2/3} + \xi_K \mathrm{Re}\mathcal{V}_{K^-}(\rho)$, where $\mathcal{V}_{K^-} = V_{K^-} + V_C$, and $\xi_{N(K)} = m_{N(K)}/(m_N + m_K)$ [7]. It is seen that the subthreshold extrapolation $\sqrt{s}$ is crucial for the depth of V_{K^-} both in the 'no SE' and '+SE' case. While at threshold the depth of ReV_{K^-} in the '+SE' case is about half of the depth in the 'no SE' case, for $\sqrt{s}$ both in-medium versions give a similar depth $-\mathrm{Re}V_{K^-}(\rho_0) \approx 80 - 90$ MeV. The imaginary parts of V_{K^-} and consequently the widths which represent only $K^-N \to YN$ decays, are considerably reduced in the self-consistent calculations of the subthreshold amplitudes owing to the proximity of the $\pi\Sigma$ threshold. This reduction of Γ_K which appears model independent, confirms conclusions of Wycech [10].

In Fig. 3, we present binding energies B_K of $1s$ K^- nuclear states obtained by solving Eq. 1 self consistently for several nuclei, using in-medium 'no SE' and '+SE' subthreshold amplitudes with argument $\sqrt{s}$. These values of B_K are compared to those calculated using threshold amplitudes (E_{th}). The figure again illustrates the peculiar role of energy dependence of the K^-N scattering amplitudes. It is worth noting that the self consistent calculations of B_K using 'no SE' and '+SE' in-medium amplitudes give very similar results. These B_K values are remarkably close to those

Fig. 3 Binding energies B_K of 1s K^-—nuclear quasibound states in several nuclei, calculated using static RMF nuclear densities and TW1 chiral amplitudes at threshold (E_{th}) and with $\sqrt{s}$, in both in-medium versions: 'only Pauli blocking' (no SE) and including also hadron selfenergies (+SE). Results of static RMF calculations, with a K^-—nuclear interaction mediated by vector mesons only, are shown for comparison

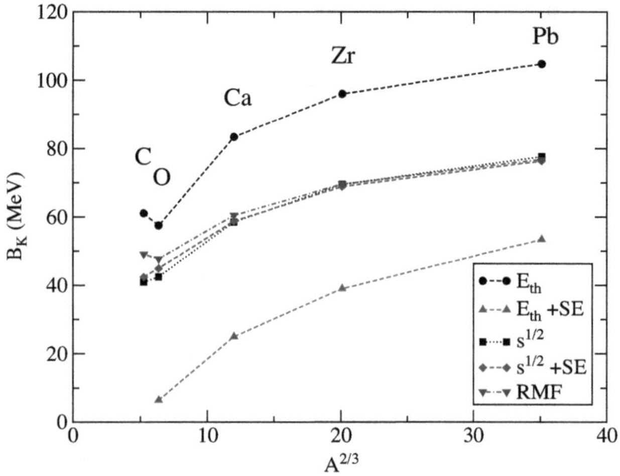

calculated within a static RMF approach, when K^- nucleus interaction is mediated exclusively by vector mesons with purely vector SU(3) F-type couplings (denoted RMF). Last, when phenomenological energy depedent imaginary ρ^2 terms are added self consistently to simulate two-nucleon $K^-NN \to YN$ absorption modes and their available phase space [11], the resulting widths of order $\Gamma_K \approx 50$ MeV become comparable in light nuclei to the binding energies B_K.

Acknowledgements I wish to thank my colleagues Ales Cieply, Eli Friedman, Avraham Gal and Daniel Gazda.

References

1. Ramos, A., Oset, E.: Nucl. Phys. A **671**, 481 (2000)
2. Borasoy, B., Nissler, R., Weise, W.: Eur. Phys. J. A **25**, 79 (2005); Phys. Rev. Lett. **94**, 213401 (2005)
3. Cieplý, A., Smejkal, J.: Eur. Phys. J. A **43**, 191 (2010)
4. Waas, T., Kaiser, N., Weise, W.: Phys. Lett. B **365**, 12 (1996); ibid. **379**, 34 (1996)
5. Cieplý, A., Friedman, E., Gal, A., Mareš, J.: Nucl. Phys. A **696**, 173 (2001)
6. Cieplý, A., Friedman, E., Gal, A., Gazda, D., Mareš, J.: Phys. Lett. B **702**, 402 (2011)
7. Cieplý, A., Friedman, E., Gal, A., Gazda, D., Mareš, J.: Phys. Rev. C **84**, 045206 (2011)
8. Lutz, M.: Phys. Lett. B **426**, 12 (1998)
9. Bazi M., et al.: [SIDDHARTA Collaboration], Phys. Lett. B **704**, 113 (2011)
10. Wycech, S.: Nucl. Phys. A **450**, 399c (1986)
11. Mareš, J., Friedman, E., Gal, A.: Nucl. Phys. A **770**, 84 (2006)

Hyperfine Interact (2012) 210:97–101
DOI 10.1007/s10751-011-0533-z

Recent results on the $K^-_{stop} + A \rightarrow \Sigma^\pm + \pi^\mp + A'$ reaction with FINUDA

Stefano Piano on behalf of the FINUDA collaboration

Published online: 30 December 2011

Abstract A study of the $K^-_{stop}A \rightarrow \Sigma^\pm\pi^\mp A'$ reaction in some p-shell nuclei, i.e., ^{6,7}Li, ^{9}Be, ^{13}C and ^{16}O, performed by the FINUDA experiment, is reported.

Keywords Kaon absorption reactions

1 Introduction and experimental procedure

The reaction $K^-_{stop}\ A \rightarrow \Sigma^\pm\pi^\mp A'$ (A: ^{6}Li, ^{7}Li, ^{9}Be, ^{13}C and ^{16}O), was studied by means of the FINUDA magnetic spectrometer, operating at the DAΦNE ϕ-factory (LNF-INFN, Italy). In this study the charged Σ-hyperons are reconstructed through their decay channels $\Sigma^\pm \rightarrow n\pi^\pm$, whose full details are given in Ref. [1]. This approach allows for the selection of the elementary process $K^-_{stop}\ p \rightarrow \Sigma^\pm\pi^\mp$, with the absorbing proton being embedded in A. In fact, charged Σ's (and Λ's) from the two-body absorption process $K^-_{stop}\ 2N \rightarrow N\Sigma(\Lambda)$ have no access to the selected final state. However, the request for a final Σ excludes the detection of the Σ conversion reaction, $\Sigma N \rightarrow \Lambda N$, and thus the identification of possible Σ-hypernuclear states.

All the particles involved in the $A(K^-_{stop}, \Sigma^\pm\pi^\mp)A'$ reaction, i.e. $K^\pm$'s, $\pi^\mp$'s and $\Sigma^\pm$'s, are fully reconstructed in the apparatus. The negative kaons from the decays of the ϕ meson (B.R.∼ 50%), after traversing the innermost layers of the spectrometer, stop in ∼0.25 gr/cm^2 thick targets. The targets are mostly manufacted as solid slabs except for ^{13}C and ^{16}O, the first being carbon powder and the second liquid D_2O.

The FINUDA spectrometer consists of an inner (TOFINO [2]) and an outer (TOFONE [3]) sensitive layers of plastic scintillator, which constitute the time-of-flight

S. Piano (✉)
INFN Trieste, via A. Valerio, 2, 34127 Trieste, Italy
e-mail: stefano.piano@ts.infn.it

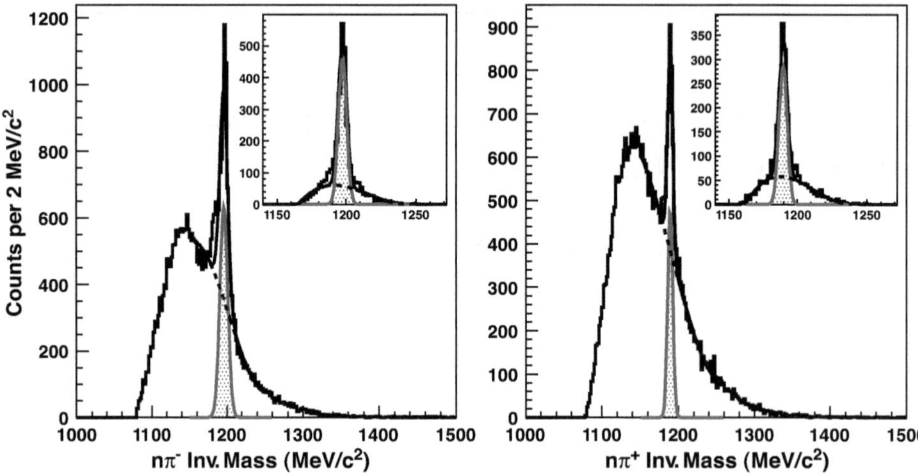

Fig. 1 Unconstrained $n\pi^-$ and $n\pi^+$ invariant mass distributions for the $^6Li(K^-_{stop}, n\pi^\pm)A'$ reaction. The *grey-filled* peaks represent the Σ^- (left) and Σ^+ signals. In the insets the constrained $n\pi^\mp$ invariant mass distributions are shown

(TOF) system of the apparatus. The TOF system is used to determine the momentum of neutrons as well as for triggering purposes. In this measurement, the neutron momentum resolution is obtained from the distribution width of monokinetic neutrons from $\Sigma^+_{stop} \rightarrow n\pi^+$: at 187.6±0.2 MeV/c, $\sigma = 8.4 \pm 0.2$ MeV/c. The tracking system of FINUDA is located between TOFINO and TOFONE. It consists of a vertex detector [4], two intermediate layers of low-mass drift chambers [5] and an outer array of straw tubes [6]. The vertex detector is composed by two layers of double-sided microstrip silicon sensors [4] which surrounds the targets stack. For the FINUDA magnetic field set at 1 T, the spectrometer is capable of analyzing 184.5±0.1 MeV/c positive pions with a resolution at σ of 1.72±0.06 MeV/c, which only slightly worsens for pions of lower momenta. The primary condition for processing a $n\pi^-\pi^+$ event requires a vertex in one of the targets between the tracks of the K^- and one of the two final pions. The vertex position is reconstructed with an average spatial resolution of 0.7 mm, mostly due to the kaon straggling inside the target. The particle identification is performed through the particle energy deposit in the two layers of the vertex detector as well as in the two drift chambers. The dE/dX technique ensures an overall pion ID efficiency above 98%.

Figure 1 shows the $n\pi^-$ and $n\pi^+$ invariant mass spectra for ^{6}Li. For these spectra, the background (B) is mostly due to γ's which may emulate neutrons in TOFONE and by neutrons which may scatter before being detected. The background due to misidentification of pions being negligible. A clear background reduction is obtained by requiring the kinematic observables of the $n\pi^\pm$ events to belong to the phase space volume of the $\Sigma^\pm \rightarrow n\pi^\pm$ decays. In this way, the $n\pi^-$ and $n\pi^+$ invariant mass distributions are dominated by the Σ^- and Σ^+ peaks, see the insets of Fig. 1. When peak fitting these distributions, $(S/B)_{\Sigma^-} = 4.5$ and $(S/B)_{\Sigma^+} = 3.6$, where S is the peak strength taken at 2σ.

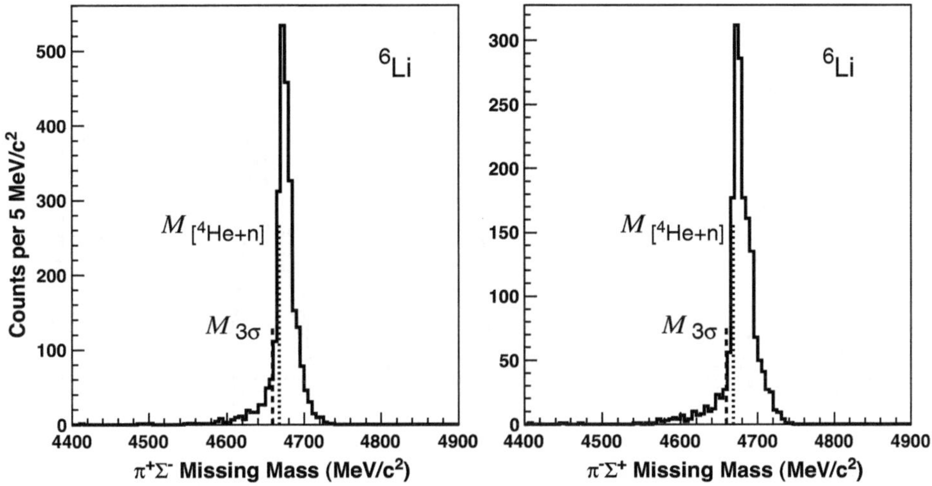

Fig. 2 Missing mass distribution for the ^{6}Li(K^-_{stop}, $\Sigma^{\pm}\pi^{\mp}$)A' reactions

2 Results

Figure 2 shows the missing mass (M) spectra of the (K^-_{stop}, $\Sigma^{\pm}\pi^{\mp}$) reaction on ^{6}Li, for the constrained events; i.e., for the events in the inset of Fig. 1.

In the figure, the dotted line denotes the physical threshold of the missing mass ($A' \equiv [^4He_{g.s.} + n]$), the dashed line points to the missing mass threshold ($M_{3\sigma}$) used in the analysis. The overall uncertainty in measuring M is $\sigma = 3.1$ MeV/c^2. Both distributions are peaked at about 10 MeV/c^2 above $M_{[^4He+n]}$ and have a FWHM in the range 22–24 MeV/c^2, which indicates clearly the quasi-exclusive character of the measurement. The observed particles carry away the 96% of the available kinetic energy, thus no other particles can be produced in the reaction.

Figure 3 shows the measured momentum distributions of prompt pions, which are corrected for the apparatus acceptance. The arrows denote the condition p[$B_{\Sigma^{-(+)}} =$ 0], where B_{Σ} is the binding energy of possible Σ-hypernuclei ($_{\Sigma}A$): beyond this threshold, there are no peaks indicating the presence of Σ-hypernuclei, in spite of a negligible background. The pion momentum spectra, which extend down to 100 MeV/c, are slightly affected by distortions due to the traversed media.

For ^{6}Li, the $\Sigma^{\pm}$ momentum distributions are shown in Fig. 4 (grey-filled histograms). For Σ^-, the distribution resembles a rather symmetric bump peaked at about 165 MeV/c with a FWHM of 155 MeV/c and shows no strength at around 0 MeV/c. The Σ^+ momentum distribution is peaked at around 0 MeV/c and a 160 MeV/c broad-bump arises at $\sim$ 174 MeV/c, the mass difference between the central values of the two bumps being basically due to the $\Sigma^{\pm}$ mass difference. At these momenta, positive and negative Σ's behave differently when traveling through nuclear targets of finite thickness. To account for this, a Monte-Carlo simulation was developed based on the $K^-_{stop}A \rightarrow \Sigma^{\pm}\pi^{\mp}A'$ quasi-free reaction. Events, related to such a reaction, were then filtered out by the geometry of FINUDA to reproduce the interaction of $\Sigma^{\pm}$'s with the target media. The resulting momentum spectra (open histograms in the Fig. 4) show a reasonable overall agreement with the experimental

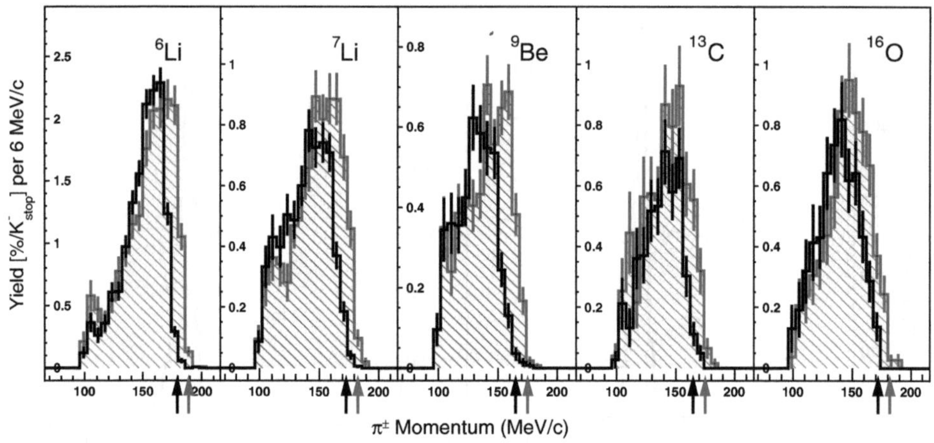

Fig. 3 Measured momentum distributions of prompt pions from the $A(K^-_{stop}, \Sigma^{\pm}\pi^{\mp})A'$ reactions. The open (*grey-filled*) histograms describe the behavior of positive (negative) pions and the black (*grey*) arrows point to the p[$B_{\Sigma^{-(+)}} = 0$] threshold

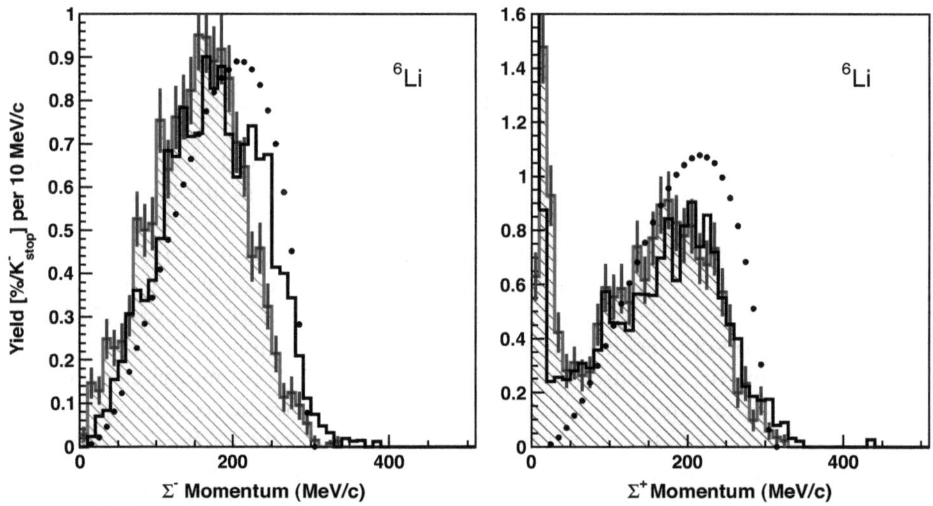

Fig. 4 Momentum distributions of Σ's from the $^6Li(K^-_{stop}, \Sigma^{\pm}\pi^{\mp})A'$ reactions. The grey-filled histograms are the measured distributions. The distributions of Monte-Carlo generated sigmas are depicted by *full dots* and the open diagrams are the M-C generated Σ's reconstructed by FINUDA

distributions. The generated $\Sigma^{\pm}$ momentum spectra are displayed in Fig. 4 with full dots. For other nuclei, the results are similar: as a whole, for p-shell nuclei the $K^-_{stop}A$ absorption reaction depends rather weakly on A.

The K^-_{stop} A emission rate (R_A), leading to final $\Sigma^{\pm}\pi^{\mp}$ pairs, is also evaluated and the values are reported in Fig. 5 left together with the few existing older measurements [7–9]. The R_A values are corrected for the amount of the unobserved Σ^-'s, as deduced from Σ^+ and Σ^- measured momenta distributions.

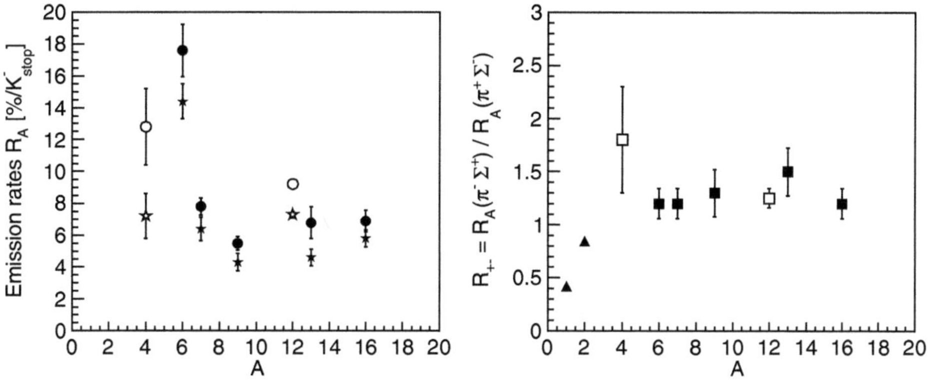

Fig. 5 *Left*: emission rates of $K^-_{stop} + A \to \Sigma^+ + \pi^- + A'$, circles, and $K^-_{stop} + A \to \Sigma^- + \pi^+ + A'$, stars. The *black symbols* are from FINUDA [1], the open ones are from Refs. [7] (^{4}He) and [8, 9] (^{12}C). Right: Ratio between the $\Sigma^+\pi^-$ and $\Sigma^-\pi^+$ yields: the *black squares* are from FINUDA [1], the open ones are from Refs. [7] (^{4}He) and [8, 9] (^{12}C), the *black triangles* are from Ref. [10] (H, D_2)

The ratio $R_{+-} = R_A(\pi^-\Sigma^+)/R_A(\pi^+\Sigma^-)$ is compared for different nuclei (Fig. 5 right). When the $K^-_{stop}\,p \to \Sigma^\pm\pi^\mp$ absorption occurs on a bound proton, $R_{+-} > 1$. On the contrary, when the same process occurs on a free proton (H) this relationship is reversed [10]. Such a clear increase of R_{+-}, when moving the interaction from a free proton to a bound proton, can be related to the sub-threshold modification of the $\bar{K}N$ interaction.

For the light nuclei examined, R_{+-} features a flat behavior indicating that the kaon absorption occurs preferentially at the surface of a nucleus. The average value, $R_{+-} = 1.3 \pm 0.1$, can be compared directly with the ratio calculated in Ref. [11]. The agreement found is remarkable.

The absorption of stopped kaons involves only surface protons when the final state particles are $\Sigma^\pm\pi^\mp$ pairs. Nevertheless, Fig. 5 left shows that when a neutron is added to a nucleus R_A drops for both channels, which is likely due to the excess of surface neutrons over surface protons. As an example, compare ^{7}Li and ^{13}C to ^{6}Li and ^{12}C, respectively. As a final note, the large values of R_A's found for ^{6}Li reflect the previously measured large value for ^{4}He since this nucleus accounts for the ^{6}Li core.

References

1. FINUDA Collaboration, Agnello, M., et al.: Phys. Lett. **B704**, 474 (2011)
2. Filippini, V., Marchesotti, M., Marciano, C.: Nucl. Instrum. Methods **A424**, 343 (1999)
3. Pantaleo, A., et al.: Nucl. Instrum. Methods **A545**, 593 (2005)
4. Bottan, P., et al.L: Nucl. Instrum. Methods **A427**, 423 (1999)
5. Agnello, M., et al.: Nucl. Instrum. Methods **A385**, 58 (1997)
6. Benussi, L., et al.: Nucl. Instrum. Methods **A361**, 180 (1995), **A419**, 648 (1998)
7. Katz, P.A., et al.: Phys. Rev. **D1**, 1267 (1970)
8. Vander Velde-Wilquet, C., et al.: Nucl. Phys. **A241** 511 (1975)
9. Vander Velde-Wilquet, C., et al.: Nuovo Cim. **39 A**, 538 (1977)
10. Nowak, R.J., et al.: Nucl. Phys. **B139**, 61 (1978)
11. Wycech, S.: Nucl. Phys. **B28**, 541 (1971)

Hyperfine Interact (2012) 210:103–106
DOI 10.1007/s10751-011-0509-z

Neutron spectra from the low-energy $K^- + d \to \pi + \Sigma + n$ reaction and the shape of the $\Lambda(1405)$ resonance

János Révai

Published online: 10 February 2012

Abstract Neutron spectra from the low-energy $K^- + d \to \pi + \Sigma + n$ reaction were calculated. This quantity should carry information on the $\Lambda(1405)$ resonance. We checked the possibility of extracting this information.

Keywords Kaonic nuclear cluster · $\Lambda(1405)$ resonance

The $\Lambda(1405)$ resonance plays a central role in low-energy kaon-nuclear physics. Being a manifestation of the assumed attraction between negative antikaons and nucleons, its observability and properties are crucial for the possible existence of antikaonic nuclear states (clusters). Its structure and origin are subject of sometimes passionate discussions among representatives of different opinions on this issue. This state can not be reached in two-body reactions due to the fact, that it lies below the K^-p threshold. Therefore it can be observed only in reactions involving $n \geq 3$ particles, the simplest of which is probably the following one:

$$\begin{aligned} & \nearrow K^- + d \\ K^- + d & \to (\pi + \Sigma)_{I=0,1} + n \\ & \searrow (\pi + \Sigma)_{I=1} + p \end{aligned} \tag{1}$$

We investigated reaction (1) trying to find a signature of $\Lambda(1405)$ resonance in the neutron spectrum.

The treatment is based on the coupled particle-channels Faddeev-AGS equations for the $\bar{K}NN \leftrightarrow \pi\Sigma N$ three-body system. The equations for the system were discussed in detail earlier [1], however, it was necessary to write additional formulae for

The support from the OTKA grant T71989 is gratefully acknowledged.

J. Révai (✉)
BLTP, JINR, 141980 Dubna, Russia
e-mail: revai@rmki.kfki.hu

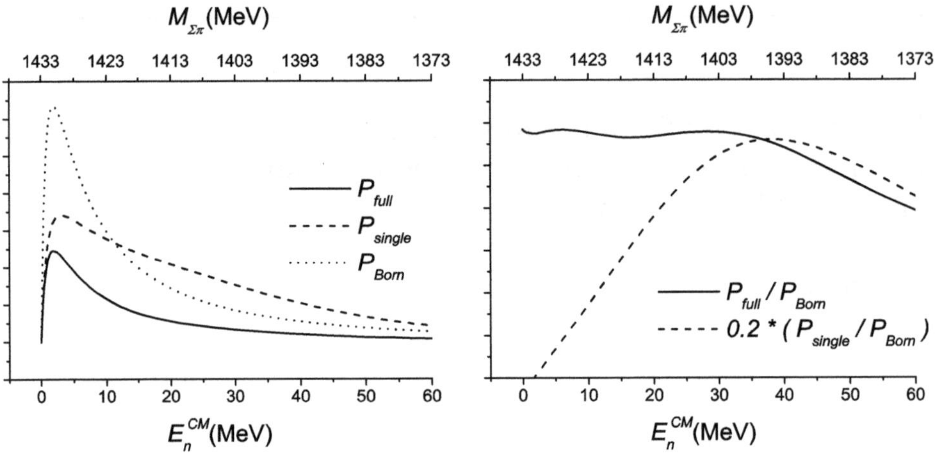

Fig. 1 Total neutron spectra for different approaches. *Left panel* $P_{\rm full}$, $P_{\rm sing}$ and $P_{\rm Born}$ spectra. *Right panel* $P_{\rm full}/P_{\rm Born}$ and $P_{\rm sing}/P_{\rm Born}$ (normalized) spectra

the break-up processes in terms of the matrix elements for elastic and re-scattering processes. A further extension was the account for the K^-, $\bar{K}^0$ mass difference, due to which the free Green's operator becomes non-diagonal in isospin indices, and states with three-body isospins $I = 1/2$ and $I = 3/2$ are mixed.

For a fixed energy of the incident kaon the amplitude of $K^- + d \to \pi + \Sigma + n$ reaction for a given isospin or particle state of the $\pi\Sigma$ subsystem depends on the neutron energy E_n and on the cosine of the angle between the outgoing π and n. As an input for the AGS equations two-body T-matrices in s-wave separable form were taken from [1]. In particular, one- and two-pole models of $\bar{K}N - \pi\Sigma$ interaction, describing all existing experimental data on K^-p scattering, were used. One-term spin-dependent ΣN model describes all measured cross-sections, while the two-term NN model gives proper deuteron binding energy and reproduces Argonne V18 phase shifts. The remaining πN interaction was neglected, the calculation was restricted to total orbital momentum $L = 0$. The total energy of the reaction was taken below the threshold of deuteron break-up, therefore, at very low energies, where the laboratory momentum of antikaon is less than 50 MeV/c.

It turned out, that the neutron spectra do not show a clear signature of the $\Lambda(1405)$ resonance. This can be seen in Fig. 1 (left panel), where $P_{\rm full}$ denotes the total (angular integrated) spectrum, summed over isospin indices, for initial antikaon (c.m.) energy 1 MeV. The $M_{\pi\Sigma}$ region of $\Lambda(1405)$ resonance corresponds to emitted neutron energy of the order $20 - 30$ MeV. In the considered low-energy $K^- + d$ reaction energetic neutrons are unlikely to be emitted by two kinematical reasons. First, the dominant quasi-free mechanism, which roughly speaking means, that slow collisions tend to produce mainly slow particles. Second, the momentum distribution of the neutron in the deuteron is also dominated by the low-energy part.

We also checked the dependence of the differential and total neutron spectra on mass difference of K^- and $\bar{K}^0$ particles, incident (c.m.) energy of antikaon (calculations were done for 0, 1 and 2 MeV) and on one- or two-pole versions of $\bar{K}N - \pi\Sigma$ interaction. It turned out, that the effect of all the above factors remains within a few per cent.

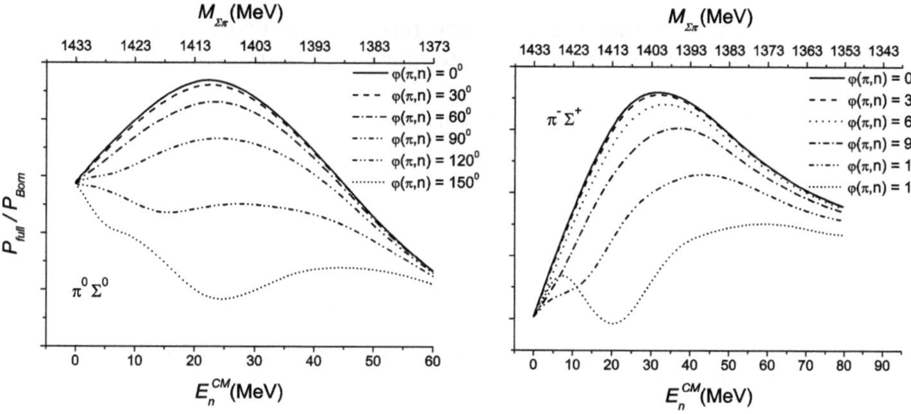

Fig. 2 Differential DEV neutron spectra for $n\pi^0\Sigma^0$ (left) and $n\pi^-\Sigma^+$ final states for different angles

Recently, a method of eliminating the disturbing kinematical effects in order to reveal the dynamical ones was proposed [2]. The authors suggested to consider the deviation spectrum

$$P_{DEV} = \frac{P(E_n)}{P_{\text{nonres}}(E_n)}, \tag{2}$$

where P_{nonres} is a non-resonant spectrum of the reaction, containing the reaction kinematics; it can be chosen as the Born approximation ($P_{\text{nonres}} = P_{\text{Born}}$). In principle, the numerator of (2) should correspond to the full spectrum of $K^- + d \to \pi + \Sigma + n$ reaction P_{full}, containing both three-body kinematical and dynamical effects. However, the authors of [2] used their own approach for calculation of the numerator, which, in fact, corresponds to the zero-order iteration of the Faddeev equations, generally called single scattering approximation. Since in this case the transition operator is replaced by a two-body T-matrix, the corresponding spectrum P_{single} contains three-body kinematics, but only two-body dynamics. All three spectra P_{Born}, P_{full} and P_{single} are shown in Fig. 1 (left). The right panel of the same figure shows proper DEV spectrum $P_{\text{full}}/P_{\text{Born}}$ together with $P_{\text{single}}/P_{\text{Born}}$ (the second one is normalized). It is seen, that the clear resonant structure, seen in $P_{\text{single}}/P_{\text{Born}}$, comes from the two-body T-matrix (containing the $\Lambda(1405)$ pole), while the DEV formula with full spectrum, obtained by Faddeev calculation shows no definite sign of the $\Lambda(1405)$ resonance. However, for certain charge combinations of $\pi\Sigma$ subsystem and angles between π and n the corresponding DEV spectra show a more pronounced trace of the $\Lambda(1405)$ resonance. It can be seen in Fig. 2 for $n\pi^0\Sigma^0$ (left) and $n\pi^-\Sigma^+$ (right) final states for certain angles.

Our conclusions can be summarized as

- No sign of the $\Lambda(1405)$ resonance in any (differential, total or summed over $\pi\Sigma$ charge states) direct neutron spectra from the low-energy $K^- + d \to \pi + \Sigma + n$ reaction.
- Kinematical effects mask the dynamical ones.

- Using the DEV spectrum method some resonance structure can be revealed, however, it is not straightforward, how to relate the obtained maxima to the pole position of the $\Lambda(1405)$.
- Calculations (or experiments) for higher incident kaon energies might improve the situation.

References

1. Shevchenko, N.V.: One- versus two-pole $\bar{K}N - \pi\Sigma$ potential: $K^{-}d$ scattering length. arXiv:1103.4974 [nucl-th] (2011)
2. Esmaili, J., Akaishi, Y., Yamazaki, T.: Resonant formation of $\Lambda(1405)$ by stopped-K^{-} absorption in deuteron. Phys. Rev. C **83**, 055207 (2011)

Hyperfine Interact (2012) 210:107–110
DOI 10.1007/s10751-011-0523-1

Studies of the $\bar{K}N$ interaction at DAΦNE

O. Vázquez Doce · M. Bazzi · C. Berucci · L. Bombelli · A. M. Bragadireanu · M. Cargnelli · A. Clozza · C. Curceanu (Petrascu) · A. d'Uffizi · C. Fiorini · F. Ghio · C. Guaraldo · M. Iliescu · T. Ishiwatari · P. Levi Sandri · A. Longoni · J. Marton · S. Okada · D. Pietreanu · K. Piscicchia · M. Poli Lener · T. Ponta · A. Rizzo · A. Romero Vidal · E. Sbardella · A. Scordo · H. Shi · D. L. Sirghi · F. Sirghi · H. Tatsuno · E. Widmann · J. Zmeskal

Published online: 20 December 2011

Abstract AMADEUS is an experiment planned to be performed at the DAΦNE e^+e^- collider of the Frascati National Laboratories (Italy) of INFN, to investigate the antikaon-nuclei interaction at low energies. AMADEUS will perform, for the first time, full-acceptance studies of antikaon interaction in light nuclei, including a complete experimental program for the case of the kaonic nuclear clusters. The study of the absorption of antikaon by the nucleus will provide information concerning the $\bar{K}N$ interaction and the modification of the kaon mass in the nuclear medium. The experiment is being preceded by the study of the hadronic interactions of K^- in the ^{4}He of the drift chamber from the KLOE experiment data.

Keywords Kaonic nuclear clusters · Nuclear matter · $\bar{K}N$ potential

O. Vázquez Doce (✉) · M. Bazzi · C. Berucci · A. Clozza · C. Curceanu (Petrascu) · A. d'Uffizi · C. Guaraldo · M. Iliescu · P. Levi Sandri · S. Okada · K. Piscicchia · M. Poli Lener · A. Rizzo · A. Romero Vidal · E. Sbardella · A. Scordo · D. L. Sirghi · F. Sirghi · H. Tatsuno
Laboratori Nazionali di Frascati, INFN, Frascati, Roma, Italy
e-mail: oton.vazquezdoce@lnf.infn.it

L. Bombelli · C. Fiorini · A. Longoni
Politechno di Milano, Sez. di Elettronica, Milano, Italy

A. M. Bragadireanu · D. Pietreanu · T. Ponta
IFIN-HH, Magurele, Bucharest, Romania

M. Cargnelli · T. Ishiwatari · J. Marton · E. Widmann · J. Zmeskal
Stefan-Meyer-Institut für subatomare Physik, Vienna, Austria

F. Ghio
INFN Sez. di Roma I and Inst. Superiore di Sanita, Roma, Italy

H. Shi
University of Tokyo, Tokyo, Japan

1 The $\bar{K}N$ potential and kaonic nuclear clusters

The studies of the low-energy Kaon-nucleon interaction have a rather long tradition at the DAΦNE collider, with the DEAR [1] and the SIDDHARTA [2] experiments, measuring the strong interaction in the kaonic hydrogen and other kaonic atoms.

The approach proposed by the AMADEUS experiment deals with one of the current hot topics in low-energy QCD, namely how deep is the antikaon-nucleon potential in the nuclear environment, and the possible existence of the kaonic nuclear clusters, where an antikaon plays glue-type role to bind nucleons forming deeply bound states with a narrow width.

These states are supposed to be formed, favoured by the strongly attractive $\bar{K}p$ potential. The intense theoretical debate, a comparison between predictions based on phenomenological optical potentials included in the original prediction [3] of these states and the more recent calculations leading to larger widths and shallower potentials, parallelled by the diverse interpretations of the experimental results obtained so far indicate the strong need of a dedicated experiment to study this scientific case.

2 The AMADEUS experiment

In order to confirm or deny the existence of the kaonic nuclear clusters the AMADEUS collaboration will study the K^- hadronic interactions in light nuclei. Moreover, AMADEUS aims to perform other types of measurements as elastic and inelastic kaon interactions on various nuclei, obtaining important information for a better understanding of the undergoing processes.

DAΦNE is an e^+e^- collider tuned to be a Φ meson factory, where kaons coming from the decay of the Φ are copiously produced. After its recent upgrade [4], it has reached a luminosity as high as 5×10^{32} cm^{-2} s^{-1}.

The AMADEUS first phase program foresees the investigation of the most basic antikaon-mediated clusters, namely:

- kaonic dibaryon state ppK^-, produced via ^{3}He (stopped K, n) reaction;
- kaonic 3-baryon states $ppnK^-$ and $pnnK^-$, produced via ^{4}He (stopped K^-, n/p) reactions.

If such deeply bound kaonic nuclear clusters exist, the physics program will be completed performing a systematic spectroscopy study including the determination of binding energies, decay widths and quantum numbers of all states. The AMADEUS experiment plans to use the KLOE detector [5] that has been successfully taking data at DAΦNE since 1999, with its excellent 4π drift chamber and calorimeter, implementing it in the central region with a dedicated setup.

Three main components for the AMADEUS setup are presently under study, including the high gaseous density target, a trigger system made of scintilliating fibers, and a tracking device placed internally to the KLOE drift chamber.

The R&D work has already started for the construction of the target and of the trigger system, in order to optimize the kaon stopping efficiency and background suppression, taking advantage of the 50 cm diameter gap present around the interaction point inside the drift chamber.

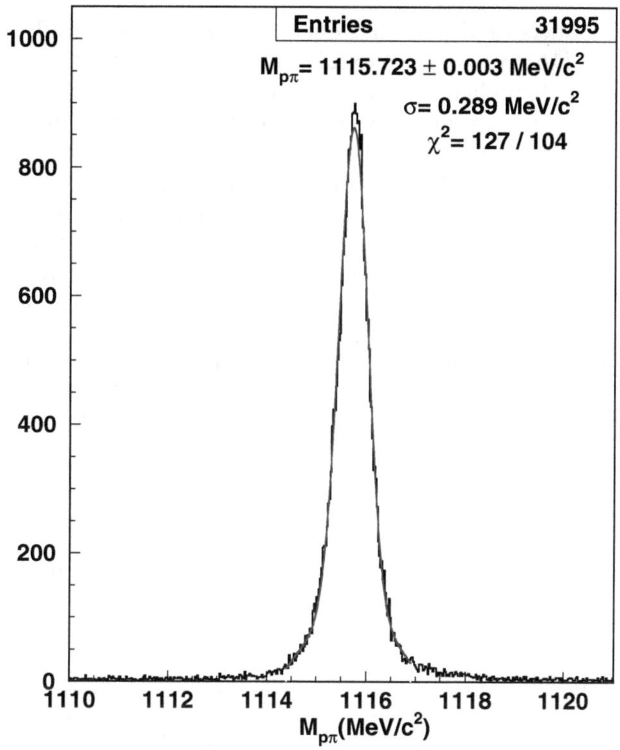

Fig. 1 Λ invariant mass reconstruction (events/20 eV) for the KLOE 2005 data

A toroidal or half-toroidal gaseous target filled with different gases (d, ^{3}He, ^{4}He) will be used surrounding the beam pipe around the interaction region.

Two layers of scintilliating fibers surrounding the beam pipe will be used to trigger the passage of kaons. This detector is essential, delivering an optimal trigger condition by making use of the back-to-back topology of the charged kaons generated from the Φ-decay. A prototype was already built, with fibers read by silicon photomultipliers and successfully tested in DAΦNE. Further studies and optimizations are undergoing.

3 KLOE data analysis in search for $K^{-4}He$ interactions

As a preliminary search for signals of kaonic nuclear clusters inside the KLOE setup, and as first output of the fruitful collaboration between the AMADEUS and KLOE groups, the hadronic interactions of K^- in the KLOE setup in the collected data from previous KLOE runs is under analysis. Being the drift chamber filled mainly with helium, from Monte Carlo simulations we know that 0.1% of the K^- stop in the gas, giving a unique opportunity to study the developed hadronic interactions in such an "active target".

Preliminary results of the analysis of a sample of the 2005 KLOE data has shown the capabilities in performing nuclear physics measurements with the KLOE detector [6]. The strategy is focused on the identification of possible specific decay products of the kaonic nuclear clusters, specifically into channels containing the $\Lambda(1116)$ hyperon. An excellent result has been achieved with a precise determination of the lambda mass, as can be seen in the Fig. 1, where the signal shape in the proton-pion invariant mass spectrum for the selected events is shown.

Vertices produced by these lambdas with protons or deuterons are searched for along the K^- decay path as direct signals of the formations of the kaonic nuclear clusters, or absorptions of K^- by the nucleons of the gas nuclei.

The KLOE detector capability to reconstruct hyperons with a very good resolution, ideal for kaonic clusters studies, has been proven, opening the door for studies of many hadronic physics hot topic items, to be performed by AMADEUS.

Acknowledgements We acknowledge the support of the European Community-Research Infrastructure Integrating Activity "Study of Strongly Interacting Matter" (aconym HadronPhysics2, Grant Agreement n. 227431) under the Seventh Framework Protramme of EU.

References

1. Beer, G., et al.: Measurement of the kaonic hydrogen X-ray spectrum. Phys. Rev. Lett. **94**, 212302 (2005)
2. Bazzi, M., et al.: A new measurement of kaonic hydrogen X rays. Phys. Lett. B **704**, 113–117 (2011)
3. Akaishi, Y., Yamazaki, T.: Nuclear anti-K bound states in light nuclei. Phys. Rev. C **65**, 044005 (2002)
4. Boscolo, M., et al.: Luminosity and background measurements at the e+e- DAΦNE collider upgraded with the crab waist scheme. Nucl. Instrum. Methods A **621**, 121–129 (2010)
5. Adinolfi, M., et al.: The tracking detector of the KLOE experiment. Nucl. Instr. Methods A **488**, 51–73 (2002)
6. Cargnelli, M., Petrascu, C., Vazquez Doce, O.: KLOE K-charged group, KLOE memo 337 (2007)

Hyperfine Interact (2012) 210:111–114
DOI 10.1007/s10751-012-0595-6

Neutral kaon production in p+p and p+Nb collisions

**Jia-Chii Berger-Chen · Laura Fabbietti ·
Kirill Lapidus for the HADES collaboration**

Published online: 6 March 2012

Abstract The kaon nucleus (KN) interaction in dense nuclear matter is predicted to be repulsive and increasing with density. However, determined values for this potential are not yet consistent with each other (Benabderrahmane et al., Phys Rev Lett 102:182501, 2009; Agakishiev et al., Phys Rev C 82:044907, 2010; Büscher et al., Eur Phys J A 22:301–317, 2004). We analyze K^0_S mesons identified with the HADES detector in p+p and p+^{93}Nb reactions at 3.5 GeV kinetic beam energy. To determine the KN potential at normal nuclear density we propose to compare the K^0_S differential distributions in p+^{93}Nb and p+p collisions. High statistics of low p_t-kaons (p_t < 100 MeV/c) ensure the sensitivity of our measurements to the nuclear matter effects. We present the data analysis method and first results.

Keywords Neutral kaon · Cold nuclear matter · Kaon nucleus potential

1 Introduction

The modification of kaon properties in nuclear medium is a permanent subject of interest in related theoretical and experimental research [4]. In particular it is important to extract the in-medium KN potential as a function of density. At normal nuclear density it can be studied employing γ, π or proton induced reactions [5]. Theory predicts a slight repulsive potential, which can be observed in yields and

J.-C. Berger-Chen (✉) · L. Fabbietti · K. Lapidus
Excellence Cluster Universe, Technische Universität München, Boltzmannstr. 2,
85748 Garching b. München, Germany
e-mail: jia-chii.chen@tum.de

L. Fabbietti
e-mail: laura.fabbietti@ph.tum.de

K. Lapidus
e-mail: kirill.lapidus@ph.tum.de

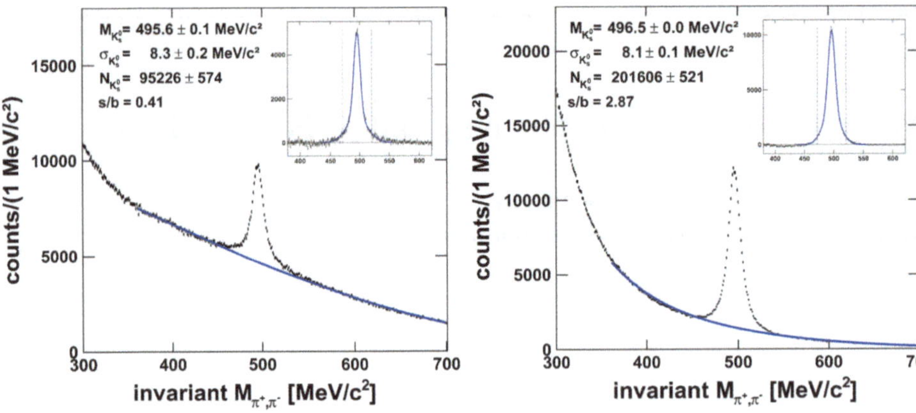

Fig. 1 Invariant mass $\pi^+\pi^-$ for p+p (*left*) and p+^{93}Nb (*right*)

kinematical distributions. One may conclude from theoretical model predictions that momentum distributions, in particular in the low momentum region, are sensitive to the influence of the KN potential. Moreover, it is favorable to analyze neutral kaons as they are not distorted by Coulomb interaction. In different experiments this aspect has been studied at normal nuclear densities with pion and proton induced reactions and at higher densities by means of heavy ion collisions. The results are not conclusive: the data obtained in pion/proton induced reactions are best described by models employing $U_{KN}(\rho_0) \approx 20$ MeV [1, 3], whereas heavy ion collisions are reproduced with $U_{KN}(\rho_0) \approx 40$ MeV [2]. This discrepancy motivates us to remeasure the K^0's in cold nuclear matter with the HADES setup [6]. The advantage of our experiment is a high acceptance for low p_t-kaons. Around $4\cdot 10^9$ events were collected for the p+^{93}Nb and $1.2\cdot 10^9$ for the p+p reaction, both at 3.5 GeV kinetic beam energy. As no effect from the finite nuclear density is expected in the latter system, a direct indication for the KN potential can be deduced from the ratio $\sigma(pNb)/\sigma(pp)$ in the momentum distribution. The quantitative strength can then be extracted by a comparison to theoretical models. Due to a larger data sample, the uncertainty of our measurement is expected to be smaller as reported in [2].

2 K^0_S analysis

Similar analyses were carried out for both data sets. The K^0_S was reconstructed using the HADES detector by its charged decay particles π^+ and π^- (BR = 69.20%), which were identified via graphical cuts on the energy loss versus momentum distribution. The contribution of $\bar{K}^0$ has been neglected, as the $\bar{K}^0/K^0$ ratio in both data samples is less than 4% (estimated from HSD [7] simulations). To increase the signal-to-background ratio, following secondary vertex cuts were applied: (1) distance between the two pion tracks ($d_{\pi^+-\pi^-} < 7$ mm), (2) distance between the primary reaction vertex and the secondary decay vertex ($d(K^0_S - V) > 25$ mm), (3) distance of closest approach to the primary vertex for the two pion tracks ($\mathrm{DCA}_{\pi^+} > 7$ mm, $\mathrm{DCA}_{\pi^-} > 7$ mm). These cuts were optimized to maximize the

Table 1 Three K^0 production channels with highest cross sections used as simulation input

Reaction	$\sigma[\mu b]$
$p + p \rightarrow \Sigma^+ + p + K^0$	21.3
$p + p \rightarrow \Lambda + p + \pi^+ + K^0$	18.4
$p + p \rightarrow \Sigma^0 + p + \pi^+ + K^0$	12.4

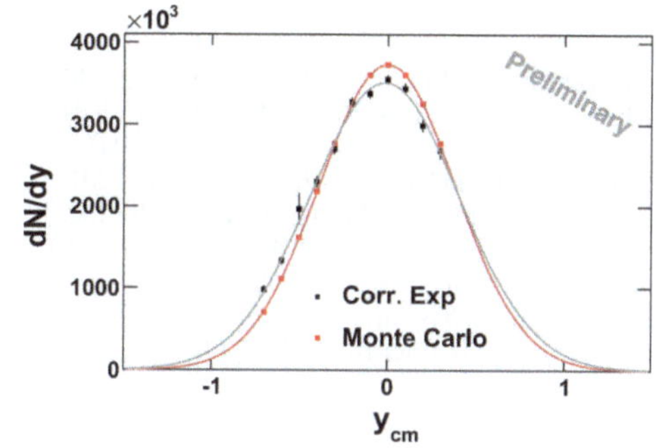

Fig. 2 Comparison of the K^0_S rapidity distribution of corrected experimental data (p+p) and simulated data fitted with a Gaussian ($\mu_{exp} = -0.02 \pm 0.01$)

ratio S^2/B. The resulting $\pi^+\pi^-$ invariant mass spectra for the p+p and p+^{93}Nb colliding systems are shown in Fig. 1. Both spectra are described by a fit of the signal (two Gaussians[1]) and the combinatorial background (Landau function and polynoms). High statistics allow differential yield analyses in rapidity ($\Delta y = 0.1$) and transverse momentum ($\Delta p_t = 75$ MeV/c). For acceptance and efficiency corrections of the p+p reaction the Pluto [8] event generator based on Monte Carlo was used. Fourteen K^0 production channels were simulated with isotropic angular distributions. Each channel was weighted with its corresponding cross section and processed through the full analysis chain. The cross sections from the Landolt–Börnstein database were recalculated with updated experimental cross sections using a phasespace fit from [9]. Three of the channels with the highest contributions are listed in Table 1. In case of the p+^{93}Nb reaction simulated tracks embedded into real data were used.

3 Preliminary results

Transverse momentum distributions obtained by the Pluto [8] event generator were compared to acceptance and efficiency corrected experimental data from p+p collisions. The spectra were normalized using a single scaling factor obtained at mid-rapidity. Overall an agreement within 10% deviation was observed. Moreover the rapidity distributions were compared (Fig. 2). To extract the yields the measured points in the p_t-spectra were summed up in the region where measurements are available. A Boltzmann function was fitted to the tail of the p_t-spectra to extrapolate the yield to the not measured region. The simulation is underestimating the experimental data in

[1]Two Gaussians are needed to take multiple scattering effects into account.

backward and forward rapidity areas. This might be explained by angular anisotropy of K^0 production channels, which is not yet included into the simulation. Finally the goal will be to determine absolute normalized yields and the in-medium KN potential from the ratio $\sigma(pNb)/\sigma(pp)$.

References

1. Benabderrahmane, M.L., et al.: Phys. Rev. Lett. **102**, 182501 (2009)
2. Agakishiev, G., et al.: Phys. Rev. C **82**, 044907 (2010)
3. Büscher, et al.: Eur. Phys. J. A **22**, 301–317 (2004)
4. Hartnack, C., et al.: arXiv:nucl-th/1106.2083v2 (2011)
5. Korpa, et al.: Acta Physiol. Hung. A **22**, 21–28 (2005)
6. Agakishiev, et al.: Eur. Phys. J. A **41**, 243–277 (2009)
7. Cassing, W., et al.: Phys.Rep. **308**, 65 (1999)
8. Fröhlich, I., et al.: arXiv:nucl-ex/0708.2382v2 (2007)
9. Sibirtsev, et al.: arXiv:nucl-th/9802019 (1998)

Hyperfine Interact (2012) 210:115–118
DOI 10.1007/s10751-011-0546-7

Investigating in-medium lambda production in pion induced reactions

Ivana Carević · Olaf Hartmann · Mile Dželalija

Published online: 30 December 2011

Abstract We present the status of the evaluation of in-medium production cross sections for Lambda baryons and K_0^S mesons in pion induced reactions at 1.15 GeV/c beam momentum. A systematics of five target nuclei, from carbon to lead, has been measured. The inclusive in-medium cross section is studied as well as possible conclusions on in-medium potential by comparing to transport model calculations. The data have been recorded by the FOPI detector at the GSI.

Keywords In-medium effects · Production cross section · Pion induced reactions · Strangeness · Transport model calculations

1 Introduction and motivation

The presence of in-medium effects which manifest in the change of basic hadron properties is one of the central questions in analysing the properties of nuclear matter [1–3]. For example, the suppression of K^0 production on heavy target nucleus (Pb) comparing to the light one (C) was shown through the comparison of the ratio of yields as the function of the momentum of K^0 mesons [4]. Furthermore, the HSD[1] transport model [1] suggests a repulsive KN potential (20 ± 5 MeV) at normal nuclear matter density due to K^0 interactions with the nuclear medium.

Strange particles are essential probes to study in-medium effects for hadrons produced in nucleus-nucleus collisions at energies close to the nucleon-nucleon production thresholds. The direct method to investigate the threshold behaviour is

[1]Hadron String Dynamics

I. Carević (✉) · Mile Dželalija
Faculty of Science, University of Split, Split, Croatia
e-mail: ivana@pmfst.hr

O. Hartmann
Stefan Meyer Institute, Vienna, Austria

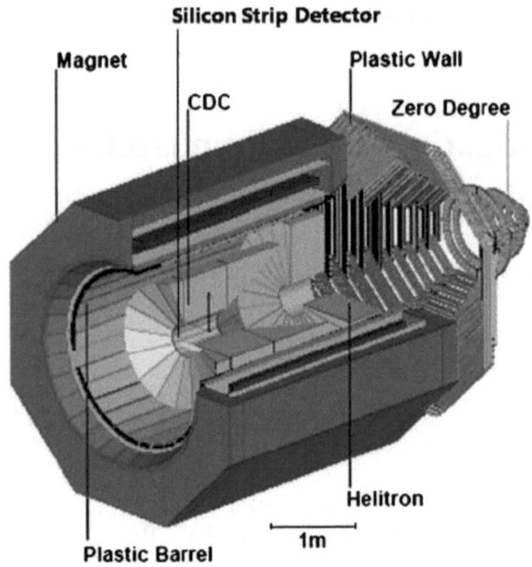

Fig. 1 Schematic drawing of the FOPI detector with its sub-detectors as used in this experiment (distances and dimensions do not show the real case). The z-axis is the beam axis

the use of pion beams to study the reactions like $\pi\ N \rightarrow YK$, involving a hyperon Y and a K meson. The investigation of inclusive Λ production in the $\pi^- p \rightarrow \Lambda K^0$ reaction in medium should contribute to better understanding of in-medium effects with the benefit of extra constraints to transport model parameters. It is also worth to do the comparison of the Λ inclusive cross section to one obtained for the K^0 and to investigate the relevance of the associated strangeness production.

2 Data analysis

At the SIS[2] of the GSI[3], a secondary pion beam in the momentum range from 0.6 to 2.8 GeV/c is available [5]. The data discussed here were taken by the FOPI detector in the experiment with a secondary pion beam of 1.15 GeV/c momentum (which corresponds to the energy of about 1.75 GeV in the system of π^- mesons colliding with protons at rest), performed in August 2004 using C, Al, Cu, Sn and Pb nuclear targets. The FOPI detector system (covering nearly the full solid angle 4π – Fig. 1), assembled at the SIS accelerator, measures charged particles and, among other, investigates the production of strangeness in heavy-ion collisions, especially close to the threshold [4, 6–8].

The main goal of this experiment was to measure the in-medium cross section of the pion induced reaction $\pi^- p \rightarrow \Lambda K^0$. K^0_S and Λ particles are reconstructed from their charged decay products: $K^0_S \rightarrow \pi^+\pi^-$ with 69% branching ratio, and $\Lambda \rightarrow \pi^- p$ with 64% branching ratio.

[2]Schwer-Ionen-Synchrotron

[3]Helmholtzzentrum für Schwerionenforschung GmbH, Darmstadt, Germany

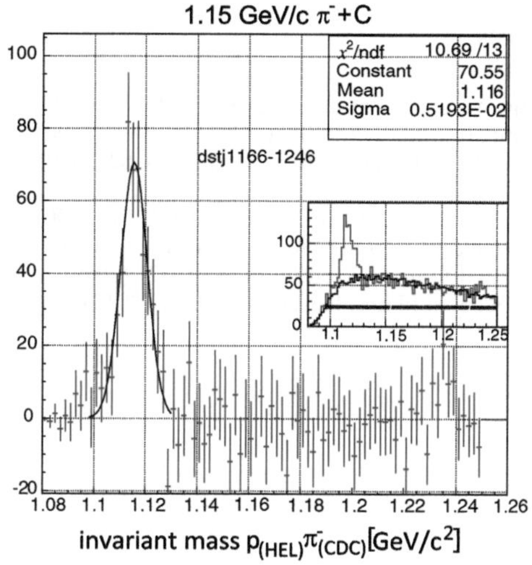

Fig. 2 Reconstructed invariant mass spectrum of Lambda particles for pion induced reaction on Carbon target at 1.15 GeV/c, after background subtraction (reconstructed $\sim 10^{-4}$ Lambda particles per event). The inset shows both signal and mixed-event background. The π^- was measured by the CDC and the proton by the Helitron (the forward drift chamber of the FOPI detector)

The analysis for the K^0_S channel, measured by the CDC[4] of the FOPI detector and the results for the measurement of in-medium production of K^0_S mesons have been presented and published in [4]. The current investigation of the second channel with the Λ reconstruction includes also the data from the forward parts of the FOPI detector, because for the reactions like $\pi\ N \rightarrow Y\quad K$ a preferential emission of hyperons in forward direction is expected [2]. Exemplarily, in Fig. 2 the invariant mass spectrum of Λ reconstructed for the reaction on Carbon target is shown, with the proton measured in the forward region.

In this experiment two silicon strip detectors were placed in front of the target to track the beam particles with a precision in the order of mm. The so-achieved information on the primary vertex should provide an improvement of the momentum resolution and the reconstruction of secondary vertices.

3 Summary and outlook

For the main aim of the analysis, the study of in-medium production, the reconstruction of Λ particles from reactions on C and Pb for all available data of the FOPI S273 pion experiment and the determination of a statistical limit on co-produced Λ-K^0 pairs will be done. The obtained results will be compared with theoretical model predictions in order to extract hadronic properties in medium and to get more complete picture of the in-medium properties of $\pi\ N \rightarrow YK$ reactions. The favour of two silicon strip detectors used in the experiment will be investigated. The analysis will also be extended for the data acquired in the recent FOPI pion experiment (June 2011) at a higher beam momentum (1.7 GeV/c).

[4]Central Drift Chamber

References

1. Cassing, W., et al.: Phys. Rep. **308**, 65 (1999); Nucl. Phys. **A614**, 415 (1997)
2. Tsushima, K., et al.: Phys. Rev. **C62**, 0649904 (2000)
3. Hayano, R., Hatsuda, T.: Rev. Mod. Phys. **82**, 2949 (2010)
4. Benabderrahmane, M.L., et al.: Phys. Rev. Lett. **102**, 182501 (2009)
5. Diaz, J., et al.: Nucl. Instr. Methods **A478**, 511 (2002)
6. Merschmeyer, M., et al.: Phys. Rev. **C76**, 024906 (2007)
7. Hartmann, O.: Int. J. Mod. Phys. **A24**, 271 (2009)
8. Lopez, X., et al.: Phys. Rev. **C81**, 061902 (2010)

Hyperfine Interact (2012) 210:119–121
DOI 10.1007/s10751-011-0528-9

Energy dependence of $\bar{K}N$ interaction in nuclear medium

Aleš Ciepý

Published online: 3 January 2012

Abstract When the $\bar{K}N$ system is submerged in nuclear medium the $\bar{K}N$ scattering amplitude and the final state branching ratios exhibit a strong energy dependence when going to energies below the $\bar{K}N$ threshold. A sharp increase of $\bar{K}N$ attraction below the $\bar{K}N$ threshold provides a link between shallow $\bar{K}$-nuclear potentials based on the chiral $\bar{K}N$ amplitude evaluated at threshold and the deep phenomenological optical potentials obtained in fits to kaonic atoms data. We show the energy dependence of the in-medium K^-p amplitude and demonstrate the impact of energy dependent branching ratios on the Λ-hypernuclear production rates.

Keywords Kaon-nucleon amplitude · Nuclear medium · Hypernuclei

A key issue in studying in-medium K^- meson interactions concerns the strength of the attractive K^- nuclear potential. Chirally based coupled channels calculations lead to K^- nuclear optical potentials that are about 80 MeV deep at nuclear densities $\rho = \rho_0 = 0.17\ \text{fm}^{-3}$. The incorporation of kaon selfenergy reduces the depth of the potential to approximately $40 - 50$ MeV while much deeper potentials, in the range $\text{Re}\ V_{K^-}(\rho_0) \sim -(150\text{–}200)$ MeV are obtained in comprehensive global fits to kaonic atoms data [1]. In our recent paper [2] we reported on a new self consistent treatment of chirally motivated K^-N amplitudes that lead to deep K^- nuclear potentials, considerably deeper than the 'shallow' potentials deduced in earlier models based on the chiral amplitude [3, 4].

The elementary $\bar{K}N$ interaction at threshold and low energies is well understood within chiral models combined with the coupled channels re-summation techniques.

A. Cieplý (✉)
Nuclear Physics Institute, 250 68 Řež, Czech Republic
e-mail: cieply@ujf.cas.cz

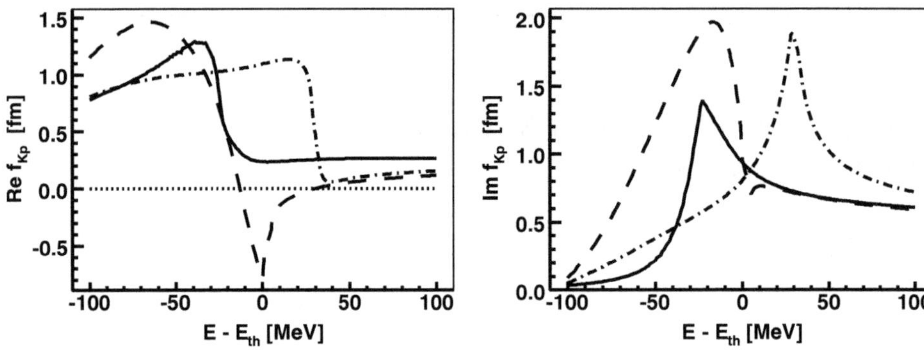

Fig. 1 Energy dependence of the K^-p scattering amplitude. The *left and right panels* refer to the real and imaginary parts of the amplitude, respectively, and E_{th} denotes the K^-p threshold energy. *Dashed curves*: free space, *dot-dashed*: Pauli blocking in nuclear medium at $\rho = \rho_0$, *solid curves*: combined effect of Pauli blocking and hadron selfenergies at $\rho = \rho_0$

In our approach we employ chirally motivated coupled-channel potentials that are taken in a separable form [5]. The meson-baryon channels are composed from the $\pi\Lambda$, $\pi\Sigma$, $\bar{K}N$, $\eta\Lambda$, $\eta\Sigma$, and $K\Xi$ states (taken with all appropriate charge combinations). The parameters of the model, the inverse ranges that define the off-shell Yamaguchi form factors and low energy constants of the SU(3) chiral Lagrangian that contribute to the inter-channel couplings, are fitted to the kaonic hydrogen and low energy K^-p reactions data. A detailed description of our model can be found in Refs. [5] and [2].

When the $\bar{K}N$ system is submerged in nuclear medium the $\bar{K}N$ scattering amplitude and the final state branching ratios exhibit a strong energy dependence when going to energies below the $\bar{K}N$ threshold. In Fig. 1 we show an energy dependence of the elastic scattering amplitudes f_{K^-p} in free space and for two versions of in-medium modifications, performed with a leading order chiral model labeled as TW1 in Ref. [2]. The free space K^-p amplitude exhibits a typical structure related to an $I = 0$ $\bar{K}N$ quasi-bound state which is assigned to a well known $\Lambda(1405)$ resonance. When the K^-p system is submerged in nuclear medium the Pauli blocking shifts the structure to higher energies, about 30 – 40 MeV above the $\bar{K}N$ threshold. However, the addition of kaon selfenergies (we also included baryon and pion selfenergies in our model, though their impact is minute) brings it back below the threshold as demonstrated by the solid curves in the figure. In this regime the most striking feature is a sharp increase in the real part of the amplitude when going to subthreshold energies. Consequently, the K^-p interaction becomes much stronger at energies about 30 MeV below the K^-p threshold with respect to its strength at threshold. We have shown [2] that this is exactly the region of energies probed by kaons at the lowest K-atomic orbits, so the K^--nuclear optical potential becomes much deeper than when it were constructed from the amplitudes taken at the $\bar{K}N$ threshold.

The impact of the in-medium $\bar{K}N$ dynamics and energy dependence of the $\bar{K}N$ amplitudes on the characteristics of kaonic atoms and kaon-nuclear states was investigated in Ref. [2]. Here we mention another application to the Λ-hypernuclear production in the $(K^-_{\text{stop}}, \pi^-)$ reactions. Recently, the production rates were established by the FINUDA collaboration [6] for five p-shell nuclear targets from ^{7}Li to

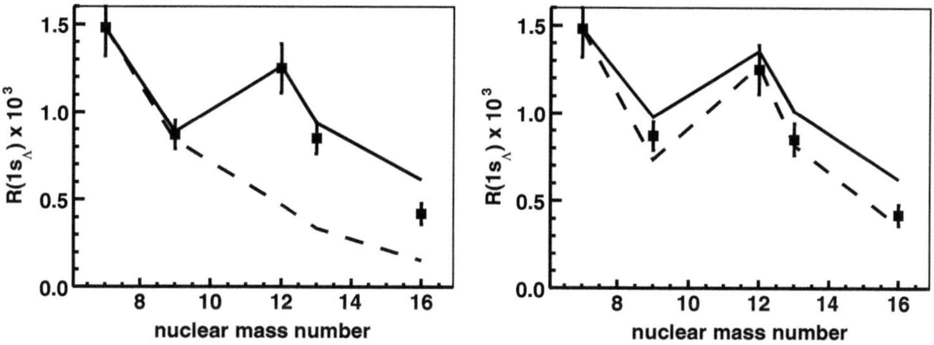

Fig. 2 The A-dependence of the $1s_\Lambda$ hypernuclear formation rates, experimental data from Ref. [6]. The theoretical rates are normalized to the ^{7}Li experimental value and were calculated with a phenomenological density dependent kaon-nuclear optical potential (*dashed lines*) and with the chirally motivated K^--nuclear optical potential (*solid lines*). *Left panel*: elementary branching ratio $\text{BR}(K^-N \to \pi\Lambda)$ fixed at the threshold value for nuclear density $\rho = \rho_0/2$, *right panel*: energy and density dependent BR

^{16}O. Since the absolute normalization of the experimental capture rates is a delicate matter and the calculated rates are generally much lower than the measured ones, we have focused on the A-dependence of the $1s_\Lambda$ formation rates [7]. The Fig. 2 shows the effect of energy dependent branching ratios $\text{BR}(K^-n \to \pi^-\Lambda)$ on the rates computed within the framework of the distorted wave impulse approximation. There, the K^--nuclear optical potential is used to generate the K-atomic initial state wave function. In the left panel (which was not published earlier) of Fig. 2 we present the rates calculated with the BR fixed at a value obtained at the $\bar{K}N$ threshold while the theoretical rates in the right panel (taken from Ref. [7]) were calculated with energy dependent BR averaged over the region of relevant subthreshold energies, specific for each nuclear target. With the BR fixed at the $\bar{K}N$ threshold the A-dependence is reproduced much better by the chirally motivated optical potential which is shallow at the $\bar{K}N$ threshold. Since both the phenomenological and chirally motivated optical potentials are sufficiently deep at the subthreshold energies relevant for the evaluation of the energy dependent BR, they lead to similar $1s_\Lambda$ formation rates.

Acknowledgements The author acknowledges a fruitful collaboration with E. Friedman, A. Gal, D. Gazda, J. Mareš, V. Krejčiřík and J. Smejkal who coauthored the papers the report is based on. This work was supported by the GACR Grant No. 202/09/1441.

References

1. Friedman, E., Gal, A., Batty, C.J.: Nucl. Phys. **A579**, 518 (1994)
2. Cieplý, A., Friedman, E., Gal, A., Gazda, D., Mareš, J.: Phys. Rev. **C84**, 045206 (2011)
3. Waas, T., Kaiser, N., Weise, W.: Phys. Lett. **B365**, 12 (1996); *ibid.* **B379**, 34 (1996)
4. Ramos, A., Oset, E.: Nucl. Phys. **A671**, 481 (2000)
5. Cieplý, A., Smejkal, J.: Eur. Phys. J. **A43**, 191 (2010)
6. Agnello, M., et al.: [FINUDA Coll.], Phys. Lett. **B698**, 219 (2011); *ibid.* **B622**, 35 (2005)
7. Cieplý, A., Friedman, E., Gal, A., Krejčiřík, V.: Phys. Lett. **B698**, 226 (2011)

Hyperfine Interact (2012) 210:123–126
DOI 10.1007/s10751-011-0507-1

Kaon production at finite temperature and baryon density in an effective relativistic mean field model

Felice Iazzi · Andrea Lavagno · Daniele Pigato

Published online: 9 December 2011

Abstract We investigate the kaon production at finite temperature and baryon density by means of an effective relativistic mean-field model with the inclusion of the full octet of baryons. Kaons are considered taking into account of an effective chemical potential depending on the self-consistent interaction between baryons. The obtained results are compared with a minimal coupling scheme, calculated for different values of the anti-kaon optical potential.

Keywords Strangeness production · Nuclear equation of state

1 Hadronic equation of state and main results

The relativistic mean-field model (RMF) is widely successful used for describing the properties of finite nuclei as well as hot and dense nuclear matter [1–8]. In this context, the total baryon Lagrangian density can be written as $\mathcal{L}_B = \mathcal{L}_{\rm octet} + \mathcal{L}_K$, where $\mathcal{L}_{\rm octet}$ stands for the full octet of baryons ($p, n, \Lambda, \Sigma^+, \Sigma^0, \Sigma^-, \Xi^0, \Xi^-$) and $\mathcal{L}_K$ corresponds to the kaon mesons.

F. Iazzi · A. Lavagno (✉) · D. Pigato
Dipartimento di Fisica, Politecnico di Torino, 10129 Torino, Italy
e-mail: andrea.lavagno@polito.it

F. Iazzi · A. Lavagno · D. Pigato
Istituto Nazionale di Fisica Nucleare (INFN), Sezione di Torino, 10126 Torino, Italy

The quantum hadrodynamics (QHD) model for the full octet of baryons was originally studied with the following standard Lagrangian density [4]

$$\mathcal{L}_{\rm octet} = \sum_i \bar{\psi}_i \left[i\,\gamma_\mu\,\partial^\mu - (m_i - g_{\sigma i}\,\sigma) - g_{\omega i}\,\gamma_\mu\,\omega^\mu - g_{\rho i}\,\gamma_\mu\,\boldsymbol{\tau}\cdot\boldsymbol{\rho}^{\,\mu} \right] \psi_i$$
$$+\frac{1}{2}(\partial_\mu\sigma\partial^\mu\sigma - m_\sigma^2\sigma^2) - U(\sigma) + \frac{1}{2}\,m_\omega^2\,\omega_\mu\omega^\mu + \frac{1}{4}\,c\,(g_{\omega N}^2\,\omega_\mu\omega^\mu)^2$$
$$+\frac{1}{2}\,m_\rho^2\,\boldsymbol{\rho}_\mu\cdot\boldsymbol{\rho}^{\,\mu} - \frac{1}{4}F_{\mu\nu}F^{\mu\nu} - \frac{1}{4}\mathbf{G}_{\mu\nu}\mathbf{G}^{\mu\nu}\,, \qquad (1)$$

where m_i is the vacuum baryon mass of index i, $\tau = 2\,\mathbf{t}$ denotes the isospin operator and $U(\sigma)$ is the nonlinear self-interaction potential of σ meson [2, 5]. It is proper to remark that the above QHD Lagrangian is very different from genuine SU(3) models, defined in terms of baryon and meson octets [9].

Because we are going to describe finite temperature and density nuclear matter with respect to strong interaction, we have to require the conservation of three "charges": baryon number, electric charge and strangeness number. Therefore, the chemical potential of particle of index i can be written as

$$\mu_i = b_i\,\mu_B + c_i\,\mu_C + s_i\,\mu_S\,, \qquad (2)$$

where b_i, c_i and s_i are, respectively, the baryon, the electric charge and the strangeness quantum numbers of the i-th hadrons. The effective chemical potential μ_i^* of the i-th baryon is given by $\mu_i^* = \mu_i - g_{\omega i}\,\omega - g_{\rho i}\,\tau_{3i}\,\rho$.

In this context, kaons degrees of freedom are treated in two distinct approaches. In the first case, we consider the interaction between kaons and baryons by means of a direct minimal coupling scheme with the meson fields [10–12]. The kaon lagrangian density can be written as $\mathcal{L}_K = D_\mu^*\Phi^* D^\mu\Phi - m_K^{*2}\Phi^*\Phi$, where $D_\mu = \partial_\mu + ig_{\omega K}\omega_\mu + ig_{\rho K}\tau_{3K}\rho_\mu$ is the covariant derivative of the meson field, $m_K^* = m_K - g_{\sigma K}\sigma$ is the effective kaon mass and τ_{3K} is the third component of the isospin operator. The kaon-meson vector coupling constants are obtained from the quark model and isospin counting rules, setting equal to $g_{\omega K} = g_{\omega N}/3$ and $g_{\rho K} = g_{\rho N}$. Whereas the scalar $g_{\sigma K}$ coupling constant is determined from the study of the real part of the anti-kaon optical potential, at saturation nuclear density, in symmetric nuclear matter: $U(K^-) = -g_{\sigma K}\sigma - g_{\omega K}\omega$. In this investigation we set the anti-kaon optical potential equal to $U(K^-) = -50$ MeV, -100 MeV and -160 MeV, based on recent theoretical calculations and experimental measurements [13–17]. The meson-baryons couplings constant have been fixed to the parameters set marked as GM3 of [5, 18].

In the second approach, we use an alternative formulation, based on the self-consistent interaction between baryons [18]. In this scheme, kaons are treated as a quasi-ideal Bose gas with an effective chemical potential μ_K^*, obtained from the bare one given in (2) and subsequently expressed in terms of the corresponding effective baryon chemical potentials, respecting the strong interaction. More explicitly, the kaon effective chemical potential can be written as

$$\mu_{K^+}^* = (\mu_p^* - \mu_\Lambda^*) = \mu_p - \mu_\Lambda - (1 - x_{\omega\Lambda})g_{\omega N}\omega - \frac{1}{2}g_{\rho N}\rho\,, \qquad (3)$$

where $x_{\omega\Lambda} = g_{\omega\Lambda}/g_{\omega N}$. Thus, the hadronic system is still regarded as an ideal gas but with an effective chemical potential that contains the self-consistent interaction of the meson fields, related to the interaction between baryons.

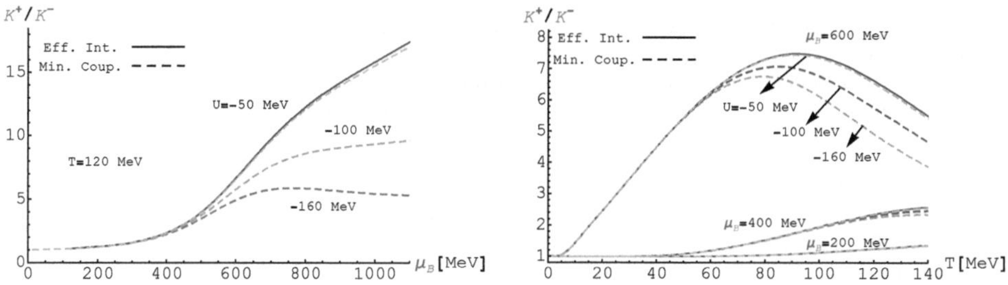

Fig. 1 Kaon to anti-kaon ratio as a function of baryon chemical potential (*left panel*) and temperature (*right panel*) in the effective relativistic mean field model (*solid lines*) and for different values of anti-kaon optical potential (*dashed lines*)

In this paper we focus our attention into a comparative study between the above two different approaches. At this scope, for simplicity, other strangeless mesons (mainly pions) are not considered in our analysis, assuming that they do not sensibly affect the strangeness production but contribute essentially to the total pressure and energy density. Heavier strange meson degrees of freedom have been also neglected.

At finite temperature the thermodynamical quantities can be obtained in the standard way from the total grand potential $\Omega_H = \Omega_B + \Omega_K$ for the two different approaches [11, 12].

In the Fig. 1, we report the major results obtained in this comparative study. In both panels we observe a good correspondence between the effective relativistic mean-field model (solid lines) and the minimal coupling scheme (dashed lines) for moderate values of the anti-kaon optical potential. In particular, recent self-consistent calculations based on coupled-channel G-matrix theory [13] and chiral Lagrangian [19] seem to suggest that the real part of the anti-kaon optical potential could be close to $U_{K^-} \cong -50$ MeV, in good agreement with our results. Moreover, it is interesting to observe that, for a very strong attractive optical potential ($U_{K^-} = -160$ MeV) and for high values of μ_B, a lowering in the K^+/K^- ratio takes place, especially at high temperature. This behavior could be considered as a relevant feature for the determination of the real part of the anti-kaon optical potential, for example in the future CBM (compressed baryonic matter) experiment of the FAIR (Facility of Antiproton and Ion Research) project [20–22].

References

1. Walecka, J.D.: Ann. Phys. **83**, 491 (1974)
2. Boguta, J., Bodmer, A.R.: Nucl. Phys. **A292**, 413 (1977)
3. Serot, B.D., Walecka, J.D.: Adv. Nucl. Phys. **16**, 1 (1986)
4. Glendenning, N.K.: Phys. Lett. **B114**, 398 (1982)
5. Glendenning, N.K., Moszkowski, S.A.: Phys. Rev. Lett. **67**, 2414 (1991)
6. Bonanno, L., Drago, A., Lavagno, A.: Phys. Rev. Lett. **99**, 242301 (2007)
7. Alberico, W.M., Lavagno, A.: Eur. Phys. J. **A40**, 313 (2009)
8. Lavagno, A., Quarati, P.: Phys. Lett. **B498**, 47 (2001)
9. Schaffner-Bielich, J., Gal, A.: Phys. Rev. **C62**, 034311 (2002) and references therein
10. Glendenning, N.K., Schaffner-Bielich, J.: Phys. Rev. **C60**, 025803 (1999)
11. Cavagnoli, R., Providencia, C., Menezes, D.P.: Phys. Rev. **C83**, 045201 (2011)
12. Banik, S., Greiner, W., Bandyopadhyay, D.: Phys. Rev. **C78**, 065804 (2008)

13. Tolos, L., Ramos, A., Polls, A.: Phys. Rev. **C65**, 054907 (2002)
14. Friedman, E., Gal, A., Batty, C.J.: Nucl. Phys. **A579**, 518 (1994)
15. Batty, C.J., Friedman, E., Gal, A.: Phys. Rep. **C60** 024314 (1999)
16. Friedman, E., Gal, A., Mares, J., Cieply, A.: Phys. Rev. **C60**, 024314 (1999)
17. Gal, A.: Nucl. Phys. **A691**, 268 (2001)
18. Lavagno, A.: Phys. Rev. **C81**, 044909 (2010)
19. Lutz, M.F.M., Korpa, C.L.: Nucl. Phys. **A700**, 309 (2002)
20. Senger, P.: J. Phys. G: Nucl. Part. Phys. **30**, S1087 (2004)
21. Senger, P., et al.: Ibid. **36**, 064037 (2009)
22. Henning, W.F.: Nucl. Phys. **A805**, 502c (2008)

Hyperfine Interact (2012) 211:1–7
DOI 10.1007/s10751-011-0501-7

Hadron physics at J-PARC

Tomofumi Nagae

Published online: 2 December 2011

Abstract At the hadron experimental hall of J-PARC, a lot of interesting experiments on strangeness nuclear physics and hadron physics are going to be carried out. Here I introduce several examples, and report the status of the J-PARC.

Keywords Hadron · J-PARC · Strangeness

1 Status of J-PARC before the earthquake on March 11, 2011

At the end of January 2009, the first proton beam from the main ring at 30 GeV was delivered to the hadron experimental hall (Fig. 1) of J-PARC, successfully. At that time, we had one secondary beam line, K1.8BR, which is a branch of the K1.8 beam line after the first electro-static separator. Secondary beams of pions and kaons were observed in February 2009, at this beam line [1].

The beam commissioning of a neutrino beam line was carried out during spring and summer of 2009. In this period, the remaining part of the K1.8 beam line was constructed; such as the second electro-static separator system and a beam spectrometer system. The superconducting kaon spectrometer (SKS) was also installed in the K1.8 beam line [2] for various experiments on strangeness nuclear physics. A neutral kaon beam line, K_L, was also constructed at the same time.

From October 2009, the beam commissioning of the K1.8 beam line was carried out. We confirmed that the double-stage electro-static separator system worked very effectively to achieve a good K/π ratio better than 1. In January and February 2010, the energy resolution of the SKS spectrometer together with the beam line

This work is partially supported by the Grant-In-Aid for Specially Promoted Research 23000003 by Japan Society for the Promotion of Science (JSPS).

T. Nagae (✉)
Department of Physics, Kyoto University, Kyoto 606-8502, Japan
e-mail: nagae@scphys.kyoto-u.ac.jp

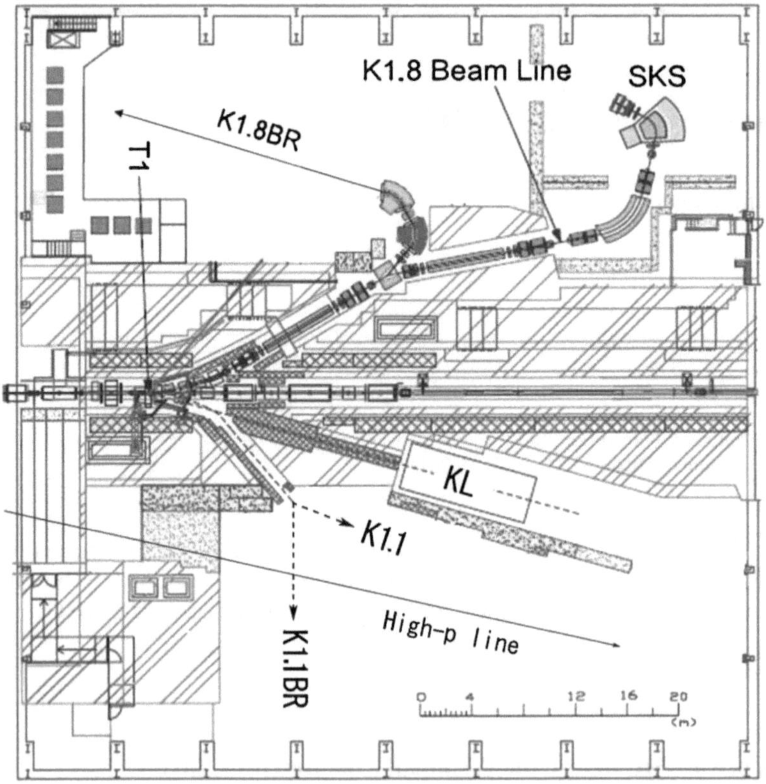

Fig. 1 A schematic layout of the hadron experimental hall. The primary proton beam at 30 GeV from the main ring comes from the left to hit a production target, T1

spectrometer was studied by using the beam straight-through runs and the reactions $\pi^{\pm}p \to K^{+}\Sigma^{\pm}$. A good performance was confirmed in a preliminary analysis. The proton beam intensity from the main ring was a few kW on a production target, which is about 1% of the design intensity of 270 kW. Even with this proton intensity, we could get enough pion beam intensity more than a few million pions per second in average. However, the time structure of the beam is not so good that we can use only about one million pions per two-second spill period.

Nevertheless, a successful physics data taking was carried out during October and November 2010, for the E19 experiment. It is a penta-quark search experiment in $\pi^{-} + p \to K^{-} + \Theta^{+}$. Around 40 people from KEK, Kyoto, Tohoku, Nara, Osaka, JAEA, USA, Italy, Korea, and Russia, participated the beam time.

In the beam time in the winter of 2011, the typical beam power in the rapid-cycling synchrotron (RCS) at 3 GeV was 200 kW and that for the neutrino beam line from the main ring was 145 kW, steadily.

2 Status of J-PARC after the earthquake

After the big earthquake in the east part of Japan on March 11, 2011, a lot of utility facilities, roads, etc. are damaged in J-PARC. Fortunately, there were no

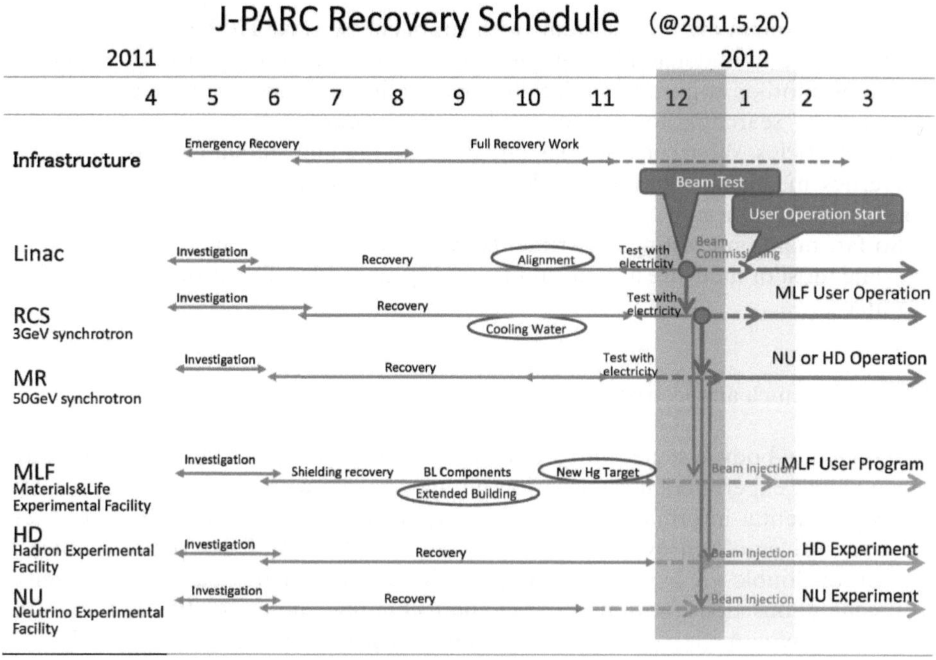

Fig. 2 A J-PARC recovery schedule announced on May 20, 2011, from the J-PARC center

direct effects of "Tsunami". There were also no serious damages on the accelerator equipment, main tunnels, and in the experimental areas, while we found a lot of minor effects to be fixed. After the careful inspection of these damages in the whole J-PARC facilities during April and May 2011, a recovery plan of J-PARC was announced to the public on May 20, 2011 (Fig. 2). According to the plan, we are going to start the beam commissioning of the proton LINAC in December 2011. Although it will depend on the commissioning conditions, the beam would be delivered to the user facilities for neutron, muon, kaon, and neutrino beams in January or February of 2012. Further, a budget request for nine-month running in the JFY2012 is submitted to the government this summer.

So, we expect to have the next beam time for physics data taking in the hadron experimental hall around March or April 2012; it was scheduled to run in the last April.

3 Nuclear physics programs at J-PARC

The hadron experimental hall, one of three experimental facilities in J-PARC (others are the materials and life sciences facility and the neutrino facility), is the place to conduct nuclear physics programs at J-PARC. One of important research subjects is strangeness nuclear physics. We aim to establish a modern picture of baryon-baryon interactions based on quantum chromo dynamics (QCD) and to explore new forms of hadron many-body systems with strangeness. Through these investigations, we would like to understand a role of strangeness in dense nuclear matter. Hadron physics is another important subject. We would like to understand the confinement of quarks by searching for new exotic hadrons, penta-quark, tetra-quark particles, hybrid particles, dibaryons and so on. Possible medium modifications of hadron properties in nuclear matter are also very interesting from a viewpoint of chiral symmetry restoration.

So far, many interesting experiments are proposed at the hadron experimental hall, and most of them are already approved [3]. Here I briefly introduce some of the examples.

3.1 Ξ Hypernuclear spectroscopy

Baryon many-body systems with strangeness are called hypernuclei. While we have accumulated experimental information on hypernuclei with strangeness(S) −1, still the experimental information of $S = -2$ hypernuclei is very much limited. There exist several emulsion events showing a sequential weak decay pattern which is typical for double−Λ hypernuclei. One of the events called "Nagara Event" gives us unambiguous identification of ${}^{6}_{\Lambda\Lambda}\mathrm{He}$ and its binding energy [4]. In addition to the double−Λ hypernuclei, Ξ hypernuclei could exist as the bound states of $S = -2$. There is experimental evidence suggesting such a bound state production with the Ξ-nucleus potential depth of ∼14 MeV [5]. However, the bound state peak was not observed because of the poor energy resolution of >10 MeV.

In the J-PARC E05(spokesperson: T. N), we will use the same reaction of ${}^{12}\mathrm{C}(K^-, K^+)$ to produce ${}^{12}_{\Xi}\mathrm{Be}$ with a much better energy resolution of 1.5–3 MeV. The high-intensity K^- beam at the K1.8 beam line will be used for the experiment at around 1.8 GeV/c. Just recently, we have started to design a new spectrometer, S-2S (Fig. 3), with a good momentum resolution of $\Delta p/p \approx 5 \times 10^{-4}$ and a large solid angle of 60 msr, to be constructed in the next four years.

3.2 Kaonic nuclei search

The first experimental evidence of a kaonic bound state, K^-pp, was reported from the FINUDA collaboration [6]. While we have observed many pairs of $\Lambda - p$ emitted in the back-to-back from the stopped K^- absorption on ${}^{6}\mathrm{Li}$, ${}^{7}\mathrm{Li}$, and ${}^{12}\mathrm{C}$ targets, the mass of the $\Lambda - p$ pairs was largely shifted suggesting a binding of the K^-pp system as much as 115 MeV. A similar binding of about 105 MeV was also observed in an exclusive measurement in the $p + p \rightarrow K^+ + \Lambda + p$ reaction at 2.85 GeV by the DISTO group [7].

Stimulated with these experimental results, several theory groups [8–11] carried out the calculations of the K^-pp system based on few-body techniques. They confirmed the system has a bound state with the binding energy of 40–90 MeV and the broad width of 60–100 MeV depending on their interaction models.

Thus, it is of vital importance to experimentally establish the existence of the lightest kaonic nucleus, K^-pp. In the J-PARC E15(spokespersons: M. Iwasaki (RIKEN) and T. N), we will use the ${}^{3}\mathrm{He}(K^-, n)$ reaction at 1 GeV/c at the K1.8BR beam line. The forward neutrons are measured with arrays of plastic scintillators.

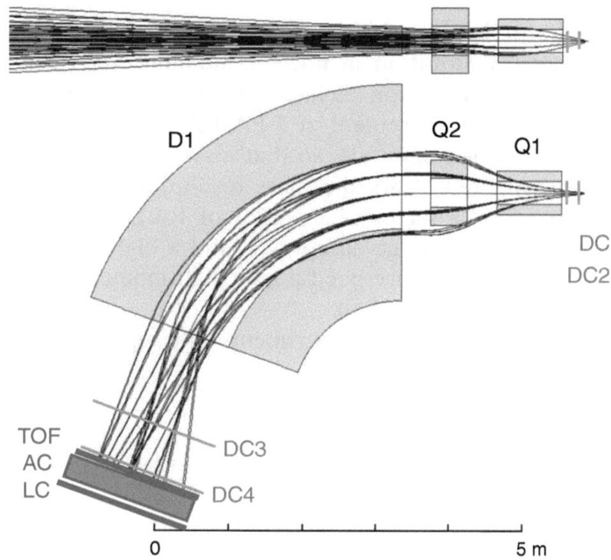

Fig. 3 A schematic design of the *S*-2*S* spectrometer for the (K^-, K^+) reactions. It is composed of *QQD* magnets

At the same time, the decay products of the K^-pp are detected with a cylindrical detector system (CDS). Thereby, the mass of the K^-pp system is measured both in the missing mass and in the invariant mass. There is another experiment, E27 (spokesperson: T. N), to look for the K^-pp in the d(π^+, K^+) reaction at ~1.5 GeV/*c*. The SKS spectrometer will be used for the missing-mass measurement together with a range counter system surrounding the target to enhance the signal to background ratio.

3.3 Penta-quark search

There are many negative results for the existence of a penta-quark $\Theta^+(1535)$, while the LEPS group still claims it in the γd reaction [12]. It is important to investigate the production mechanism by using different reactions. In this regard, there is a suggestive data in the $\pi^- + p \to K^- + X$ reaction [13], while a negative result was observed in the reverse reaction of $K^+ + p \to \pi^+ + X$ at 1.2 GeV/*c* [14]. The experimental group did not claim the peak existence, because the statistical significance of the possible peak was only 2.5σ mainly due to their poor energy resolution of 13.4 MeV(FWHM).

In the J-PARC E19 (spokesperson: M. Naruki (KEK)), we carried out the first data taking at the pion incident momentum of 1.92 GeV/*c* in October and November of 2010. While the estimated energy resolution of the spectrometer is 1.4±0.1 MeV(FWHM), no significant peak structure is observed in much better statistics. We are ready to submit a paper on this data.

3.4 Medium modification of vector mesons

In the J-PARC E16 (spokesperson: S. Yokkaichi (RIKEN)), the mass of ϕ meson produced in nuclear medium is measured through its decay into e^+e^- pairs. The change of the mass compared with that in vacuum could be a signal of chiral symmetry restoration in nuclear medium. Such a hint of mass modification was observed in the proton nucleus reactions at 12 GeV in KEK E325 experiment [15, 16]. In the new measurement at J-PARC, they are going to improve the statistics by two orders of magnitude, so that a systematic study of the mass modifications can be investigated. Also, the mass resolution of the e^+e^- pair spectrometer will be improved by a factor of 2. A lot of R&D works on new detectors are now ongoing to have a large acceptance and a high-rate capability. In order to carry out this measurement, we need a new high-momentum beam line up to 30 GeV in the hadron experimental hall.

There are also other experiments with regard to this subject on ω meson(E26) and ϕ meson(E29).

4 Summary

After the big earthquake on March 11, 2011, we expect the J-PARC beam commissioning will start in December 2011. We hope to have the next data taking of the hadron experimental hall in March or April 2012, after about one year of shutdown. There are a number of important experiments in hadron physics waiting for the beam.

References

1. Tanaka, K.H.: Construction and status of the Hadron experimental hall. Nucl. Phys. A. **835**, 81–87 (2010)
2. Takahashi, T.: Experimental apparatus in the Hadron hall at J-PARC. Nucl. Phys. A. **835**, 88–95 (2010)
3. Program Advisory Committee (PAC): All the information of the submitted proposals and the considerations by the program advisory committee is found at http://j-parc.jp/NuclPart/PAC_for_NuclPart_e.html
4. Takahashi, H., et al.: Observation of a ${}^{6}_{\Lambda\Lambda}$He double hypernucleus. Phys. Rev. Lett. **87**, 212502 (2001)
5. Khaustov, P., et al.: Evidence of Ξ hyper nuclear production in the ^{12}C(K^-, K^+)${}^{12}_{\Xi}$Be reaction. Phys. Rev. C. **61**, 054603 (2000)
6. Agnello, M., et al.: Evidence for a Kaon-bound state of K^-pp produced in K^- absorption reactions at rest. Phys. Rev. Lett. **94**, 212303 (2005)
7. Yamazaki, T., et al.: Indication of a deeply bound and compact K^-pp state formed in the $pp \to p\Lambda K^+$ reaction at 2.85 GeV. Phys. Rev. Lett. **104**, 132502 (2010)
8. Schevchenko, N.V., Gal, A., Mareš, J., Révai, J.: $\bar{K}NN$ quasibound state and the $\bar{K}N$ interaction: coupled-channels Faddeev calculation of the $\bar{K}NN - \pi\Sigma N$ system. Phys. Rev. C. **76**, 044004 (2007)
9. Ikeda, Y., Sato, T.: Resonance energy of the $\bar{K}NN - \pi YN$ system. Phys. Rev. C. **79**, 035201 (2009)
10. Yamazaki, T., Akaishi, Y.: Basic $\bar{K}$ nuclear cluster, K^-pp, and its enhanced formation in the $p + p \to K^+ + X$ reaction. Phys. Rev. C. **76**, 045201 (2007)
11. Doté, A., Hyodo, T., Weise, W.: Variational calculation of the ppK^- system based on chiral SU(3) dynamics. Phys. Rev. C. **79**, 014003 (2009)
12. Nakano, T., et al.: Evidence of the Θ^+ in the $\gamma d \to K^+K^-pn$ reaction by detecting K^+K^- pairs. Phys. Rev. C. **79**, 025210 (2009)
13. Miwa, K., et al.: Search for Θ^+ via $\pi^-p \to K^-X$ reaction near production threshold. Phys. Lett. B. **635**, 72–79 (2006)

14. Miwa, K., et al.: Search for Θ^+ via the $K^+p \to \pi^+X$ reaction with a 1.2 GeV/c K^+ beam. Phys. Rev. C **77**, 045203 (2008)
15. Muto, R., et al.: Evidence for In-Medium Modification of the ϕ Meson at Normal Nuclear Density. Phys. Rev. Lett. **98**, 042501 (2007)
16. Sakuma, F., et al.: Nuclear-Matter Modification of Decay Widths in the $\phi \to e^+e^-$ and $\phi \to K^+K^-$ channels. Phys. Rev. Lett. **98**, 152302 (2007)

Hyperfine Interact (2012) 211:9–14
DOI 10.1007/s10751-012-0584-9

Search for electric dipole moments at storage rings

C. J. G. Onderwater

Published online: 28 February 2012

Abstract Permanent electric dipole moments (EDMs) violate parity and time-reversal symmetry. Within the Standard Model (SM) they are many orders of magnitude below present experimental sensitivity. Many extensions of the SM predict much larger EDMs, which are therefore an excellent probe for the existence of "new physics". Until recently it was believed that only electrically neutral systems could be used for sensitive searches of EDMs. With the introduction of a novel experimental method, high precision for charged systems will be within reach as well. The features of this method and its possibilities are discussed.

Keywords Permanent electric dipole moment · Standard model test · New physics search

1 Motivation

The symmetry properties of fundamental processes and particles are a strong guide to understand the underlying interactions. The QED Lagrangian in the current Standard Model (SM) predicts that all electromagnetic observables are even under the discrete symmetries $\mathcal{C}$ (charge conjugation), $\mathcal{P}$ (parity) and $\mathcal{T}$ (time reversal) individually and thus under each combinations of them. Strong interaction observables, described by QCD, are also predicted to be even under $\mathcal{C}$, $\mathcal{P}$ and $\mathcal{T}$, with the exception of those proportional to $\bar{\theta}$ which are $\mathcal{P}$ and $\mathcal{T}$-odd [1]. The weak interaction violates both $\mathcal{P}$ and $\mathcal{C}$ because of the handedness of the coupling of the W and Z-bosons. Many observables are even under their combination $\mathcal{CP}$. Nevertheless, the weak interaction also predicts $\mathcal{CP}$-odd ones. These are all proportional to the Jarlskog invariant, $J \propto \sin^2\theta_{12}\sin\theta_{23}\sin\theta_{13}\sin\delta \sim 3\times10^{-5}$ [2]. Here θ_{12}, θ_{23} and θ_{13} are the quark flavor mixing angles and δ the $\mathcal{CP}$-violating complex phase associated

C. J. G. Onderwater (✉)
KVI, University of Groningen, Zernikelaan 25, 9747AA Groningen, The Netherlands
e-mail: onderwater@kvi.nl

with the Kobayashi-Maskawa mechanism [3]. The SM is built on the assumption of Lorentz invariance and hence the invariance of the combination $\mathcal{CPT}$. Consequently $\mathcal{T}$ and $\mathcal{CP}$ violation are equivalent.

In the SM there are thus two sources of $\mathcal{CP}/\mathcal{T}$ violation, $\bar{\theta}_{QCD}$ and J. The magnitude of the former is as-of-yet undetermined, $|\bar{\theta}| < \mathcal{O}(10^{-11})$, whereas the smallness of the latter guarantees that SM $\mathcal{CP}$-odd observables are generally small.

Violation of $\mathcal{CP}$ is also expected to be necessary to explain the baryon asymmetry in the universe [4]. The baryon asymmetry predicted from the SM and the Cosmological Standard Model falls short of the observed one by as much as ten order of magnitude. This suggests the presence of additional sources of $\mathcal{CP}$-violation beyond those incorporated in the SM.

Permanent electric dipole moments (EDMs) are an excellent tool to search for such additional sources of $\mathcal{CP}$-violation [5]. EDMs break both $\mathcal{P}$ and $\mathcal{T}$, which is manifest when considering the field dependent part of the interaction Hamiltonian $\mathcal{H}$ for a particle in an electric field $\mathbf{E}$ and a magnetic field $\mathbf{B}$,

$$\mathcal{H} = -(\boldsymbol{\mu} \cdot \mathbf{B} + \mathbf{d} \cdot \mathbf{E}) = -(\mu\mathbf{B} + d\mathbf{E}) \cdot \frac{\mathbf{J}}{J}. \tag{1}$$

The second equality holds because the spin $\mathbf{J}$ is the only vector in the rest frame of a fundamental particle. The electric and magnetic dipole moments must point along it; $\mu = g(e/2m)$ and $d = \eta(e/4m)$ are the respective proportionality constants (using $\hbar = c = 1$). In units of the Bohr magneton $e/2m$ they are respectively given by the dimensionless moments g and $\eta/2$. When $d \neq 0$ this Hamiltonian violates both $\mathcal{P}$ and $\mathcal{T}$.

Particles acquire non-zero EDMs through $\mathcal{CP}$ violating radiative corrections. Quark mixing, *i.e.* δ, contributes only through three or more weak-interaction loop corrections. This makes these EDMs extremely small, of order $10^{-31}\,e \cdot$ cm for hadronic systems down to $10^{-41}\,e \cdot$ cm for leptons. This is far below present detection limits [6]. Hadronic systems may also acquire an EDM through $\bar{\theta}$ without the need for multiple loops. The non-observation of the neutron EDM limits the magnitude of $|\bar{\theta}| < \mathcal{O}(10^{-11})$.

In many proposed extensions of the SM, the need for multiple loops is not present, and EDMs may occur even at first order [7]. For example, many supersymmetric (SUSY) models predict a neutron EDM

$$d_n(SUSY) \sim \sin\delta_{CP}^{SUSY} \left(\frac{1\,\mathrm{TeV}}{M_{SUSY}}\right)^2 \times 10^{-25}\,e \cdot \mathrm{cm}. \tag{2}$$

Once the SUSY mass scale M_{SUSY} is determined, the $\mathcal{CP}$ violating phase δ_{CP}^{SUSY} can be determined using EDMs.

The first observation of a non-zero EDM would already unambiguously establish the presence of physics beyond the SM. Different forms of new physics manifest themselves differently already at the level of the fundamental fermions and their interactions, and propagate into increasingly larger composite systems, from hadrons to nuclei, atoms and molecules [8]. At each stage the appropriate theory needs to be applied. The most stringent limits on quark and proton EDM, as well as on $\mathcal{CP}$-violating electron-nucleus interactions are derived from the EDM search on the ^{199}Hg atom [9]. The most strict electron EDM limit stems from the YbF molecule [10]. The muon EDM is the only fundamental particle for which the EDM was obtained directly [11].

A single EDM measurement cannot be traced back to a specific source of $\mathcal{CP}$-violation. Several complementary measurements are necessary. For example, the combination of the neutron EDM combined with that for the proton, deuteron and possibly helion and triton makes it possible to distinguish new sources of $\mathcal{CP}$-violation from that introduced by $\bar{\theta}$ [12]. The uncertainties in the theory to describe light nuclei are well under control permitting reliable predictions [13–15]. Light nuclei thus offer the theoretically cleanest way to study hadronic $\mathcal{CP}$ violation.

Light nuclei cannot be probed for an EDM using atom- or molecule-based methods because of shielding effects [16]. Several experimental methods that circumvent these problems make use of the motional electric field a fast moving particle experiences when traversing a magnetic field [17–20]. These methods provide direct access to the very interesting realm of light nuclei, which so far have not been examined for EDMs. Also the muon can be probed sensitively, offering the unique possibility to explore the flavor structure of fundamental particles.

2 Storage ring techniques

A charged particle with magnetic moment anomaly a and EDM η moving in an electromagnetic field will exhibit spin precession. The evolution of the spin $\mathbf{S}$ is described by the (simplified) BMT equation [21],

$$\frac{d\mathbf{S}}{dt} = \frac{e}{m}\mathbf{S} \times \left[a\mathbf{B} + \left(\frac{1}{\gamma^2 - 1} - a \right) \boldsymbol{\beta} \times \mathbf{E} + \frac{\eta}{2} \left(\mathbf{E} + \boldsymbol{\beta} \times \mathbf{B} \right) \right] \equiv \mathbf{S} \times \boldsymbol{\Omega}. \tag{3}$$

The first two terms arise from the interaction of the magnetic moment with the magnetic field, whereas the last term is due to the interaction of the EDM with the electric field. The EDM interacts with the electric field in the rest-frame of the particle, which may have a strength $E^{CM} \sim E + vB \sim$ GV/m far in excess of those attainable in the laboratory.

In a purely magnetic storage ring the spin precesses about $\boldsymbol{\Omega}$ which is tilted with respect to $\mathbf{B}$ by an angle $\psi \simeq \eta\beta/2a$. The precession rate increases to $\Omega = \sqrt{1+\psi^2}\Omega_0$ with $\Omega_0 = a(e/m)B$. This quadratic sensitivity of the precession rate precludes a sensitive measurement of η. Because $\boldsymbol{\Omega}$ is tilted with respect to $\mathbf{B}$ the spin component along the magnetic field oscillates with an amplitude that depends linearly on the EDM. In the muon g-2 experiment at BNL this was used to limit the muon EDM [11]. It will be used again in the new muon g-2 experiment at Fermilab [22]. The muon EDM limit of $10^{-19}\, e \cdot$ cm corresponds to a tilt in the precession plane of order 1 microradian.

The statistical power of an EDM experiment is proportional to $PE\sqrt{NT}A$ with polarization P, effective electric field E, number of particles N, characteristic time scale T and analyzing power A. The sensitivity of the $\mathbf{B}$-only method is limited by the short precession cycle, which determines T.

Reducing the precession rate will prolong T. This can be done by applying a suitably chosen combination of radially oriented electric and vertically oriented magnetic fields. For

$$\frac{E}{B} = \frac{a\beta}{1-(1+a)\beta^2}. \tag{4}$$

the first two terms in (3) cancel, so that $\mathbf{\Omega} = \eta/2(e/m)(\mathbf{E} + \boldsymbol{\beta} \times \mathbf{B})$. The spin precession rate is now entirely determined by the EDM. The spin precesses about the electric field in the rest frame of the particle, which is oriented radially in the laboratory frame. The signature of an EDM is the change of the polarization component out of the particle orbit plane, generally along the magnetic field.

For particles with positive a a "magic" momentum $p_{\text{magic}} = \frac{m}{\sqrt{a}}$ exists. At this momentum the particles can be stored in an all-electric setup with $B = 0$. The bending radius ρ of a particle with mass m moving in an electric field E is

$$\rho = \frac{1}{\sqrt{a(a+1)}} \frac{m}{E}. \tag{5}$$

For a proton $p_{\text{magic}} \simeq 700\,\text{MeV}/c$ and $\rho \simeq 42\,\text{m}$, assuming $E = 10\,\text{MV/m}$. The spin precesses at a rate of $\Omega/d = 2E \simeq 10^{20}\,\text{rad/s}/(e \cdot \text{cm})$. At the expected sensitivity for Ω of 1 nrad/s a proton EDM of $d_p = 10^{-29}\,e \cdot \text{cm}$ can be measured. Efforts are ongoing to realize such an experiment [23].

An all-electric setup is not feasible for particles with small a and impossible for those with a negative one. A combination of electric and magnetic fields is necessary to "freeze" the spin. Expressed in E the EDM-induced spin precession rate and corresponding bending radius are given by

$$\Omega = 2d(E + vB) = \frac{a+1}{a\gamma^2} 2dE \quad \text{and} \quad \rho = \frac{a\beta^2\gamma^3}{a+1} \frac{m}{E} \tag{6}$$

Both from the point of view of the spin precession rate and the size of the setup low momenta are preferred. The electric field is effectively amplified by $(a+1)/a\gamma^2$. This becomes sizable for particles with a small a (see [24]). For the electron or muon with $a \simeq 0.00116$ this is $(a+1)/a \simeq 860$.

For muons with $p = 500\,\text{MeV}/c$ and $E = 2.2\,\text{MV/m}$ as proposed in [25] $\rho = 7\,\text{m}$. A considerably smaller setup with $\rho = 0.42\,\text{m}$ is proposed in [26] with $p = 125\,\text{MeV}/c$ and $E = 0.64\,\text{MV/m}$. The projected sensitivities are about the same at $d_\mu \simeq 10^{-16}\,e \cdot \text{cm}/\sqrt{N}$, with N the number of detected muon decays. At existing muon facilities $N = 10^{12}$ could be collected yielding a statistics limited sensitivity of $d_\mu \simeq 10^{-22}\,e \cdot \text{cm}$. This improves the current limit by three orders of magnitude. This experiment could well serve as a small-scale low-cost demonstration of this novel technique. At a future high-intensity muon facility this can be further improved by several orders of magnitude.

At the Forschungszentrum Jülich the possibilities for a light-ion EDM facility are explored [27]. Several options for an "all-in-one" storage ring were presented to search for EDMs on protons, deuterons and ^{3}He [28]. For a $p_p = 435\,\text{MeV}/c$, $p_d = 1000\,\text{MeV}/c$ and $p_{^3\text{He}} = 765\,\text{MeV}/c$ a single ring with a bending radius of 10 m can be constructed requiring $B < 0.5\,\text{T}$ and $E < 17\,\text{MV/m}$.

3 Outlook

A sensitive EDM search using a storage ring requires besides statistical also systematic precision. Current R&D efforts address several aspects that affect both. A system is being developed to reliably generate the electric field strengths of 10 MV/m planned for the proton EDM search. Such systems have been employed on a much smaller scale as electrostatic separators at e.g. the AGS at BNL and the Tevatron at FNAL. A critical aspect is the alignment of the electric field with respect to the magnetic field. Spin and beam dynamics must be understood at an unprecedent level of precision to guarantee optimal statistical precision through a long spin coherence time and to exclude or reduce systematic uncertainties to an acceptable level. An active research program is underway using the COSY facility at the Forschungszentrum Jülich. A lower limit on the spin coherence time of 75 s was demonstrated already, just one order of magnitude below the goal for the proton and deuteron EDM experiments [29]. Also at COSY a scheme to efficiently measure deuteron polarization and to correct for systematic errors was demonstrated [30]. The demonstrated sensitivity is sufficient to reach the proposed sensitivity of $d_d = 10^{-29}\, e \cdot \text{cm}$.

In conclusion storage rings make it possible to enter new territory in the search for EDMs. It is expected that an experiment on the proton and deuteron can be realized in the near future. A first small-scale implementation of a storage ring could be realized already now to search for a muon EDM. They make it possible to directly probe charged particles with competitive sensitivity. Such systems have a complementary sensitivity to new sources of CP-violation and may help to pin-down the last unconfirmed source of CP-violation in the Standard Model, $\bar{\theta}$.

Acknowledgements The author is indebted to his colleagues at KVI, PSI, RCNP and the storage ring-EDM collaboration for valuable discussions. The research described in this paper was partly financed by an Innovative Research grant (#680-47-203) from the Dutch Organization for Fundamental Research (NWO).

References

1. Cheng, H.Y.: The strong CP problem revisited. Phys. Rep. **158**, 1 (1988)
2. Jarlskog, C.: A basis independent formulation of the connection between quark mass matrices, CP violation and experiment. Phys. Z. **C29**, 491 (1985)
3. Kobayashi, M., Maskawa, T.: CP violation in the renormalizable theory of weak interaction. Prog. Theor. Phys. **49**, 652 (1973)
4. Sakharov, A.D.: Expanding universe and the appearance of a nonuniform distribution of matter. Pis'ma Z. Eksp. Teor. Fiz. **5**, 32 (1967) [JETP Lett. **5**, 24 (1967)]
5. Pospelov, M., Ritz, A.: Electric dipole moments as probes of new physics. Ann. Phys. **318**, 119 (2005)
6. Khriplovich, I.B., Lamoreaux, S.K.: CP Violation Without Strangeness. Springer (1997)
7. Sandars, P.G.H.: Electric dipole moments of charged particles. Contemp. Phys. **42**, 97 (2001)
8. Gould, C.R., Davis, E.D.: Time reversal invariance in nuclear physics: from neutrons to stochastic systems. In: Beyer, M. (ed.) CP Violation in Particle, Nuclear and Astrophysics, p. 207. Springer (2002)
9. Griffith, W.C., et al.: Improved limit on the permanent electric dipole moment of Hg-199. Phys. Rev. Lett. **102**, 101601 (2009)
10. Hudson, J.J., et al.: Improved measurement of the shape of the electron. Nature **473**, 493 (2011)
11. Bennett, G.W., et al.: An improved limit on the muon electric dipole moment. Phys. Rev. **D80**, 052008 (2009)
12. de Vries, J., et al.: Electric dipole moments of light nuclei from chiral effective field theory. Phys. Rev. **C84**, 065501 (2011)

13. Liu, C.-P., Timmermans, R.G.E.: P- and T-odd two-nucleon interaction and the deuteron electric dipole moment. Phys. Rev. **C70**, 055501 (2004)
14. Afnan, I.R., Gibson, B.F.: Model dependence of the (2)H electric dipole moment. Phys. Rev. **C82**, 064002 (2010)
15. de Vries, J., Mereghetti, E., Timmermans, R.G.E., van Kolck, U.: P and T violating form factors of the deuteron. Phys. Rev. Lett. **107**, 091804 (2011)
16. Schiff, L.I.: Measurability of nuclear electric dipole moments. Phys. Rev. **132**, 2194 (1963)
17. Berley, D., Gidal, G.: Search for an electric dipole moment structure of the muon. Phys. Rev. **118**, 1086 (1960)
18. Bailey, J., et al.: New limits on the electric dipole moment of positive and negative muons. J. Phys. G **4**, 345 (1978)
19. Farley, F.J.M., et al.: A new method of measuring electric dipole moments in storage rings. Phys. Rev. Lett. **93**, 052001 (2004)
20. Orlov, Y.F., Morse, W.M., Semertzidis, Y.K.: Resonance method of electric-dipole-moment measurements in storage rings. Phys. Rev. Lett. **96**, 214802 (2006)
21. Bargmann, V., Michel, L., Telegdi, V.L.: Precession of the polarization of particles moving in a homogeneous electromagnetic field. Phys. Rev. Lett. **2**, 435 (1959)
22. Carey, R.M., et al.: The new (g-2) experiment: a proposal to measure the muon anomalous magnetic moment to ±0.14 ppm precision. FERMILAB-PROPOSAL-0989
23. Semertzidis, Y.K.: A storage ring proton electric dipole moment experiment: most sensitive experiment to CP-violation beyond the standard model. http://arxiv.org/abs/1110.3378 (2011)
24. Khriplovich, I.B.: Nuclear electric dipole moments at ion storage rings. Hyperfine Interact. **127**, 365 (2000)
25. Aoki, M., et al.: J-PARC letter of intent: search for a permanent muon electric dipole moment at the 10^{-24} $e \cdot$ cm level (2003)
26. Adelmann, A., Kirch, K., Onderwater, C.J.G., Schietinger, T.: Compact storage ring to search for the muon electric dipole moment. Phys, J. **G37**, 085001 (2010)
27. http://www2.fz-juelich.de/ikp/en/future_projects.shtml
28. Lehrach, A.: 485. WE-Heraeus-Seminar. Bad Honnef, Germany. http://www2.fz-juelich.de/ikp/edm/en/talks/A.Lehrach.pdf, 4–6 Jul 2011
29. Stephenson, E.J., Onderwater, C.J.G., Lenisa, P.: Proc. 8th International Conference on Nuclear Physics at Storage Rings (Stori11) Laboratori Nazionali di Frascati, Italy, 9–14 Oct 2011
30. Brantjes, N.P.M., et al.: Correcting systematic errors in high-sensitivity deuteron polarization measurements. Nucl. Instrum. Methods **A664** 49, (2012)

Hyperfine Interact (2012) 211:15–20
DOI 10.1007/s10751-012-0580-0

The $\overline{\mathsf{P}}$ANDA detector at FAIR

Thomas Würschig
on behalf of the $\overline{\mathsf{P}}$ANDA collaboration

Published online: 9 February 2012

Abstract The $\overline{\mathsf{P}}$ANDA experiment at the future FAIR facility in Darmstadt will perform measurements with antiprotons to explore the physics in the charm quark sector. A state-of-the-art general purpose detector is planned that will allow a variety of complementary studies from high-precision spectroscopy to hypernuclei physics. This article gives a brief introduction to the experiment and the physics program. Main detector requirements are extracted subsequently. Finally, an overview of the instrumentation and the current state of development of the main detector subsystems are given.

Keywords Hadron physics · Antiprotons · Spectrometer · Instrumentation · DAQ · Tracking · Calorimetry

1 Introduction

$\overline{\mathsf{P}}$ANDA is a dedicated antiproton experiment planned at the FAIR facility [1], which is currently built up at GSI Darmstadt. It will study annihilation reactions of antiprotons with protons and in nuclear matter. The broad scientific program focused on the charm quark sector aims on a deeper insight into the nature of the strong interaction in the non-pertubative regime of QCD.

Designed as a fixed target experiment, $\overline{\mathsf{P}}$ANDA will use internal targets. These will be introduced at one interaction point (IP) into the high-energy storage ring (HESR) [2], that provides up to 10^{11} antiprotons with an adjustable momentum range from 1.5 GeV/c to 15 GeV/c. High beam intensity and optimized beam-target overlap will yield a maximum luminosity of $2 \cdot 10^{32}$ cm^{-1}s^{-1} with a relative momentum spread

Th. Würschig (✉)
Helmholtz-Institut für Strahlen- und Kernphysik,
Rheinische Friedrich-Wilhelms-Universität Bonn,
Nussallee 14-16, D-53115, Bonn, Germany
e-mail: t.wuerschig@hiskp.uni-bonn.de

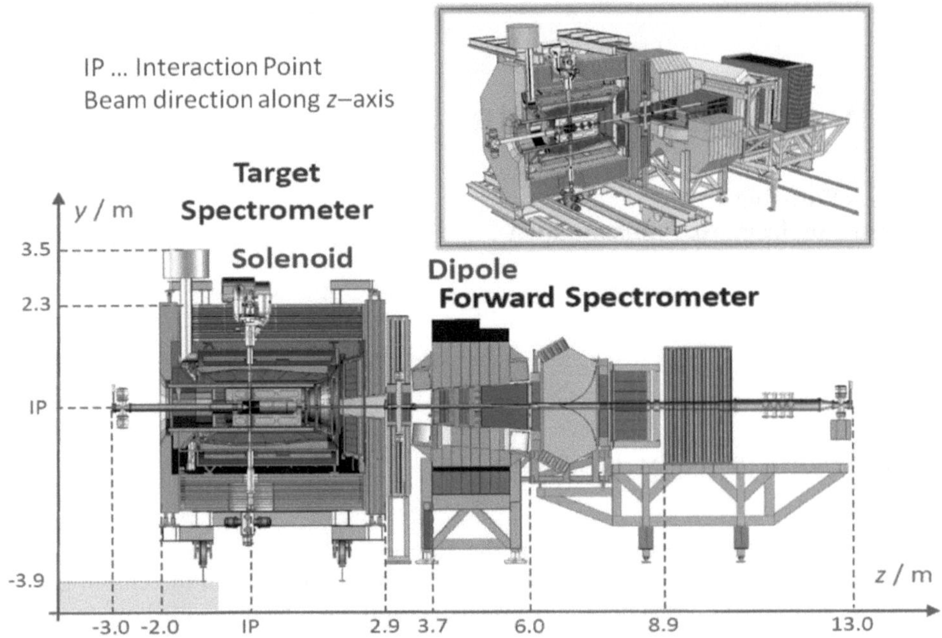

Fig. 1 Artistic side view of the $\overline{\text{P}}$ANDA detector along the beam axis. A 3D representation is shown in the framed inlay

$\delta p/p$ of only 10^{-4}. Even better values down to $4 \cdot 10^{-5}$ are expected at a reduced luminosity of $10^{31}\ \text{cm}^{-1}\text{s}^{-1}$.

Figure 1 shows the basic layout of the $\overline{\text{P}}$ANDA detector. It consists of a 4 m long, 2 T superconducting solenoid and a 2 Tm resistive dipole magnetic spectrometer. Both are instrumented to detect charged and neutral particles emitted at large and backward angles (Target Spectrometer, TS) and at angles between 0° and 20° (Forward Spectrometer, FS), respectively. The TS has an onion-shell like design surrounding the interaction point (IP), which is very similar to collider experiments. The FS will be formed by an arrangement of subsystems along the beam axis in forward direction.

2 Physics program

The copious production of particles with gluonic degrees of freedom as well as particle-antiparticle pairs in $\bar{p}p$ annihilations allow spectroscopic studies with very high statistics and precision. With the given design parameters $\overline{\text{P}}$ANDA will be a unique tool to improve the current experimental status and to further exploit the QCD spectrum in the relevant mass region. Main experiments of the rich and diversified hadron physics program are (1) High-precision charmonium and open charm spectroscopy, (2) Search for gluonic excitations, (3) Study of hadrons in nuclear matter, (4) Hypernuclear physics and (5) Electromagnetic processes. More details can be found in the $\overline{\text{P}}$ANDA physics performance booklet [3].

3 Detector requirements

To fully exploit the physics potential of the unique experimental conditions at $\overline{\text{P}}$ANDA the detector must provide high precision and reconstruction efficiency for a set of final states. Therefore, a multi-purpose detector based on modular sub-components is best suited. The design must also be compatible with different target systems. Moreover, a high flexibility of the trigger is needed. $\overline{\text{P}}$ANDA wants to achieve this goal by an autonomously signal detection of the individual sub-systems. This approach requires data reduction in the preprocessing step, sufficient buffering and powerful computing nodes that access the buffers via a high-bandwidth network fabric to set the final trigger decision.

For a complete event reconstruction the detection of both charged and neutral particles is prerequisite. Full 2π azimuthal coverage and good particle identification (PID) are mandatory to identify multi-particle final states and to study correlations within the produced particles. The fixed-target setup implies a Lorentz boost with a γ_{CM} factor ranging from 1.2 to 2.9 that corresponds to a large difference in the typical event topologies and a large dynamic range of emitted particles. Thus, the detector must be also suited to accurately track slow particles with momenta of about 100 MeV/c to 200 MeV/c and to detect soft photons. Therefore, a very low material budget is needed to reduce scattering effects.

For high-precision spectroscopy and background suppression a relative resolution of particle momentum and energy on the percent level is required. A very important task of the tracking system is the determination of secondary vertices of shorted-lived particles. In this context, D mesons with mean decay lengths $c\tau$ about 0.1 mm to 0.3 mm are of particular interest. These can be identified only by a vertex detector close to the IP.

The high intensity of the quasi-continuous beam will lead to an interaction rate of 20 million antiproton annihilations per second thus representing a technical challenge in terms of high-rate capability for the detector components, the trigger and the data acquisition systems. The expected radiation level for the detector components will stay roughly one order of magnitude below the one of the LHC experiments.

4 Main detector components

The $\overline{\text{P}}$ANDA apparatus will consist of tracking detectors (MVD, STT, GEM), electromagnetic calorimeters (EMC), a muon system, Cherenkov detectors (DIRC, RICH) and a time-of-flight (TOF) system. For the absolute measurement of total cross sections a luminosity monitor measuring elastically scattered antiprotons will be installed in very forward direction. The overall detection concept is shown schematically in Fig. 2.

In the following, the main components of the $\overline{\text{P}}$ANDA detector will be briefly described. An overview of the present state of development is given in Table 1. For further details the reader is referred to the given references.

Magnets and muon system The solenoid magnet of the TS will deliver a very homogeneous field of 2 T with fluctuations of less than $\pm 2\%$. The superconducting

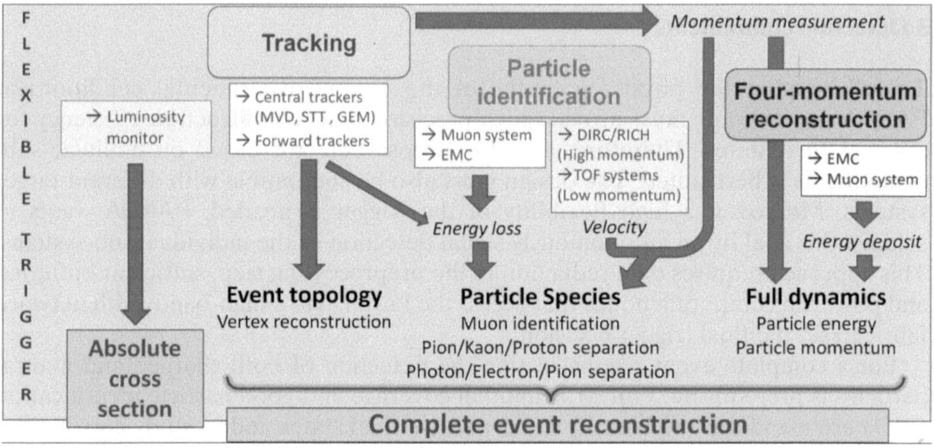

Fig. 2 Basic detection concept. The main components will be described in Section 4

Table 1 Present state of development for main detector parts. For most of them a technical design report (TDR) exists or is expected to be available in the near future

Subsystem	Status of the technical project	Hardware development
Magnets	Design finalized, TDR published [4]	First parts ordered
Muon system	Basic design studies, TDR underway	Prototypes (PT) developed
H-Targets	Design optimization, TDR in final stage	Full-scale cluster-jet PT in commision [5]
MVD	Full design study [6], TDR submitted	Full-size PTs under study
STT	Full design study [7], TDR submitted	Full-scale PT system tested
GEM	Extended design studies	Full-scale PTs developed
EMC (TS)	Full design study, TDR published [8]	First crystals delivered, prototypes evaluated [9]
DIRC	Advanced design studies [10, 11]	Extensive PT tests

coil has a length of 2.8 m and an inner diameter of 90 cm. It will use a laminated iron yoke for the flux return that will be instrumented with 2600 Mini Drift Tubes (MDT) of the muon system. For the insertion of internal targets there are two bores of 100 mm diameter positioned above and below the IP. The large aperture resistive dipole magnet of the FS has a field integral of 2 Tm and will cover an angular acceptance of ±5° in vertical and ±10° in horizontal direction, respectively. Additional muon filters with 700 MDTs will be moved between the solenoid and the dipole magnet. Another range system equipped with 900 MDTs will be installed at the downstream end of the FS.

Target systems Hydrogen targets will be used at the commissioning of the experiment. Afterwards, an extension to heavier gases and solid targets will be accomplished. Challenges for either system result from the long distance of roughly 2 m between the target injection point at the top and the dumping system at the bottom of the flux return yoke. Present developments concentrate on the hydrogen target, which will be either a cluster-jet or a pellet target. For the hyper-nuclear program a secondary target formed by a sandwiched structure of silicon detectors and carbon layers will be introduced. Therefore, a re-design of the innermost part of the TS is required.

Tracking systems There are three tracking detectors inside the TS: the micro-vertex detector (MVD), the straw tube tracker (STT) and three tracking stations with gaseous micro-pattern detectors based on GEM foils as amplification stage in forward direction. The MVD consists of four barrel layers and six wheels arranged perpendicular to the beam axis in forward direction. It will be equipped with silicon pixel and double-sided strip detectors in the inner and outer part, respectively. While the outermost radius of the MVD is restricted to 15 cm the innermost layers have a distance of only 2 cm to the IP. The STT will have a total length of 1.5 m and a maximum outer radius of 50 cm. It will consist of self-supporting aluminium Mylar tubes filled with a gas mixture and operated at a pressure of 2 bar. Each straw tube is constructed with a single anode wire in the centre. In total there are 4636 straws, which are arranged in a hexagonal shape around the MVD. Four skewed stereo double-layers will facilitate a 3D reconstruction. The tracking in the FS will be realized by three stations equipped with two skewed straw tube layers.

Electromagnetic calorimters (EMC) The EMC of the TS consists of a barrel part, a forward and a backward endcap. It will be equipped with lead tungstate ($PbWO_4$) crystals. Their length of 20 cm corresponds to roughly 22 radiation lengths and thus allows a compact design. In total there are 11,360 crystals in the barrel part and 4,200 crystals in the end-cap structures. An increase of the light yield will be achieved by cooling to −25°C. A Shashlyk-type calorimeter based on lead-scintillator sandwiches placed at a distance of 7.5 m from the IP will be used in forward direction. It will consist of 351 modules arranged in 13 rows and 27 columns.

Cherenkov detectors and time-of-flight (TOF) systems Two different Cherenkov detectors will be used to measure the velocity of charged particles above 1 GeV/c. While the FS will be equipped with a ring imaging Cherenkov detector (RICH), a modified RICH design based on the detection of internally reflected Cherenkov (DIRC) light will be applied for the TS. The DIRC system will be made of a circular arrangement of quartz slabs at radii from 45 cm to 54 cm in the barrel part and a disk-shape radiator with an outer radius of 110 mm in the forward end-cap. Fused silica will be used as radiator material. For the RICH system a dual radiator design made of silica aerogel and C_4F_{10} is favoured. The TOF system will consist of scintillator bars or pads of multi-gap resistive plate chambers, which are cylindrically arranged around the outer tracker, and a wall of plastic scintillator slabs at about 7 m in downstream direction. Additional detectors of the same type will placed inside the dipole magnet opening.

5 Summary

The presented $\overline{\text{P}}$ANDA setup describes a multi-purpose machine that serves a rich and diversified hadron physics program. Its main features are:

- High spatial vertex resolution in the order of 100 μm
- Precise momentum reconstruction at a level of 1%

- Photon detection in a wide energy range from 1 MeV to 10 GeV
- High-quality calorimetry with an energy resolution from 2.7% at 1 GeV to 1.5% at 10 GeV
- Good particle identification of photons, muons, pions, kaons and protons in a large dynamical range from $pc = 0.5$ GeV to 15 GeV
- Flexible data acquisition based on a triggerless readout concept for the near detector front-end electronics

The present state of development for the main sub-systems indicates that $\overline{\text{P}}$ANDA is on track to fix technical specifications and to start with the production of first detector components for the final detector assembly.

Acknowledgement This work was supported by the Bundesministerium für Bildung und Forschung (BMBF) under contract no. 06BN9005 I.

References

1. Henning, W.F.: FAIR—recent developments and status. Nucl. Instrum. Methods A **805**, 502c–510c (2008)
2. Lehrach, A., et al.: Beam dynamics of the high-energy storage ring (HESR) for FAIR. Int. J. Mod. Phys. E **18/2**, 420–429 (2009)
3. PANDA collaboration: Physics performance report for: $\overline{\text{P}}$ANDA-strong interaction studies with antiprotons. arXiv:0903.3905v1 (2009)
4. PANDA collaboration: Technical design report for the solenoid and dipole spectrometer magnets. arXiv:0907.0169v1 (2009)
5. Täscher, A., et al.: High density cluster jet target for storage ring experiments. Nucl. Instrum. Methods A **660**(1), 22–30 (2011)
6. Würschig, Th.: Design optimization of the $\overline{\text{P}}$ANDA Micro-Vertex-Detector for high performance spectroscopy in the charm quark sector. Dissertation, urn:nbn:de:hbz:5N-26230 (2011, also submitted as: ISBN 978-3-8381-3092-7)
7. Costanza, S.: Design of the central tracker of the $\overline{\text{P}}$ANDA experiment. Dissertation, ISBN 978-88-95767-38-3 (2010)
8. PANDA collaboration: Technical design report for Electromagnetic Calorimeter (EMC). arXiv:0810.1216v1 (2008)
9. Kavatsyuka, M., et al.: Performance of the prototype of the electromagnetic calorimeter for $\overline{\text{P}}$ANDA. Nucl. Instrum. Methods Phys. Res. A **648**(1), 77–91 (2011)
10. Schwarz, C., et al.: Development of the $\overline{\text{P}}$ANDA barrel. DIRC, JINST **4**, P12016 (2009)
11. Cowie, E.N., et al.: A focussing disc DIRC for $\overline{\text{P}}$ANDA. Nucl. Instrum. Methods A **639**(1), 181–184 (2011)

Hyperfine Interact (2012) 211:21–25
DOI 10.1007/s10751-012-0578-7

Startup of the high-intensity ultracold neutron source at the Paul Scherrer Institute

Bernhard Lauss
on behalf of the PSI UCN Project Team

Published online: 2 March 2012

Abstract Ultracold neutrons (UCN) can be stored in suitable bottles and observed for several hundreds of seconds. Therefore UCN can be used to study in detail the fundamental properties of the neutron. A new user facility providing ultracold neutrons for fundamental physics research has been constructed at the Paul Scherrer Institute, the PSI UCN source. Assembly of the facility finished in December 2010 with the first production of ultracold neutrons. Operation approval was received in June 2011. We give an overview of the source and the status at startup.

Keywords Ultracold neutrons · UCN · Neutron source · Ultracold neutron production and transport

1 Introduction

The fundamental properties of the neutron can serve as a window into the first minutes of the universe. The neutron lifetime is an important parameter in understanding the abundance of light elements created in the big-bang nucleo-synthesis [1]. Finding a neutron electric dipole moment (nEDM) [2] would hint at a new CP violating process necessary to understand the asymmetry of matter over antimatter in our universe. nEDM searches (e.g. [3, 4]) probing physics beyond the Standard Model are presently statistics limited and are the main driving force behind the proposals of several new high-intensity ultracold neutron sources around the globe [5]. Moreover, stored neutrons allow to search for extra forces not included in the Standard Model of particle physics or search for modifications of Newtonian gravity in the sub-millimeter range predicted by string theories [6].

The members of the PSI UCN Project Team are listed at http://ucn.web.psi.ch.

B. Lauss (✉)
Paul Scherrer Institute, 5232 Villigen-PSI, Switzerland
e-mail: bernhard.lauss@psi.ch

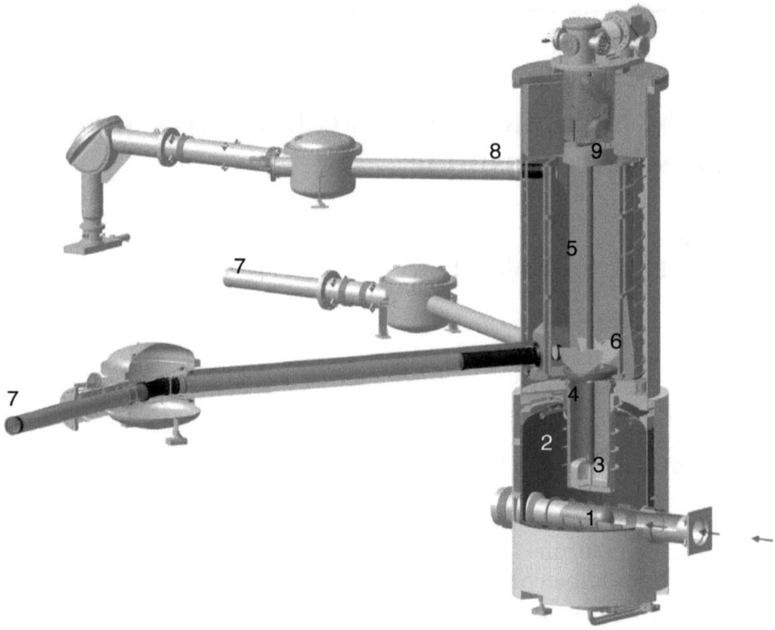

Fig. 1 Cut view of the neutron production, storage and guiding system of the PSI UCN source inside a 7 m high vacuum tank. The indicated components are described in the text

2 The ultracold neutron source

Neutrons with kinetic energies below ~350 neV, corresponding to a few milli-Kelvin, are termed ultracold neutrons (UCN). The material optical potential of certain materials (e.g. Ni, Be, steel, diamond-like carbon/DLC) is high enough—due to their high density and large neutron bound-coherent scattering length—that UCN undergo total reflection under all angles of incidence [2]. Hence, UCN can be stored in material bottles for several hundreds of seconds. They can also be contained via gravity, with an energy change of 100 neV per meter and they can be manipulated via their magnetic moment, where a 1 Tesla field change corresponds to a change of 60 neV in potential energy.

Over the last years an ultracold neutron source based on accelerator driven spallation neutron production has been constructed at the Paul Scherrer Institute (PSI), Switzerland [7, 8].

The method, experimentally pioneered at PNPI [9] and Los Alamos National Lab [10], is based on (i) neutron production via proton induced spallation on lead [11]; (ii) superthermal UCN production in solid ortho-deuterium (sD_2) [2]; and (iii) intermediate UCN storage and distribution of UCN on demand to experiments. The main experimental components covering the neutron production and transport are shown in Fig. 1. PSI's 590 MeV proton beam with up to 2.4 mA beam current is impinging on a heavy-water-cooled canneloni-type lead spallation target (see (1) in Fig. 1) [11]. UCN operation is limited by radiation safety requirements to an integrated proton flux of 20 μA/hour. This means that a 2 mA proton beam kick

Fig. 2 Parts of the UCN source during assembly. **a** The container for the solid deuterium crystal. The *donut shaped top-lid*, necessary for pressure safety, is only 0.5 mm thick to allow for a high UCN transmission. **b** The storage volume shutter during operation testing. The 2 aluminum jaws visible on each flap are used for stopping the closing flaps via an eddy-current brake working at 80 K. **c** The UCN storage volume before insertion in the vacuum tank. Visible is the vertical neutron guide on the *bottom*, the radiation shield surrounding the UCN storage vessel with the 80 K cooling tubes and the opening for the neutron guides. The steel shielding is on *top*

of 8 s is followed by a minimum waiting period of 800 s. Spallation neutrons are thermalized in the surrounding heavy water tank (2). Close to the spallation target inside a AlMg3 container (No. (3) in Figs. 1 and 2a) a pure ~30 liter ortho-deuterium crystal is kept at a temperature of 5 K. It serves to moderate neutrons first to be cold and finally to be ultracold neutrons via downscattering [2, 5, 13].

The UCN production rate strongly depends on the deuterium temperature and spin [5, 12, 13] favoring solid ortho-deuterium as production medium. Some UCN can then be emitted into the vacuum on top of the crystal, where they get a 102 neV boost from the sD_2 material potential at the crystal surface. Via a vertical UCN guide (4) (shown in Fig. 2c) they can reach the UCN storage vessel (5) coated with diamond-like carbon which has a high material optical potential of about 235 neV and a very low neutron loss probability.

After the beam kick the main UCN shutter (No. (6) and Fig. 2b) is rapidly closed and the produced UCN are stored inside the vessel. Three UCN guides (7) and (8) lead through the radiation shielding to experimental areas. Shutters can open and close these guides at the storage vessel exits. In order to provide optimal UCN transmission through more than 8 m of tubes penetrating the radiation shielding, most of the guides are made of 180 mm inner diameter Duran® glass tubes with approximately 1 nm surface roughness and sputter-coated with 500 nm thick NiMo on the inside. A 80 cm long polished steel guide connects the glass guides at

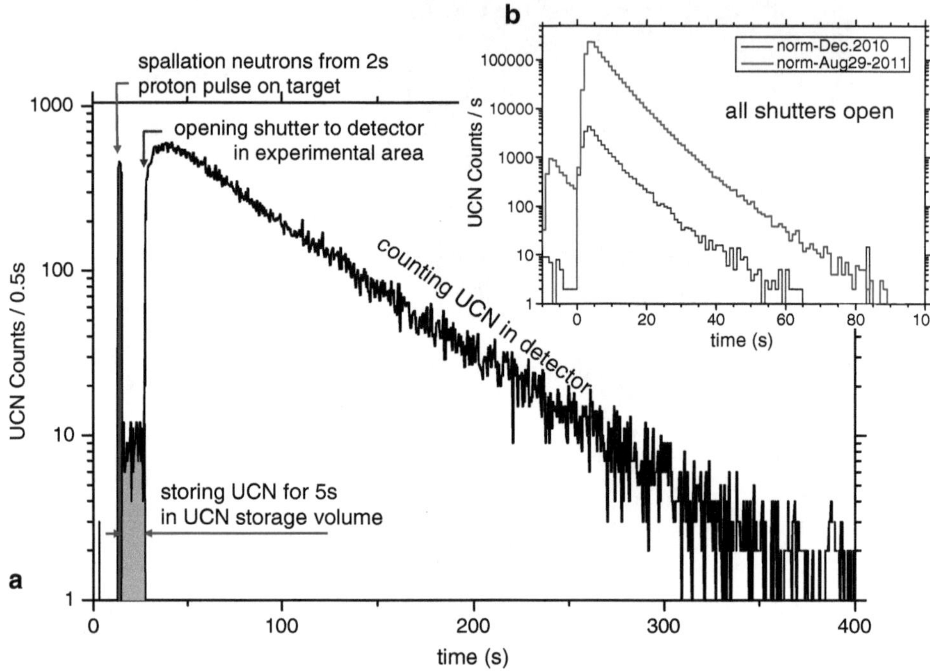

Fig. 3 **a** First UCN in December 2010: UCN counts observed in a Cascade-U detector located at the end of the UCN guide in area West, plotted versus time (seconds). The structure in the count rate reflects the initial proton beam kick onto the spallation target of 2 s which produce a neutron background in the detector. After the beam kick the UCN storage volume is closed and the UCN are stored for 5 s. A few percent of the UCN leak during that time through the guide UCN shutter. After 5 s the guide shutter towards the experimental area is opened and the UCN can arrive at the detector. The observed count rate then reflects the exponential emptying of the UCN storage vessel into the detector. **b** UCN counts versus time after beam kick observed on guide West-1 for 'normalized pulses', i.e. beam kicks with 2 s length and all shutters, including the main UCN storage vessel shutter remaining open all the time. Hence the emptying of the storage vessel is determined by the large opening towards the deuterium container. The given time distributions are for the best beam kick observed in Dec. 2010 and the startup kicks in Aug. 2011. A factor 50 improvement is obvious

room temperature to the UCN storage vessel which is operating at 80 K. Radiation protection requires 30° bends (made from polished stainless steel) to prevent direct sight onto the storage vessel. On top of the UCN storage vessel a large, activated carbon loaded cryo-pump is the coldest spot of the source vacuum in order to efficiently capture residual gas atoms and prevent them from condensing on the sD_2 container or anywhere else inside the storage vessel.

Several important components are not shown on the figure. Most prominently the cryo-system necessary for the production of the ortho-deuterium crystal. The present sD_2 production procedure [14] uses 30 m^3 pure D_2 gas. The gas is frozen into the 'condenser' volume. Then it is liquefied and transported to the nearby 'para-otho-converter' filled with a paramagnetic chromium-oxide (Oxisorb®) [15] and left boiling at 19 K for several hours. Raman spectroscopy of the rotational transitions in D_2 on extracted gas samples shows an ortho-deuterium concentration of 97±2% as expected for equilibrium conditions at 19 K. Finally, the ortho-D_2 is slowly frozen inside the crystal container. This process takes several days and, being a very important part of the source setup, still needs to be optimized.

Final assembly of all parts and commissioning of the PSI source finished in 2010. December 16, 2010 saw the first UCN production during testing the radiological and cryogenic safety. Figure 3a shows one of the first production fills with a UCN count-rate structure as expected. Beam kicks of up to 8 s length and full beam power were tested.

Operation approval of the Swiss Federal Authorities was received on June 27 and the source operation started in August, 2011. The source performance is continuously improving since then with presently already a factor 50 higher yield in comparison to Dec. 2010 as demonstrated in Fig. 3b. This can be mainly attributed to increasing the ortho-deuterium content from 66 to ~97% and a different freeze-out technique. A factor of about 30 lies ahead to full design intensity.

The experimental search for a neutron EDM is starting in area South with an installed apparatus [3]. The UCN user facility at PSI is open for experiment proposals.

Acknowledgements Cordial thanks to the more than two hundred colleagues contributing to the UCN source project at the Paul Scherrer Institute who are indispensable for the realization of this project. Support of our colleagues at PF2—ILL and the Mainz TRIGA UCN Source during component testing and of Jagellonian University Cracow and LPSC Grenoble is gratefully acknowledged. PNPI contributed in the early planning of the project.

References

1. Particle Data Group, Nakamura, K., et al.: JPG **37**, 075021 (2010)
2. Golub, R., Richardson, D., Lamoreaux, S.K.: Ultra-Cold Neutrons. Adam Hilger, Bristol, Philadelphia and New York (1991)
3. Altarev, I., et al.: Nucl. Instrum. Methods A **611**, 133 (2009)
4. Baker, C.A., et al.: Phys. Rev. Lett. **97**, 131801 (2006)
5. Kirch, K., et al.: Nucl. Phys. News **20/1**, 17 (2010)
6. Abele, H.: Prog. Part. Nucl. Phys. **60**, 1 (2008)
7. Blau, B.: In: Proceedings of the ICANS XIX Meeting. ISSN-Nr.1019–6447, Grindelwald, Switzerland, 8–12 Mar 2010
8. Lauss, B.: J. Phys.: Conf. Ser. **312**, 052005 (2011)
9. Serebrov, A.P., et al.: JETP Lett. **66**(12), 802 (1997)
10. Morris, C.L., et al.: Phys. Rev. Lett. **89**, 272501 (2002)
11. Wohlmuther, M., Heidenreich, G.: Nucl. Instrum. Methods A **564**, 51 (2006)
12. Liu, C.-Y., et al.: Phys. Rev. B **62**(6), R3581 (2000)
13. Atchison, F., et al.: Phys. Rev. C **71**, 054601 (2005)
14. Anghel, A., et al.: In: Proc. of the 10th IIR International Conference Cryogenics 2008. Prage, Czech Republic, 21–25 Apr 2008
15. Bodek, K., et al.: Nucl. Instrum. Methods A **533**, 491 (2004)

Hyperfine Interact (2012) 211:27–31
DOI 10.1007/s10751-011-0549-4

CP violation studies at Belle

Christoph Schwanda

Published online: 29 December 2011

Abstract CP-violation in the Standard Model (SM) is described by a single parameter and the pattern of possible CP-violating phenomena is thus extremely constrained. The Belle experiment at the KEK laboratory in Tsukuba (Japan) is testing if this pattern actually occurs by analyzing a sample of about 770 million B-meson decays. CP-violation observed at Belle is consistent with the Cabibbo-Kobayashi-Maskawa (CKM) mechanism, the theory of CP-violation within the SM. We briefly describe how the unitarity triangle allows to probe the CKM theory and review recent measurements of its angles and sides, namely determinations of the CKM angles ϕ_1 and ϕ_3, and the CKM magnitudes $|V_{cb}|$ and $|V_{ub}|$.

Keywords CP violation · Cabibbo-Kobayashi-Maskawa theory · Belle B factory

1 Introduction

The CP-transformation combines charge conjugation (C) with the inversion at the origin (P). E.g., a left-handed electron e_L^- is transformed under CP into a right-handed positron e_R^+. In 1964, CP was found to be violated in neutral kaon decays by Cronin and Fitch [1].

The Standard Model (SM) of particle physics contains a single source of CP-violation, the non-trivial complex phase in the unitary 3×3 quark mixing matrix, introduced by Cabibbo, Kobayashi and Maskawa [2, 3]. Given this single parameter, it is obvious that the pattern of CP-violating phenomena in the SM is extremely constrained. The main objective of the B-factory experiments Belle [4] and BaBar [5] is to test, if this pattern actually occurs.

C. Schwanda (✉)
Institute of High Energy Physics, Austrian Academy of Sciences,
Nikolsdorfer Gasse 18, 1050 Vienna, Austria
e-mail: Christoph.Schwanda@oeaw.ac.at

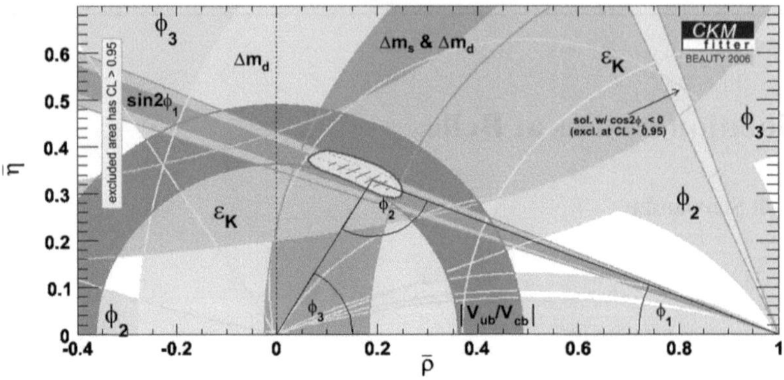

Fig. 1 The CKM unitarity triangle, analyzed by the CKM fitter group [6]. The angles ϕ_1 and ϕ_3, and the magnitudes $|V_{cb}|$ and $|V_{ub}|$ discussed in this writeup are shown

Table 1 Measurements at the B-factories used to determine the different observables in the CKM unitarity triangle. This table also lists recent Belle results

Obs.	Measurement	Recent Belle results
ϕ_1	Mixing-induced CP-violation in $B^0 \to (c\bar{c})K^0$	$B^0 \to (c\bar{c})K^0$ [winter 2011] $B^0 \to D^{(*)+}D^{(*)-}$ [summer 2011]
ϕ_2	Mixing-induced CP-violation in $B^0 \to \pi\pi, \rho\rho, \rho\pi$	
ϕ_3	Interference between $B^\pm \to \bar{D}K^\pm$ and $B^\pm \to DK^\pm$	Binned Dalitz plot [8] First evidence $B \to DK$ ADS [9] $B \to D^*(D\pi^0, D\gamma)K$ ADS [summer 2011] $B \to D_{CP}K$ GLW [summer 2011]
$\|V_{cb}\|$	$B \to X_c\ell\nu$	$B^0 \to D^{*-}\ell^+\nu$ [10]
$\|V_{ub}\|$	$B \to X_u\ell\nu$	$B^0 \to \pi^-\ell^+\nu$ [11]

This is done by measuring the angles and sides of unitarity triangle (Fig. 1), which represents the equation

$$V_{ud}V_{ub}^* + V_{cd}V_{cb}^* + V_{td}V_{td}^* = 0 \tag{1}$$

in the complex (ρ, η) plane. This relation is a consequence of the unitarity of the Cabibbo-Kobayashi-Maskawa (CKM) matrix. Analyses by the CKM fitter [6] and UTfit groups [7] show that current experimental constraints (mainly from the B-factories) are consistent with the CKM mechanism.

Table 1 lists the B-meson decay modes used to determine the CKM angles and sides at the B-factories.

2 The Belle apparatus

The Belle detector is a large-solid-angle magnetic spectrometer, located in the interaction region of the KEKB machine which collides 3.5 GeV positrons and 8 GeV electrons [4]. The center-of-mass (c.m.) energy of the collisions corresponds to the $\Upsilon(4S)$-resonance (10.58 GeV) which decays to a pair of B-mesons (either B^+B^- or $B^0\bar{B}^0$). Belle consists of a silicon vertex detector, a 50-layer central drift chamber, an array of aerogel threshold Cherenkov counters, a barrel-like arrangement of time-of-flight scintillation counters, and an electromagnetic calorimeter comprised of CsI(Tl) crystals, located inside a super-conducting solenoid coil that provides a 1.5 T magnetic field. The iron flux return is instrumented to detect K_L mesons and to identify muons.

Belle operated from 1999 to 2010 and accumulated an integrated luminosity of about 711 fb^{-1} on the $\Upsilon(4S)$-resonance. This data sample corresponds to about 772 million $B\bar{B}$ events.

3 Measurements of the angle ϕ_1

CP-violation in the interference of mixing and decay in $B^0 \to (c\bar{c})K^0$ allows to determine the angle ϕ_1 with high precision. This mixing-induced CP-violation is measured by reconstructing Belle events of the form $\Upsilon(4S) \to B^0_{CP}B^0_{\text{tag}}$, where B^0_{CP} is a neutral B-meson decaying to a $(c\bar{c})K^0$ CP-eigenstate and B^0_{tag} is a B decaying to a state with defined flavor (B^0 or $\bar{B}^0$). If CP-violation occurs, the decay distributions for B^0 and $\bar{B}^0$ tags are different,

$$P(\Delta t, q; S, A) = \frac{e^{-|\Delta t|/\tau_B}}{4\tau_B}\left[1 + q\,(A\cos\Delta m\Delta t + S\sin\Delta m\Delta t)\right] . \tag{2}$$

Here, $\Delta t = t_{CP} - t_{\text{tag}}$ is the difference in the decay time of B^0_{CP} and B^0_{tag}, and q is +1 for $B^0_{\text{tag}} = B^0$ and -1 for $B^0_{\text{tag}} = \bar{B}^0$. τ_B is the B^0 lifetime and Δm is the B^0 oscillation frequency. The parameters A and S depend on the CP-eigenmode. For $B^0 \to (c\bar{c})K^0$, theory predicts A to vanish and S to be equal to $-\eta_{CP}\sin 2\phi_1$, where η_{CP} is the CP-eigenvalue of the mode.

Experimentally, the decay of B^0_{CP} is fully reconstructed while the charge signs of leptons and charged pions and kaons among the tracks not associated with B^0_{CP} are used to identify the flavor of B^0_{tag}. To determine the decay time difference Δt, the decay vertices of B^0_{CP} and B^0_{tag} are reconstructed. As the c.m. system moves in the laboratory frame with defined boost $\beta\gamma = 0.425$ in the direction of the beam axis, Δt can be calculated from the difference in the decay vertex positions along the beam axis Δz, $\Delta t = \Delta z/\beta\gamma c$.

Belle has updated its measurement of $\sin 2\phi_1$ from $B^0 \to (c\bar{c})K^0$ using the entire 711 fb^{-1} dataset in winter 2011: With 10041 ± 154 signal events (including both $CP = -1$ and $CP = +1$ modes) we obtain $\sin 2\phi_1 = S = 0.668 \pm 0.023(\mathit{stat}) \pm 0.013(\mathit{syst})$ and $A = 0.007 \pm 0.016(\mathit{stat}) \pm 0.0.13(\mathit{syst})$.

4 Measurements of the angle ϕ_3

The relative weak phase between the diagrams for $B^- \to D^0K^-$ and $B^- \to \bar{D}^0K^-$ is to a good approximation equal to the angle $\phi_3 = \arg(V_{ud}V^*_{ub}/V_{cd}V^*_{cb})$. If D^0 and $\bar{D}^0$ are reconstructed in the same final state, ϕ_3 can be determined from the interference of these two diagrams. This is the basic idea underlying three different approaches for determining ϕ_3:

- Gronau-London-Wyler (GLW) [12]: uses D^0-decays to a CP-eigenstate,
- Atwood-Dunietz-Soni (ADS) [13]: uses Cabibbo-favored and doubly Cabibbo-suppressed D^0-modes,
- Dalitz plot analysis [14]: uses three-body D^0-decays, e.g., $D^0 \to K_S\pi^+\pi^-$.

Regarding the latter, the interference term in the $B^\pm \to D(K_S\pi^+\pi^-)K^\pm$ rate is

$$2|A\left(m_+^2, m_-^2\right)||A(m_-^2, m_+^2)|r\cos\left(\delta + \theta_{(m_+^2, m_-^2)} \pm \phi_3\right), \tag{3}$$

where $m_\pm^2 = m^2(K_S\pi^\pm)$ are the coordinates and $|A(m_+^2, m_-^2)|^2$ is the density of the $D^0 \to K_S\pi^+\pi^-$ Dalitz plot, and r is the ratio of the $B \to DK$ to the $B \to \bar{D}K$ amplitude. δ is the strong phase related to the B-decay and can be determined easily by comparing B^+ and B^- decays. The strong phase related to the D^0-decay, $\theta_{(m_+^2, m_-^2)}$, is more problematic as it depends on the position in the Dalitz plot and typically introduces a dependence on the $D^0 \to K_S\pi^+\pi^-$ model.

A recent Belle analysis (using also the full 711 fb^{-1} dataset) [8] uses a measurement of the charm-related strong phase difference at CLEO-c in $\psi(3770) \to D\bar{D}$ events [15] to overcome this model limitation: We obtain $\phi_3 = (77.3^{+15.1}_{-14.9}(stat) \pm 4.2(syst) \pm 4.3(CLEO))^\circ$. This can be compared to the previous model-dependent Belle result: $\phi_3 = (78 \pm 11(stat) \pm 4(syst) \pm 9(model))^\circ$ [16].

5 Measurements of the CKM magnitudes $|V_{cb}|$ and $|V_{ub}|$

The CKM matrix elements $|V_{cb}|$ and $|V_{ub}|$ are measured using semileptonic B-decays to charmed, $B \to X_c\ell\nu$, and charmless final states, $B \to X_u\ell\nu$, as these decays have large branching fractions and hadronic uncertainties are under control. Inclusive measurements are sensitive to all $X\ell\nu$ final states in a given region of phase space. Here, the decay amplitudes can be calculated in perturbative QCD using the Operator Product Expansion (OPE), which factorizes perturbative and non-perturbative contributions. Exclusive measurements search for a specific final state, e.g., $B \to D^*\ell\nu$, $B \to \pi\ell\nu$, ..., and lattice QCD is needed to normalize the decay rate. As theoretical and experimental systematics are completely different in these two approaches, obtaining consistency between inclusive and exclusive is a very important objective.

Belle has recently measured the exclusive decay $B^0 \to D^{*-}\ell^+\nu$ using 711 fb^{-1} of $\Upsilon(4S)$-data and about 120,000 reconstructed signal decays [10]. The CKM element $|V_{cb}|$ is determined with only 2.9% experimental uncertainty and we obtain $|V_{cb}| = (38.0 \pm 0.2(stat) \pm 1.1(syst) \pm 0.7(LQCD)) \times 10^{-3}$, where we assumed the form-factor normalization from Ref. [17]. Another recent Belle analysis measures the decay $B^0 \to \pi^-\ell^+\nu$ in bins of $q^2 = (p_\ell + p_\nu)^2$ [11]. Fitting these partial branching fractions together with a lattice QCD prediction of the differential shape [18] allows to reduce the form factor uncertainty. We obtain $|V_{ub}|$ with less than 10% experimental uncertainty, $|V_{ub}| = (3.43 \pm 0.33) \times 10^{-3}$.

6 Summary and outlook

The pattern of CP-violating effects in the Standard Model (SM) is strongly constrained by the Cabibbo-Kobayashi-Maskawa (CKM) mechanism. The B-factory experiments Belle and BaBar have confirmed that this pattern actually occurs by measuring the angles and sides of the unitarity triangle. The next generation of flavor physics experiments, at hadron and lepton colliders, will further improve the B-factory measurements and might see deviations from the CKM paradigm. This would be a clear hint for new CP-violating phases and thus physics beyond the SM.

References

1. Christenson, J.H., Cronin, J.W., Fitch, V.L., Turlay, R.: Phys. Rev. Lett. **13**, 138 (1964)
2. Cabibbo, N.: Phys. Rev. Lett. **10**, 531 (1963)
3. Kobayashi, M., Maskawa, T.: Prog. Theor. Phys. **49**, 652 (1973)
4. Abashian, A., et al.: Nucl. Instrum. Methods A **479**, 117 (2002)
5. Aubert, B., et al. [BABAR Collaboration]: Nucl. Instrum. Methods A **479**, 1. arXiv:hep-ex/0105044 (2002)
6. Charles, J., et al. [CKMfitter Group]: Eur. Phys. J. C **41**, 1. arXiv:hep-ph/0406184 (2005)
7. Ciuchini, M., et al.: JHEP **0107**, 013. arXiv:hep-ph/0012308 (2001)
8. Adachi, I., Adamczyk, K., Aihara, H. [Belle Collaboration]: arXiv:1106.4046 [hep-ex] (2011)
9. Horii, Y., et al. [Belle Collaboration]: Phys. Rev. Lett. **106**, 231803. arXiv:1103.5951 [hep-ex] (2011)
10. Dungel, W., et al. [Belle Collaboration]: Phys. Rev. D **82**, 112007. arXiv:1010.5620 [hep-ex] (2010)
11. Ha, H., et al. [BELLE Collaboration]: Phys. Rev. D **83**, 071101. arXiv:1012.0090 [hep-ex] (2011)
12. Gronau, M.: Phys. Lett. B **557**, 198. arXiv:hep-ph/0211282 (2003)
13. Atwood, D., Dunietz, I., Soni, A.: Phys. Rev. D **63**, 036005. arXiv:hep-ph/0008090 (2001)
14. Giri, A., Grossman, Y., Soffer, A., Zupan, J.: Phys. Rev. D **68**, 054018. arXiv:hep-ph/0303187 (2003)
15. Libby, J., et al. [CLEO Collaboration]: Phys. Rev. D **82**, 112006. arXiv:1010.2817 [hep-ex] (2010)
16. Poluektov, A., et al. [The Belle Collaboration]: Phys. Rev. D **81**, 112002. arXiv:1003.3360 [hep-ex] (2010)
17. Bailey, J.A., et al. [Fermilab Lattice and MILC Collaborations]: PoS **LATTICE2010**, 311. arXiv:1011.2166 [hep-lat] (2010)
18. Bailey, J.A., et al.: Phys. Rev. D **79**, 054507. arXiv:0811.3640 [hep-lat] (2009)

Hyperfine Interact (2012) 211:33–38
DOI 10.1007/s10751-012-0568-9

Physics topics at KLOE-2

Michal Silarski on behalf of the KLOE-2 Collaboration

Published online: 28 January 2012

Abstract The goal of the KLOE–2 experiment operating at the upgraded DAΦNE e^+e^- collider is to collect an integrated luminosity of about 20 fb^{-1} over 3-4 years of running. The KLOE–2 apparatus is now equipped with an inner tracker, new scintillation calorimeters and tagging detectors for $\gamma\gamma$ physics. These will allow measurements to refine and extend the KLOE programme on kaon physics and tests of fundamental symmetries as well as quantum interferometry. Here the latest results from the KLOE data analysis are presented and the perspectives at KLOE-2 outlined.

Keywords KLOE-2 · Rare kaon decays · Kaon interferometry · CP symmetry

1 Introduction

In the 2000–2006 data-taking period the KLOE detector, operating at the DAΦNE accelerator in the Laboratori Nazionali di Frascati, acquired a total integrated luminosity of 2.5 fb^{-1} at the ϕ mass peak (corresponding to about 10^{10} ϕ decays)

The KLOE- 2 Collaboration: F. Archilli, D. Babusci, D. Badoni, I. Balwierz, G. Bencivenni, C. Bini, C. Bloise, V. Bocci, F. Bossi, P. Branchini, A. Budano, S. A. Bulychjev, L. Caldeira Balkeståhl, P. Campana, G. Capon, F. Ceradini, P. Ciambrone, E. Czerwiński, E. Dané, E. De Lucia, G. De Robertis, A. De Santis, G. De Zorzi, A. Di Domenico, C. Di Donato, D. Domenici, O. Erriquez, G. Fanizzi, G. Felici, S. Fiore, P. Franzini, P. Gauzzi, G. Giardina, S. Giovannella, F. Gonnella E. Graziani, F. Happacher, B. Höistad, L. Iafolla, E. Iarocci, M. Jacewicz, T. Johansson, A. Kowalewska, V. Kulikov, A. Kupsc, J. Lee-Franzini, F. Loddo, G. Mandaglio, M. Martemianov, M. Martini, M. Mascolo, M. Matsyuk, R. Messi, S. Miscetti, G. Morello, D. Moricciani, P. Moskal, F. Nguyen, A. Passeri, V. Patera, I. Prado Longhi, A. Ranieri, C. F. Redmer, P. Santangelo, I. Sarra, I. Sarra, B. Sciascia, A. Sciubba, M. Silarski, C. Taccini, L. Tortora, G. Venanzoni, R. Versaci, W. Wiślicki, M. Wolke, J. Zdebik

M. Silarski (✉)
Institute of Physics, Jagiellonian University, 30-059 Cracow, Poland
e-mail: Michal.Silarski@lnf.infn.it

and 250 pb^{-1} at $\sqrt{s} = 1$ GeV (giving about 10^8 η mesons). This allowed precise studies to be carried out on charged and neutral kaon physics, low energy QCD, as well as tests of CP and CPT conservation [1]. In 2008 the Accelerator Division of the Frascati Laboratory tested a new interaction scheme to allow the beam size to be reduced and the luminosity increased. The test was successful and presently DAΦNE can reach a peak luminosity of 5×10^{32} cm^{-2}s^{-1}, which is a factor of three better than previously obtained [2]. Following these achievements, the data-taking campaign using the upgraded KLOE detector on the improved machine will start in Autumn 2011. The goal is to collect an integrated luminosity of about 20 fb^{-1} over 3–4 years of running.

2 The KLOE experiment at DAΦNE

DAΦNE is a e^+e^- collider operating near the ϕ meson mass peak, at a centre-of-mass energy $\sqrt{s} = 1019.45$ MeV [1]. The heart of DAΦNE consists of two rings in which 120 bunches of electrons and positrons are stored. Electrons are accelerated in the Linac, stored and cooled in the accumulator, and then transferred as a single bunch to the ring. Positrons are first created at an intermediate station in the linac using 250 MeV electrons and then follow the same procedure as for electrons. Electrons and positrons collide with small transverse momenta and produce ϕ mesons almost at rest ($\beta_\phi \approx 0.015$). The ϕ decay mainly into K^+K^- (49%), K_SK_L (34%), $\rho\pi$ (15%) and $\eta\gamma$ (1.3%). The decay products are registered using the KLOE detection setup. This consists of a ≈ 3.3 m long cylindrical drift chamber, with a diameter of ≈ 4 m, which is surrounded by the electromagnetic calorimeter. The detectors are placed in the axial magnetic field of a superconducting solenoid of strength $B = 0.52$ T. The KLOE drift chamber is constructed out of a carbon fibre composite with low-Z and low density and uses a gas mixture of helium (90%) and isobutane (10%). It provides tracking in three dimensions, with a resolution in the transverse plane of about 200 μm, a resolution in the z-coordinate measurement of about 2 mm and 1 mm in the decay vertex position. The momentum of a particle is determined from the curvature of its trajectory in the magnetic field with a fractional accuracy $\sigma_p/p = 0.4\%$ for polar angles greater than 45° [1]. The KLOE electromagnetic calorimeter consists of a barrel built out of 24 trapezoidally shaped modules and side detectors (so called endcaps) read out from both sides by a set of photomultipliers [1]. Each of the modules is constructed out of 1 mm scintillating fibres embedded in 0.5 mm lead foils to speed up the showering processes. This detector allows measurements of particle energies and flight times with accuracies of $\sigma_E = 5.7\% E/\sqrt{E[\mathrm{GeV}]}$ and $\sigma(t) = 57$ ps$/\sqrt{E[\mathrm{GeV}]} \oplus 140$ ps, respectively. Analysis of the signal amplitude distributions allows one to determine the place where the particle hit the calorimeter module with accuracy of about 1 cm in the plane transverse to the fibre direction. The longitudinal coordinate precision is energy dependent: $\sigma_z = 1.2$ cm$/\sqrt{E[\mathrm{GeV}]}$. Since the ϕ mesons are produced almost at rest, kaons arising from the decay move at low speed with their relative angle being close to 180°. As a consequence, the decay products are registered in well separated areas of the detector, which allows identification of K_L mesons using reconstructed decays of K_S (so called K_L tagging) and vice versa. This is a special feature of the DAΦNE accelerator which, together with the KLOE detector, is a unique laboratory for kaon physics [3].

3 From KLOE to KLOE–2

In recent years, a new scheme for the machine based, on Crab-waist optics and a large Piwinsky angle [2], has been proposed to increase the DAΦNE luminosity. This has been tested successfully and it has motivated the start of a new KLOE run with an improved detector setup, named KLOE-2. This aims to complete the KLOE physics programme and perform a new set of interesting measurements [3]. For the forthcoming run we have improved the performance of KLOE by adding new sub-detector systems: the tagger system for $\gamma\gamma$ physics studies, the Inner Tracker based on the Cylindrical GEM technology, a tile calorimeter surrounding the inner quadrupoles (QCALT), and a calorimeter between the interaction point (IP) and the first inner quadrupole (CCALT). The tagging system is made up of two different detectors which are already installed and ready for data taking. The Low Energy Tagger (LET) is a small calorimeter placed inside KLOE near the IP, consisting of LYSO crystals read out by silicon photomultipliers. This sub-detector will serve to measure electrons and positrons from $\gamma\gamma$ interactions within a wide energy range centred around 200 MeV with an accuracy $\sigma_E \sim 10\%$. The second tagger, which is called the High Energy Tagger (HET), provides a measurement of the displacement of the scattered leptons with respect to the main orbit. This position detector consists of 30 small BC418 scintillators $3 \times 3 \times 5$ mm^3, which provide a spatial resolution of 2 mm (corresponding to a momentum resolution of ~ 1 MeV/c). The output light is collected by light guides with SiPM sensors. The HET allows measurements of particle energies with an accuracy of $\sigma_E \sim 2.5$ MeV and time with a resolution of $\sigma_t \sim 200$ ps. To improve the acceptance for low momentum tracks, and the vertex reconstruction near the interaction point, we are building the inner tracking chamber. This employs a novel technology with cylindrical GEM (Gas Electron Multiplier) detectors. It will be composed of four concentric layers that will provide 2-D points on a cylinder of known radius. Each layer is a triple-GEM chamber with cathode and anode made of thin polyamide foils. We will also install two additional calorimeters, named QCALT and CCALT. QCALT will be a 1 m long dodecagonal structure covering the region of the new quadrupoles. It is composed of a sampling of five layers of 5 mm thick scintillator plates alternated with 3.5 mm thick tungsten plates, for a total depth of 4.75 cm. The crystal calorimeter CCALT will cover the low polar angle region to increase the acceptance for very forward photons down to 8°. The basic layout consists of two small barrels of LYSO crystals that are read out with APD photosensors. A timing resolution between 300 and 500 ps is expected for 20 MeV photons.

4 Recent KLOE results and the ongoing analysis in kaon physics

In addition to the preparation for the first KLOE-2 data-taking period (so-called Step0) and the activities for the design and construction of the new sub-detectors, there are still several ongoing physics analyses of the KLOE data. In kaon physics, apart from the recently published results of a precise K_S lifetime measurement [4], there are several studies of neutral kaon interferometry and rare kaon decays, for example $K_S K_L \to \pi^+\pi^-\pi^+\pi^-$ interferometry or the $K_S \to \pi^0\pi^0\pi^0$ branching ratio measurement. These will be described briefly here. The decay of K_S meson into three pions has not yet been observed, and the best limit on the branching ratio $BR(K_s \to 3\pi^0) < 1.2 \times 10^{-7}$ [5] is about two orders of magnitude larger than predictions based on the Standard Model. Moreover, this process violates CP symmetry and, assuming CPT invariance, allows one to investigate direct CP violation. At KLOE this decay is reconstructed by searching for events with a K_L interaction in the calorimeter (so called K_L–crash), six photon clusters and no tracks from the interaction point. The background originates mainly from $K_S \to 2\pi^0$ events with two spurious clusters from splittings or accidental activity or is due to false K_L–crash tags from $\phi \to K_S K_L \to \pi^+\pi^-\pi^0\pi^0\pi^0$ events. In the latter case, charged pions from K_S decays interact in the low-beta insertion quadrupoles, ultimately simulating the K_L–crash signal, while K_L decays close to the interaction point produce six photons [5]. To reduce the background, we first perform a kinematic fit with 11 constraints: energy and momentum conservation, the kaon mass and the velocities of the six photons. Cutting at a reasonable χ^2 value reduces considerably the background from false K_L–crash events with little signal loss. In order to gain a good rejection of background from events with split or accidental clusters, we look at the correlation between the following two χ^2–like discriminating variables, $\chi^2_{3\pi}$ and $\chi^2_{2\pi}$. $\chi^2_{3\pi}$ is the quadratic sum of the residuals between the nominal π^0 mass and the invariant masses of three photon pairs formed from the six clusters present. $\chi^2_{2\pi}$ is based instead on energy and momentum conservation in the $\phi \to K_S K_L$, $K_S \to \pi^0\pi^0$ decay hypothesis, as well as on the invariant masses of two photon pairs. Both variables are evaluated with the most favorable cluster pairing in each case [5]. In addition, in order to improve the quality of the photon selection using $\chi^2_{2\pi}$, we cut on the variable $\Delta E = (m_\Phi c^2/2 - \sum E_{\gamma_i})/\sigma_E$, where γ_i is the i–*th* photon from the four chosen in the $\chi^2_{2\pi}$ estimator and σ_E is the appropriate resolution. For $K_S \to 2\pi^0$ decays plus two background clusters, we expect $\Delta E \sim 0$ while, for $K_S \to 3\pi^0$, $\Delta E \sim m_{\pi^0} c^2/\sigma_E$. At the end of the analysis we cut also on the minimal distance between the photon clusters to refine the rejection of events with split clusters (see left panel of Fig. 1). Applying the preliminary selection cuts, we find zero candidates in 1.7 fb^{-1} of real data with zero events expected from Monte Carlo, corresponding to an effective statistics of two times that of the data. This results in a new preliminary upper limit on the branching ratio $BR(K_S \to 3\pi^0) < 2.9 \times 10^{-8}$, which suggests that a first observation of the decay might be feasible at KLOE-2. A unique feature of the Φ-factory is the production of neutral kaon pairs in a pure quantum state so that we can study quantum interference effects and tag pure monochromatic K_S and K_L beams. The decay rate of the system, for example, to the $\pi^+\pi^-\pi^+\pi^-$ final state is proportional to:

$$I(\pi^+\pi^-, \pi^+\pi^-, \Delta t) \propto e^{-\Gamma_L \Delta t} + e^{-\Gamma_S \Delta t} - 2e^{-\frac{\Gamma_L + \Gamma_S}{2}\Delta t} \cos(\Delta m \Delta t) , \qquad (1)$$

where Δt is the time difference between the decays of the two kaons. If the neutral kaon system evolves in time as a pure quantum state, kaons cannot decay at the same time due to the destructive interference. But there are several potential mechanisms leading to decoherence of the state which may result from fundamental modifications of Quantum Mechanics or CPT violation induced, e.g.,by quantum gravity [3]. Thus, by measuring the Δt distribution, we can test the foundations of Quantum Mechanics as well as different phenomenological models of Quantum Gravity. At KLOE, the selection of a $\phi \to K_S K_L \to \pi^+\pi^-\pi^+\pi^-$ signal requires two vertices, each with two opposite-curvature tracks inside the drift chamber, with invariant mass and

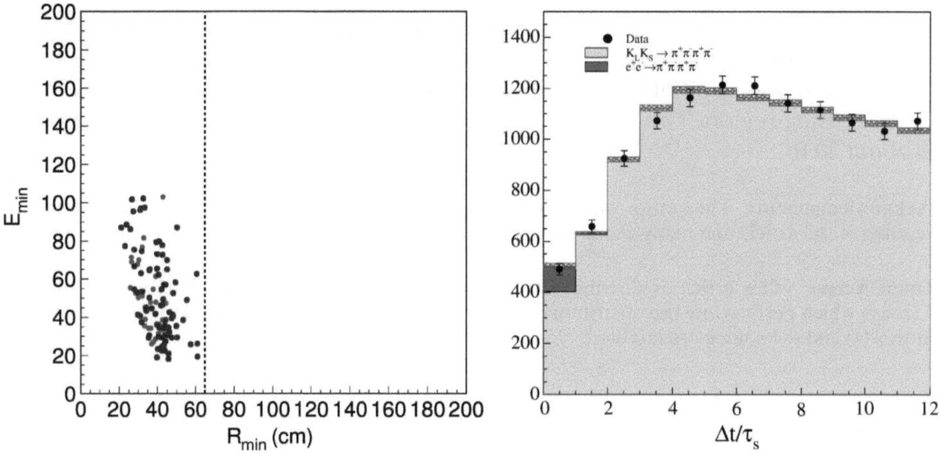

Fig. 1 *Left panel* The distribution of the minimal energy of the cluster versus minimal distance ($R_{\min}$) between clusters in the event for data (*red*) and MC (*blue*). The *dashed line* corresponds to the $R_{\min}$ cut used. *Right panel* Number of events as a function of the difference in the decay time of the two $\pi^+\pi^-$ vertices in K_S lifetime units. The measured $I(\pi^+\pi^-, \pi^+\pi^-, \Delta t)$ distribution is fitted: black points are data and the fit result is shown by the histogram

total momentum compatible with two neutral kaon decays. The resolution in Δt, the absolute value of the time difference of two $\pi^+\pi^-$ decays, benefits from the precise momentum measurements and from the completeness of the kinematics of the events [6]. The experimental Δt distribution is fitted with (1), modified by parameters expressing decoherence in the different models described in [3]. The fit is performed taking into account the resolution and detection efficiency, the background from coherent and incoherent K_S regeneration on the beam pipe wall, and the small contamination from the non-resonant $e^+e^- \to \pi^+\pi^-\pi^+\pi^-$ channel. The resulting distributions are shown together with the fit results in Fig. 1. On the basis of the 1.7 fb^{-1} of data, useful estimates could be made of several decoherence and CPT violating parameters. The details of the analysis and the results obtained can be found in [7]. It is worth mentioning that we have almost finished a refined analysis of the KLOE data with several improvements in the methodology. Moreover, KLOE-2 with increased statistics ($\times 4$ in Step0) and improved resolution in the vertex reconstruction ($\times 3$ with the Inner Tracker) will be able to achieve the best experimental sensitivity in some observables, hopefully reaching the level of the Planck scale [3].

5 Summary

KLOE is a high precision experiment which allows detailed studies of both kaons and light scalar mesons, as well as tests of the conservation of CP, CPT and low energy QCD. The success of the DAΦNE upgrade motivated a new experiment, KLOE-2, which aims at completing and extending the KLOE physics programme. We have started design and construction of new sub-detectors, which will improve the detection performance. The tagging system for $\gamma\gamma$ physics is installed and ready for the first phase of the experiment in which we expect to accumulate about 5 fb^{-1}. The next data-taking campaign during 2013–15 will be conducted with the Inner Tracker and improved photon acceptance brought about by the calorimeters in the final focusing region. The total integrated luminosity expected in this second phase is about 20 fb^{-1}.

Acknowledgement The author would like to express his gratitude to prof. Colin Wilkin for proof reading of the article and many useful comments.

References

1. Bossi, F., et al.: Precision Kaon and Hadron physics with KLOE. Riv. Nuovo Cim. **31**, 531–623 (2008)
2. Zobov, M., et al.: Test of "Crab-Waist" collisions at the DAΦNE ϕ–Factory. Phys. Rev. Lett. **104**, 174801–174806 (2010)
3. Amelino-Camelia, G., et al.: Physics with the KLOE-2 experiment at the upgraded DAΦNE. Eur. Phys. J. C **68**, 619–681 (2010)
4. Ambrosino, F., et al.: Precision measurement of the K_S meson lifetime with the KLOE detector. Eur. Phys. J. C **71**, 1604–1610 (2011)
5. Ambrosino, F., et al.: A direct search for the CP-violating decay $K_S \to 3\pi^0$ with the KLOE detector at DAΦNE. Phys. Lett. B **619**, 61–70 (2005)
6. Ambrosino, F., et al.: First observation of quantum interference in the process $\phi \to K_S K_L \to \pi^+\pi^-\pi^+\pi^-$: a test of quantum mechanics and CPT symmetry. Phys. Lett. B **642**, 315–321 (2006)
7. Di Domenico, A., et al.: Search for CPT violation and decoherence effects in the neutral kaon system. J. Phys. Conf. Ser. **171**, 012008–012023 (2009)

Hyperfine Interact (2012) 211:39–43
DOI 10.1007/s10751-012-0603-x

Trapped radioactive isotopes for fundamental symmetry investigations

The TRIμP Facility

Lorenz Willmann · Klaus Jungmann · Cornelis J. G. Onderwater · Rob G. E. Timmermans · Hans W. Wilschut

Published online: 3 April 2012

Abstract Discrete symmetries tested in high precision atomic physics experiments provide guidance to model building beyond the Standard Model (SM). Here experimental opportunities arise for searches for permanent electric dipole moments (EDMs) and measurements of atomic parity violation (APV). Heavy atoms are favorable for such experiments since symmetry violating effects in atoms increase faster than the third power of the nuclear charge Z. Of special interest are isotopes of the heavy alkaline earth element radium (Z=88) since they offer large enhancement factors for EDMs and provide a new experimental road towards high precision measurements of atomic parity violation. These opportunities are exploited at the TRIμP facility at KVI, Groningen.

Keywords Fundamental symmetries · Permanent electric dipole moments · Atomic parity violation · Radioactive beam facilities

1 Introduction

The Standard Model (SM) comprises the knowledge of physics over a wide range of energies and different interactions. Its success is visible through predictions which were verified by experimental observations. All experimental observed phenomena are accommodated by the theory, but the SM does not provide explanations for them, e.g. the origin of parity violation or the matter-antimatter asymmetry in the universe [1]. Here, experimental limits on violations of discrete symmetries, i.e. parity (P), particle-antiparticle symmetry (charge conjugation, C) and time reversal (T) provide sensitive tests of the range of validity of the SM which guide theoretical model building beyond the SM.

L. Willmann (✉) · K. Jungmann · C. J. G. Onderwater · R. G. E. Timmermans · H. W. Wilschut
KVI, University of Groningen, Groningen, Netherlands
e-mail: willmann@kvi.nl

The heavy alkaline earth element radium is at the focus of this paper. This system is preferred since symmetry violating effects scale with higher powers of the nuclear charge Z due to the larger number of coherently contributing particles and relativistic effects [2, 3]. In addition, some isotope specific enhancement factors are particularly large for neutral radium isotopes. In the case of permanent electric dipole moments (EDMs) which violate P and T at the same time, these can be as large as 4–5 orders of magnitude. The nuclear [4] and unique atomic structure [5, 6] of radium lead to this enhancements compared to ^{199}Hg in which the strongest experimental limit for a nuclear EDM of $3.1 \cdot 10^{-29}$ ecm has been determined [7]. Further interest in radium isotopes arises from the possible quantitative measurement of atomic parity violation (APV) in a single trapped ion [8] and in particular for radium isotopes [9, 10]. An experiment on a single trapped Ra^+ ion at KVI aims at a 5-fold improved determination of the weak interaction in atoms compared the previous best measurement Cs [11].

At the TRIμP Facility, KVI, Groningen, The Netherlands, a wide range of short lived isotopes are available by online production [12]. The range of $^{209-214}$Ra isotopes is produced with ^{204,206}Pb beams using the AGOR cyclotron. The primary beam power of up to 200 W impinges on a rotating target with pyrolytic graphite foils of 2 mg/cm^2 thickness. After the magnetic separator the Ra isotopes are stopped and re-ionized to Ra^+ in a Thermal Ionizer (TI) [13]. They are separated from other ions in a Wien Filter and cooled in a gas-filled Radio Frequency Quadrupole (RFQ). The rate of Ra^+ after the TI is up to $5\times10^4/s$ depending on the isotope. These isotopes are exploited in our searches for permanent electric dipole moments and measurements of atomic parity violation.

2 Electric dipole moments

A wide range of different experimental searches for EDMs from elementary particles up to complex systems like molecules are currently underway [14, 15]. EDMs exhibit a robust discovery potential for physics beyond the SM since their values within the SM are well below current experimental sensitivities. However, their sizes in a number of proposed extensions of the SM attempting to explain features like parity violation or baryon number asymmetry appear in the range of ongoing experimental efforts. Electron and nuclear EDMs are strongly enhanced in radium isotopes. The nuclear octupole deformations of isotopes near the valley of stability result in an enhancement of 2–3 orders of magnitude for nuclear EDMs [4]. In addition, the unique atomic structure of Ra gives rise to enhancements of 4–5 orders of magnitude for a nuclear EDM in the metastable 7s6d $^3D_2$ state and a factor of 5000 for the electron EDM in the 7s6d $^3D_1$ state [5]. Radium EDM experiments with an anticipated level of sensitivity of 10^{-30} ecm are currently prepared at the Argonne National Laboratory (ANL), Il, USA on the ground state of ^{225}Ra [16] and at KVI where the focus is on the exploitation of the large atomic enhancement factors in the metastable D-states for electron and nuclear EDMs.

A crucial step toward the EDM searches is the efficient capture of radium in atomic traps. At KVI we have developed the laser cooling and trapping of heavy alkaline earth elements on the strong singlet-singlet $^1S_0$—$^1P_1$ transition with barium [17], the chemical homologue of radium. The efficient laser cooling scheme involves

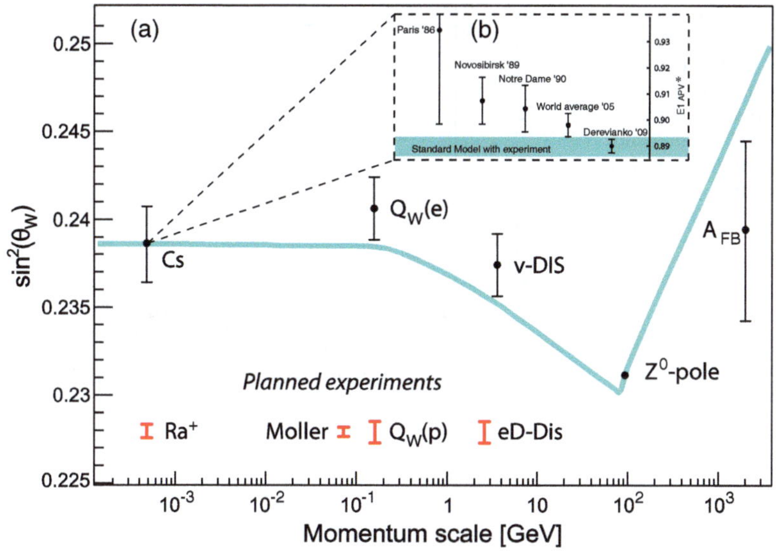

Fig. 1 (**a**) The weak mixing angle Θ_W in the Standard Model shows an energy scale dependent behavior due to radiative corrections. There are experiments planned at different energy scales to measure this dependence. Atomic parity violation provides a measurement at the lowest energy scale. Up to now a measurement in Cs [11, 20] provides the best results. APV measurements in a single trapped Ra^+ have the potential for a 5-fold improvement over the Cs results. (**b**) The inset shows the development of accuracy and precision of theoretical calculations of APV in Cs [20] over time. The anticipated precision of planned experiments is also given

seven laser at a time. The lifetime of the trapped barium atoms is about 1.5 s. The capture efficiency of barium atoms from a thermal beam into the magneto optical trap was determined to 0.5% which was limited by laser power for transitions from the metastable D-states significantly. Another approach for laser cooling and trapping of radium has been pursued by the group at ANL. However, the weak $^1S_0$—$^3P_1$ intercombination line provides a 150 times smaller optical force than the strong $^1S_0$—$^1P_1$ transition resulting in a capture efficiency of 10^{-6} [18]. The application of the laser cooling method developed at KVI with Ba promises thus a significant gain in sensitivity for Ra EDM searches due to the selection of a new sensitive system and a new type of experimental approach. The combination of the availability of the radium isotopes at online facilities and experimental techniques like laser cooling and optical dipole trapping for radium [19] opens the path to sensitive EDM searches.

3 Atomic parity violation in a single trapped radium ion

A single radium ion in an ion trap opens the path to a high precision measurement of atomic parity violation (APV). The experiment at KVI aims at a determination of the nuclear spin independent part of the electroweak interaction at low energy scales in order to extract the weak mixing (Weinberg) angle Θ_W. Such a measurement can be compared to other determinations of Θ_W at different energy scales and to the predictions of the SM (Fig. 1). The best measurement of weak interactions in atomic

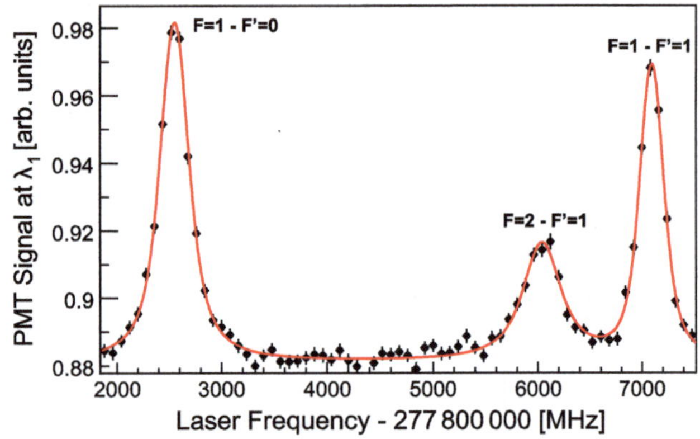

Fig. 2 Hyperfine structure of the 6d $^2D_{3/2}$—7p $^2P_{3/2}$ transition in ^{213}Ra at a wavelength of 1079 nm. The signal is obtained with a few hundred ions trapped and buffer gas cooled ions in a linear Paul trap [9]

systems has been performed with Cs [11]. The exploitation of the precision towards the weak mixing angle required an extensive treatment of the atomic theory [20] over the years. A single trapped radium ion aims at a 5-fold improvement over the Cs result. This becomes possible due to a new signature of APV due to differential light shifts [8] and the large weak charge due to the large Z of radium [10].

Spectroscopic information on singly charged radium was limited until recently. However, the exploitation of the sensitivity of a radium ion relies on high precision atomic structure calculations which require experimental input to evaluate their accuracy. We have determined the hyperfine structure in the $6d\,^2D_{3/2}$ state for ^{209,211,213}Ra, the lifetime of the $6d\,^2D_{3/2}$ state [9, 21] and isotopes shifts [22] for the $^{209-214}$Ra isotopes. The hyperfine structure of ^{213}Ra obtained with buffer gas cooled ions in a linear Paul-trap is given in Fig. 2. A comparison of the laser spectroscopy results with the atomic structure calculations permits an estimate on the current accuracy of the theoretical description. The hyperfine structure splittings, which are a sensitive probe of the atomic wave functions in the nuclear region, are in agreement with theory on the level of a few % [9]. This tests the predictions of APV matrix elements to about the same accuracy [10, 23]. Several groups are working on the improvement of the calculations [24]. The trapping of a single radium ion and its laser cooling to the motional ground state of the trap is currently underway.

Progress in the field of experimental investigations of fundamental symmetries benefits from new systems and experimental approaches. The single trapped radium ion can also be exploited as an optical clock [25] which underlines the fruitful cross-fertilization between different areas in physics. The Ra ion clock exhibits different sensitivity to temporal changes of the fine structure constant α than other available atomic clocks [25] which makes it an ideal candidate for measurements of time variation of fundamental constants.

References

1. Sakharov, A.D.: Pisma Zh. Eksp. Teor. Fiz. **5**, 32–35 (1967); translation in JETP Lett. **5**, 24–27 (1967)
2. Sandars, P.G.H.: Phys. Lett. **14**, 194 (1965)
3. Bouchiat, M.A., Bouchiat, C.C.: Phys. Lett. B **48**, 111 (1974)
4. Engel, J., Bender, M., Dobaczewski, J., Jesus, J. Hd., Olbratowski, P.: Phys. Rev. C **68**, 025501 (2003); Dobaczewski, J., Engel, J.: Phys. Rev. Lett. **94**, 232502 (2005)
5. Flambaum, V.V.: Phys. Rev. A **60**, R2611 (1999)
6. Dzuba, V.A., Flambaum, V.V., Ginges, J.S.M.: Phys. Rev. A **61**, 062509 (2000)
7. Griffith, W.C., Swallows, M.D., Loftus, T.H., Romalis, M.V., Heckel, B.R., Fortson, E.N.: Phys. Rev. Lett. **102**, 101601 (2009)
8. Fortson, N.: Phys. Rev. Lett. **70**, 2383 (1993)
9. Versolato, O.O., Giri, G.S., Wansbeek, L.W., Kruithof, W., van der Hoek, D.J., Santra, B., Shidling, P., Onderwater, C.J.G., Willmann, L., Wilschut, H.W., Timmermans, R.G.E., Jungmann, K.: Phys. Rev. A **82**, 010501(R) (2010)
10. Wansbeek, L.W., Sahoo, B.K., Timmermans, R.G.E., Jungmann, K., Das, B.P., Mukherjee, D.: Phys. Rev. A **78**, 050501(R) (2008)
11. Bennett, S.C., Wieman, C.E.: Phys. Rev. Lett. **82**, 2484 (1999)
12. Berg, G.P.A., Dermois, O.C., Dammalapati, U., et al.: Nucl. Inst. Meth. A **560**, 169 (2006)
13. Shidling, P.D., Giri, G.S., van der Hoek, D.J., et al.: Nucl. Instr. Meth. A **622**, 11 (2010)
14. Commins, E.D.: J. Phys. Soc. Jpn. **76**, 111010 (2007)
15. Onderwater, C.J.G.: this volume
16. Holt, R.J., Ahmad, I., Bailey, K., Graner, B., Greene, J.P., Korsch, W., Lu, Z.-T., Mueller, P., O'Conner, T.P., Sulai, I.A., Trimble, W.L.: Nucl. Phys. A **844**, 53 (2010)
17. De, S., Dammalapati, U., Jungmann, K., Willmann, L.: Phys. Rev. A **79**, 041402R (2009)
18. Guest, J.R., Scielzo, N.D., Ahmad, I., Bailey, K., Greene, J.P., Holt, R.J., Lu, Z.-T., O'Connor, T.P., Potterveld, D.H.: Phys. Rev. Lett. **98**, 093001 (2007)
19. Dammalapati, U., Santra, B., Willmann, L.: J. Phys. B: At. Mol. Opt. Phys. **45**, 025001 (2012)
20. Porsev, S.G., Beloy, K., Derevianko, A.: Phys. Rev. Lett. **102**, 181601 (2009)
21. Versolato, O.O., Giri, G.S., van den Berg, J.E., Boll, O., Dammalapati, U., van der Hoek, D.J., Hoekstra, S., Jungmann, K., Kruithof, W.L., Muller, S., Nuñez Portela, M., Onderwater, C.J.G., Santra, B., Timmermans, R.G.E., Wansbeek, L.W., Willmann, L., Wilschut, H.W.: Phys. Lett. A **375**, 3130 (2011)
22. Giri, G.S., Versolato, O.O., vd Berg, J.E., Dammalapati, U., vd Hoek, D.J., Jungmann, K., Kruithof, W., Nuñez Portela, M., Onderwater, C.J.G., Santra, B., Timmermans, R.G.E., Wansbeek, L.W., Willmann, L., Wilschut, H.W.: Phys. Rev. A **84**, 020503R (2011)
23. Sahoo, B.K., Das, B.P., Chaudhuri, R.K., Mukerjee, D., Timmermans, R.G.E., Jungmann, K.: Phys. Rev. A **76**, 040504 (2007)
24. Dzuba, V.A., Flambaum, V.V.: arXiv:1104.0086 [physics.atom-ph]; Pal, R., Jiang, D., Safonova, M.S., Safronova, U.I.: Phys. Rev. A **79**, 062505 (2009); Sahoo, B.K., Mandal, P., Mukkerjee, M.: Phys. Rev. A **83**, 030502(R) (2011)
25. Versolato, O.O., Wansbeek, L.W., Jungmann, K., Timmermans, R.G.E., Willmann, L., Wilschut, H.W.: Phys. Rev. A **83**, 043829 (2011)

Hyperfine Interact (2012) 211:45–48
DOI 10.1007/s10751-012-0593-8

Performance of a GEM-TPC prototype

Martin Emanuel Berger · Laura Fabbietti for the GEM-TPC collaboration

Published online: 6 April 2012

Abstract A GEM-TPC detector was build as a prototype of the inner tracker for the future $\overline{\text{P}}$ANDA experiment at FAIR and as an updgrade for the FOPI experiment at GSI. A GEM-TPC can exploit the intrinsic suppression of back drifting (Lyashenko A et al., Nucl Instrum Methods A 598:116, 2009) ions from the amplification stage of the GEM foils to overcome the problem of drift-field distortions in a ungated operation. This prototype was designed to fit into the FOPI spectrometer, where the commissioning, tests with cosmic rays and beam tests were carried out (Fabbietti, Nucl Instrum Methods A 628:204–208, 2011).

Keywords GEM · TPC · FOPI · Micro-pattern gas detectors · Particle tracking · $\overline{\text{P}}$ANDA

1 Introduction

Modern high precision experiments with high particle rates demand detectors which are capable to cope with high track densities and charge up effects. A TPC (Time Projection Chamber) with GEM (Gas Electron Multiplier) foils for the amplification instead of conventional MWPC's is such a detector. TPC's are widely used in many running experiments, for example in STAR [3] and ALICE [4] where one has high track density environments and due to the high spatial resolution of a TPC a clean identification of several thousands of tracks per event is possible. These TPC's however are operated in a gated mode to prevent back drifting secondary ions from the amplification stage to distort the drift field. A TPC without the necessity of gating would be an excellent option as an inner tracker for upcoming experiments

M. E. Berger (✉) · L. Fabbietti
Excellence Cluster Universe, Boltzmannstr. 2, 85748 Garching, Muenchen, Germany
e-mail: mberger@ph.tum.de

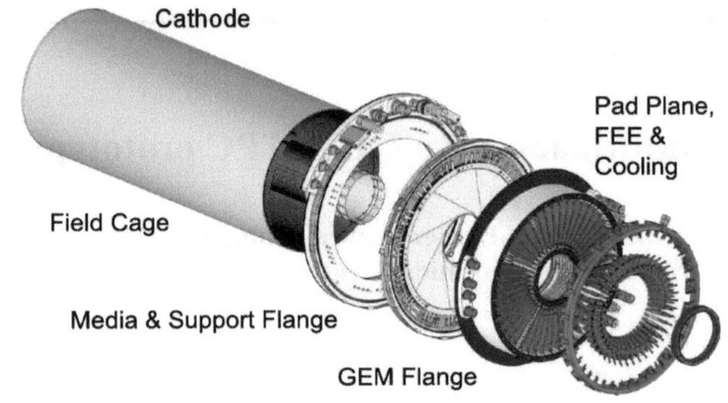

Fig. 1 Explosion view of the large GEM-TPC prototype

like $\overline{\text{P}}$ANDA which are running continuously. The amount of back drifting ions can be reduced by several orders of magnitude by using GEM foils for the amplification, exploiting their intrinsic ion back flow suppression. To investigate the performance of such a detector a prototype was built. It was designed as a fully functional detector to be used as an upgrade for the FOPI [5] experiment at GSI, Darmstadt (Germany) and for the CBELSA experiment in Bonn (Germany).

2 Mechanical design

The prototype shown in Fig. 1 in an exploded view is a cylindrical TPC with a drift length of 727.8 mm and the readout on one side only. It has an inner hole with a diameter of 104 mm and an outer diameter of 308 mm. The layout was held modular to be able to easily exchange parts either to test different designs or for maintenance. In Fig. 1 one can see the fieldcage on the left which is glued to the so called media-flange. This flange holds all necessary connectors for the detector operation e.g. gas in- and outlet, high-voltage and low-voltage supply, temperature and gas sensor connections. The next part, the GEM-flange is detachable from the media-flange and houses the triple GEM stack for the amplification. The padplane and the front-end electronics for the readout are located in the readout-flange which is also detachable. A water driven cooling was used to keep the temperature of the front-end cards stable.

2.1 Cooling and temperature stability

To avoid a overheating and a possible influence of the drift velocity it is necessary to dispose the heat of the front-end electronics. Thus an active water driven cooling was employed. To monitor the temperature and to study the influences, sensors were installed on the padplane and on the fieldcage. On the padplane temperature fluctuations of 0.5° in time and 1° between different sensors were measured. On the fieldcage temperature differences of up to 3° were recorded as visible in Fig. 2. From GARFIELD calculations one expects a linear dependence and a variation of the drift velocity of 0.024 cm/μs in this temperature regions. These measurements were done with the prototype installed inside FOPI (see Section 3.1) and the main part of the

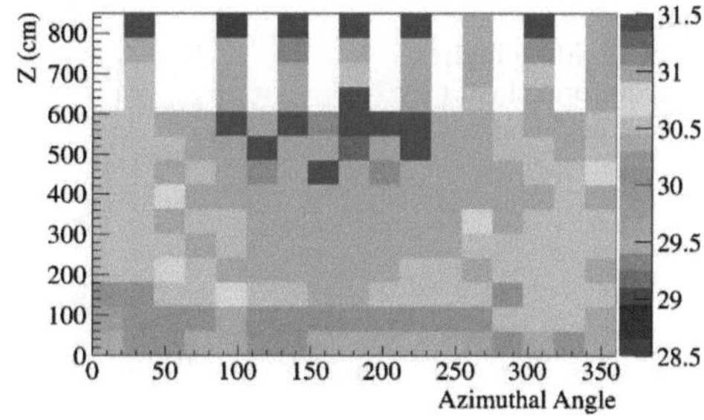

Fig. 2 Temperature distribution on the fieldcage along the z-direction as a function of the azimuthal angle. A azimuthal angle of 0° corresponds to the *lower side* ot the TPC. The bins above 600 mm represent sensors on the *backside* of the TPC (Cathode)

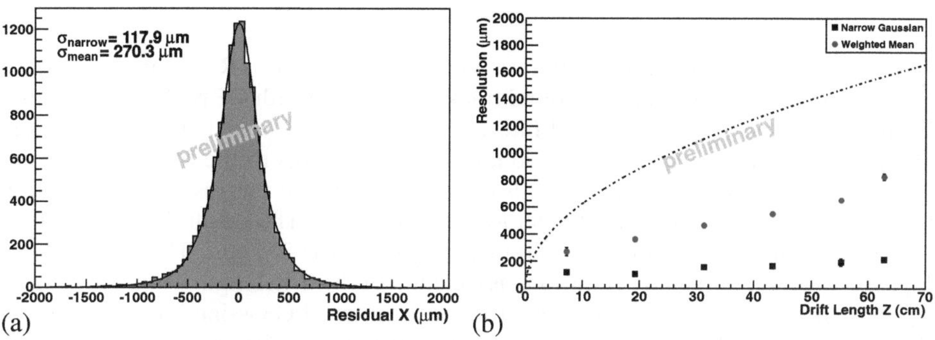

Fig. 3 The residual distribution of the first slice along the drift direction shown on the *left side* (**a**). The *right figure* (**b**) shows the obtained resolution for slices along the drift direction. The *dash-dotted line* shows the single-electron diffusion

temperature differences is caused by external heat sources which were known from a measurement done beforehand.

3 Commissioning with cosmic tracks

3.1 Experimental setup

For the commissioning the prototype was installed inside the FOPI spectrometer which was set up for triggering on cosmic particles. FOPI is a fixed target heavy ion spectrometer located at GSI in Darmstadt (Germany). It consists of an inner drift chamber which is surrounded by a scintillator and an RPC time-of-flight barrel. In forward direction is another drift chamber and a scintillator time-of-flight wall. The whole setup is surrounded by a superconducting 0.6 T solenoid magnet.

3.2 Results from cosmics tracks

Data with cosmic tracks were taken for several different detector parameters:

- Drift gases: Ar/CO_2 and Ne/CO_2, both in a 90/10 mixture
- Drift fields: 150, 200, 300 and 360 V/cm
- Different gains for the GEM stack for both gases

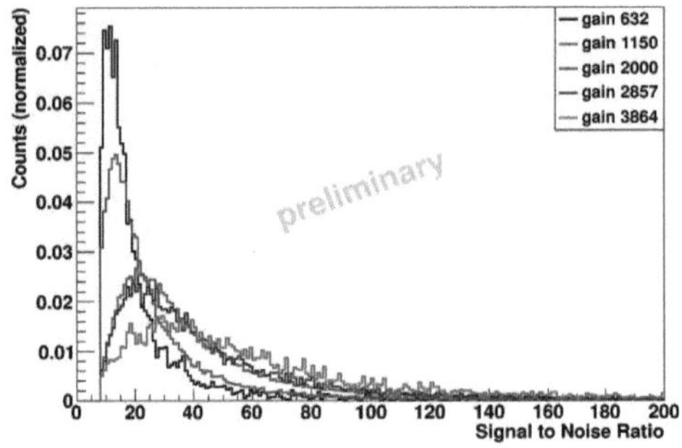

Fig. 4 Signal to noise ratio calculated by dividing highest amplitude from one cluster by corresponding pedestal

The track finding and noise suppression was done with a Riemann [6] pattern recognition after clustering. The found tracks were fitted with GENFIT [7], a generic fitting tool based on a Kalman [8] filter. The drift volume was divided into 6 slices along the drift direction and the residuals of the hit points to the fitted tracks were calculated for each slice. Figure 3a shows on the left side the residual distribution for the first slice which means the shortest drift length. The resolution one gets by fitting the distribution with three Gaussian functions is 274 μm. The right side of Fig. 3b shows the obtained resolutions as function of the drift length while the squares indicate the resolution by only taking the most centered Gaussian into account and the circles show the weighted mean of all three fitted Gaussian functions. Both plots are from a measurement with Ne/CO_2 a drift field of 360V/cm and a gain of 2,000 and are in good agreement with simulations.

To further reduce the amount of back drifting ions it is necessary to keep the gain as low as possible but still high enough to be well over the threshold of the readout. The signal-to-noise ratio is calculated by dividing the highest amplitude of a cluster by the pedestal value of the corresponding pad. One can see in Fig. 4 that the signal-to-noise ratio for a gain of 2,000 (green curve) is high enough such that the distribution is not affected by it.

References

1. Lyashenko, A., et al.: Nucl. Instrum. Methods A **598**, 116 (2009)
2. Fabbietti, L.: Nucl. Instrum. Methods A **628** 204–208 (2011)
3. Ackermann, K.H., et al.: Nucl. Instrum. Methods A **499**, 713 (2003)
4. Alme, J., et al.: Nucl. Instrum. Methods A **622**, 316 (2010)
5. Hildenbrand, K.: GSI Nachr. **91-02**, 6 (1992)
6. Strandlie, A., et al.: Comput. Phys. Commun. **131**, 95 (2000)
7. Höppner, C., Neubert, S., Ketzer B., Paul, S.: Nucl. Instrum. Methods A **620**, 518 (2010)
8. Frühwirth, R.: Nucl. Instrum. Methods A **262**, 444 (1987)

Hyperfine Interact (2012) 211:49–52
DOI 10.1007/s10751-011-0556-5

SiPM-based veto detector for the pion beam at FOPI

Gamal Ahmed · Pual Bühler · Olaf Hartmann · Johann Marton · Ken Suzuki · Johann Zmeskal

Published online: 10 January 2012

Abstract Recently the FOPI collaboration carried out an experiment to study the in medium properties of the K^+K^- system by using the pion beam interactions at 1.7 GeV/c. The experiment with a pion beam poses specific requirements to the detectors and therefore the original FOPI setup needed modifications. The new hardware developments for this experiment include the replacement of the veto detector with another more compact design. Within this report we describe the design and results of a test measurement of the new FOPI veto detector system with the pion beam.

Keywords SiPM · Veto detector · FOPI · Pion beam

1 Introduction

The FOPI detector is installed at the GSI, Darmstadt, Germany [1]. It allows identifying (event by event) light charged particles and intermediate mass fragments. Neutral hadrons (like K_s^0, Λ) can be reconstructed from their decay into charged particles. Since its assembly by an international collaboration in 1990, the experimental setup has been changed and extended several times [2]. The FOPI collaboration carried out a measurement to study the in medium properties of the K^+K^- system by using the $\pi^-\,(1.7\ \mathrm{GeV}/c) + A \rightarrow K^+K^- + X$ reaction [3]. The original FOPI setup was mainly

G. Ahmed (✉) · P. Bühler · O. Hartmann · J. Marton · K. Suzuki · J. Zmeskal
Stefan Meyer Institute for Subatomic Physics of the Austrian Academy of Sciences, Vienna, Austria
e-mail: gsmoawad@hotmail.com

G. Ahmed
Faculty of Science, Physics Department, Al-Azhar University, Cairo, Egypt

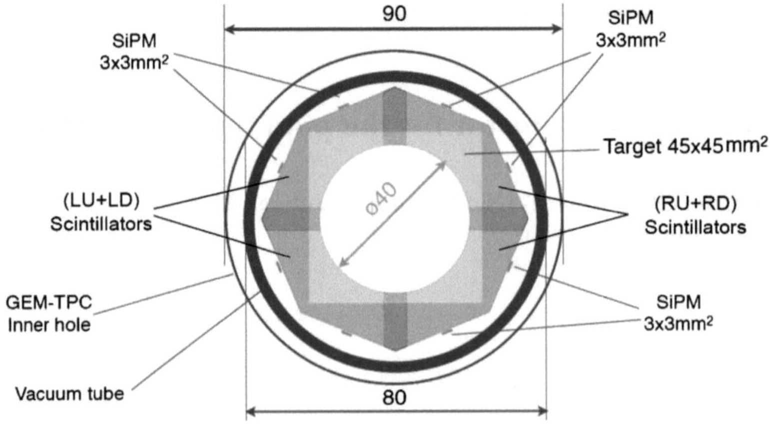

Fig. 1 The veto detector scintillator parts, where LU, LD, RD and RU refer to the positions of the veto detector scintillator parts with respect to the pion beam direction. The *dark purple* areas show overlapping of scintillators parts. The *red rectangles* define the SiPMs positions and the *light violet rectangle* shows the target area and position. The veto detector system (Scintillator + SiPMs) was housed inside an aluminum vacuum tube. The drawing shows the setup of the veto detector system inside the inner hole of the GEM-TPC

designed for heavy ion collision experiments. For experiments with a pion beam, the original FOPI setup needs modifications. In the present case a Time Projection Chambers with Gas Electron Multipliers (GEM-TPC) has been used. The GEM-TPC sits in the inner hole of the Central Drift Chamber (CDC). Around the beam axis as consequence there is no sufficient space available for the previously used photomultiplier (PMT) based veto detector and therefore the veto detector has to be exchanged by another more compact design (see Fig. 1). The main function of the veto detector is to discriminate when the pion beam hits the target. When charged particles penetrate off beam-axis it gives a veto to the data acquisition. Furthermore the veto detector is located close to the target, where a strong magnetic field is produced by the solenoid (0.6 Tesla) [4].

2 Detector design

The new veto detector prototype (Fig. 1) consists of 4 scintillator parts, designed to form together a octagon disk-like structure with an external diameter of ~8 cm. Internally it confines the cross-section of the fiducial target volume to a circle with a diameter of 4 cm. For the readout SiPMs (MPPCs, S10931–100P) have been used [5]. SiPMs have numerous advantages as compared to other photodetectors. The low operation voltage, fast timing, compactness, and moreover the insensitivity to magnetic fields make them excellent candidates for different applications [6–8] and made MPPCs the preferred choice to fit the new veto detector design requirements. Measurements of dark current, dark count rate and cross-talk and time resolution of this device have been presented in Ref. [9, 10].

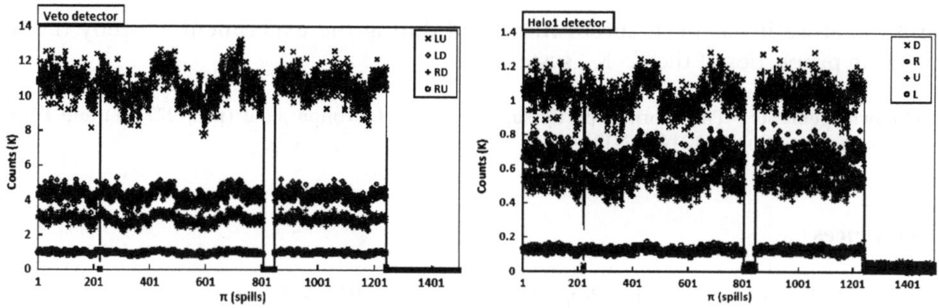

Fig. 2 Counts as a function of the pion spills for the veto detector (*Left*) and the Halol detector (*Right*)

3 Beam test and results

The FOPI detector comprises several sub-detectors. The Halol detector, sitting ~2.15 m upstream from the solenoid, is composed out of four scintillator bars read out by PMTs and arranged in a cross shape around the beam line. Thus a rectangular area of adjustable size is defined. The purpose of the Halol detector is to monitor the alignment and the focusing of the beam. The Start counter is placed ~2 m upstream the solenoid in the beam line. This detector is used to count the incoming pions and to define the reference time (start time) for all other detectors.

For testing the new veto detector, the pion beam with a momentum of 1.7 GeV/c has been used and the average beam intensity was $\sim 8 \times 10^3$ pions/spill of 2 s duration.

Figure 2 shows the test result for the response and a comparison between Halol and the veto detector. LU, LD, RD and RU refer to the positions of the veto detector scintillator parts in respect to the pion beam as described in Fig. 1. MPPCs photosensors were operated in a high temperature environment without any temperature stabilization and even at such operating conditions the achieved efficiency was >97%. The veto detector efficiency is estimated as the probability that one or both photosensors attached to one scintillator to give a signal at the same time. Efficiency has been calculated for each scintillator part, by using the following equation:

$$V_{eff} = P_{OR} = \left[1 - \left[1 - \left(\frac{2R}{1+R}\right)\right]^2\right], \quad R = \frac{P_{AND}}{P_{OR}}, \tag{1}$$

where P_{AND} is the number of cases in which both photosensor (A, B) attached to one scintillator gave a signal at the same time. P_{OR} is the number of cases in which only one photosensor gave a signal.

4 Conclusion

A new veto detector for the FOPI pion beam experiments has been built. The experimental test of the veto detector with the pion beam showed that, the veto detector has succeeded to monitor the pion beam with very good efficiency (~97%), and therefore it has been successfully used during the experiment to study the in medium properties of the K^+K^- system.

Acknowledgements This work is partly supported by Hadronphysics2 (proj 227431) and FWF Austrian Science Fund (P 21457-N16).

References

1. FOPI Collaboration: Technical Proposal GSI Report, 88–03 (1988)
2. Münzer, R., et al.: Nucl. Instr. and Meth. A **617**, 300 (2010)
3. Study of Pion-induced In-medium Production and Propagation of Strangeness., FOPI Proposal, March 2007.
4. Bühler, P., et al.: GSI scientific report, 2009.
5. http://sales.hamamatsu.com/assets/pdf/parts_S/s10362–33series_kapd1023e05.pdf
6. Suzuki, K., et al.: Nucl. Instr. and Meth. A **610**, 75 (2009)
7. Ahmed, G.S.M., et al.: Nucl. Instr. and Meth. A **628**, 393 (2011)
8. Ahmed, G., et al.: Nucl. Instr. and Meth. A **639**, 107 (2011)
9. Ahmed, G.S.M., et al.: J. Instrum. **4**, P09004 (2009)
10. Ahmed, G., et al.: Nucl. Instr. and Meth. A **652**, 528 (2011)

Hyperfine Interact (2012) 211:53–55
DOI 10.1007/s10751-011-0555-6

Energy calibration of a GEM-TPC prototype with ^{83m}Kr

Roman Schmitz for the GEM-TPC Collaboration and the CBELSA/TAPS Collaboration

Published online: 10 January 2012

Abstract A GEM-TPC has been built as a 10% scale prototype for the P̄ANDA experiment and as a full-scale prototype for the CBELSA/TAPS experiment (Fabbietti L, et al: Nucl Instrum Methods A 628:204, 2011). The intrinsic suppression of ion backflow into the drift volume makes it suitable for high rate/background environments. The GEM-TPC has been used as an inner tracking update for the FOPI experiment at GSI where cosmic and beam tests were performed. For calibration purposes a ^{83}Rb source has been produced at the Bonn HISKP Cyclotron. Using the isomeric ^{83m}Kr evaporated by the source, a relative channel-wise gain calibration as well as gain estimation for different high voltage settings could be performed.

Keywords GEM · TPC · MPGD · FOPI · Tracking · Calibration · Krypton

1 Introduction

Gain calibration of a GEM-TPC is an important factor to improve dE/dx and spatial resolution. Variations in the gain factors between channels can have various reasons besides electronic gain variations and require to be corrected. Systematic studies on the influence of the geometrical arrangement of GEM foils within a GEM stack on the effective gain of double and triple GEM detectors reveal large variations. Variations in effective gain of up to 20% due to stress on GEM frames and bending of the foils could be observed due to a change in effective transfer and induction fields between the GEMs [2]. A favourable method to perform such a calibration is by introducing radioactive ^{83m}Kr into the drift volume. A full coverage of the detector volume and the readout structure is given and channel-to-channel variations can be corrected for easily, but it is not possible to disentangle the contributions from

R. Schmitz (✉)
Helmholtz-Institut für Strahlen- und Kernphysik der Universität Bonn,
Nussallee 14–16, 53115 Bonn, Germany
e-mail: schmitz@hiskp.uni-bonn.de

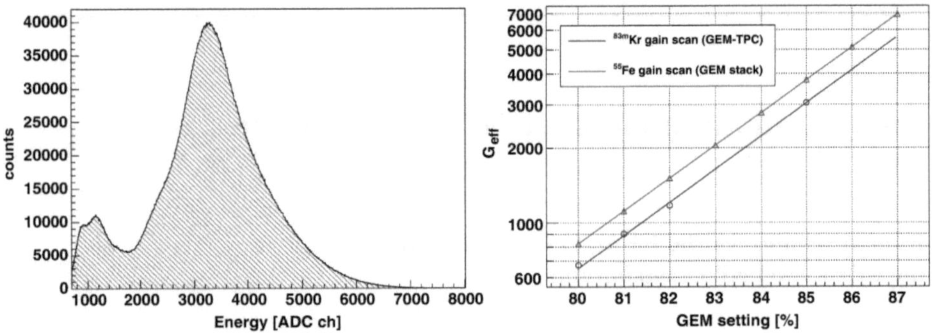

Fig. 1 Observed cluster spectrum summed over all pads and cluster sizes (*left*) and resulting gain curve for different GEM high voltage settings (*right*) in Ar/CO_2 90/10

different effects causing them. This technique has previously been used successfully in various large TPCs (e.g. [3–5]).

2 Source production and integration

A ^{83}Rb source with an activity of 2.5 MBq has been produced at the HISKP Cyclotron [6]. The radioactive source was mounted in a bypass of the TPC gas system. It decays with a half-life of 86.2 d into an isomeric 41.6 keV state of ^{83}Kr, which decays into the stable ground state with a half-life of 1.83 h via a short-lived excited state at 9.4 keV. The decay spectrum has four main peaks between 9.4 and 41.6 keV, originating from conversion electrons, that can be used for gain equalisation and calibration. For more details see [7].

3 Calibration results

A Krypton calibration was performed for both Ar/CO_2 and Ne/CO_2 with 90/10 mixture for different GEM gain settings in Ar/CO_2 and one setting in Ne/CO_2. Absolute gain estimation was performed using the peaks at 9.4 keV and 41.6 keV. A setting of 100% refers to 400 V. The observed spectra in Ar/CO_2 and Ne/CO_2 show an almost identical peak structure, as an example the spectrum for a gain setting of 81% in Ar/CO_2 90/10 is shown in Fig. 1 along with the resulting gain curve for the GEM-TPC with Ar/CO_2 90/10 and gain values from a similar 10x10 cm GEM. For a relative gain-calibration of the readout-pads a modification of the so-called "leader-pad method" was chosen. In its original form it has been developed and successfully tested for the HARP TPC as described in [3]. Clusters associated with the highest energetic decay of 41.6 keV are selected by cuts on total ADC charge, pads per cluster, and central energy deposition. The accumulated charge on each pad in a selected cluster is calculated and the energy entries of the highest 3 pads ("leader-pads", LP) that carry the main charge are filled into a histogram for each pad. The parameters and pad size of the HARP TPC are different, such that the main charge was deposited on only one single pad. Regarding the pad size of the prototype TPC

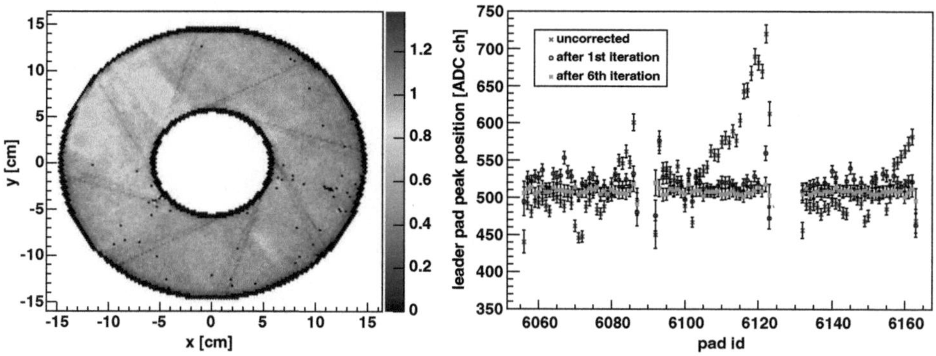

Fig. 2 Gain equalisation factors plotted for each channel over the pad plane and peak positions of central energy deposition of the 41.6 keV peak before and after performing several iterations of gain equalisation

and the width of the observed clusters the number of 3 pads was found to be optimal for this method. Normalised gain equalisation factors for the whole pad plane are calculated from the peak positions and applied as a correction factor to the raw data hits. A plot of the equalisation factors for each pad is shown in Fig. 2 (left). The sector borders of the GEM detector responsible for the higher correction factors can be clearly seen as these lines are less effective in charge collection and adjacent pads register a lower charge. The same is true for pads close to the field cage. They only register a fraction of the cluster charge. Thus the position of the LP cannot be determined clearly and the gain of such pads would be over-corrected. Such pads have been left out in this map and are shown with a correction factor of 0. Different methods using single channel spectra or single hit clusters have to be applied here. Several iterations of the LP method running the full cluster reconstruction have been used to correct for possibly wrongly reconstructed clusters in a previous iteration. After 3 iterations the relative change in the equalisation factors is as small as a few %, the resulting LP peaks before and after correction are shown on Fig. 2 (right). Gain variations show a periodic structure that has a dependence on the FE chip channels, geometrical properties and the GEM foils. The gain distribution has shown to be stable over one week taking several krypton measurements and running a calibration on each dataset.

References

1. Fabbietti, L. et al.: Nucl. Instrum. Methods A **628**, 204 (2011)
2. Hallermann, L.: Ph.D. thesis, DESY (2010)
3. Dydak, F.: HARP memo **04–103** (2004)
4. Gaertner, S.: Diploma thesis, Techn. Univ. Darmstadt (2008)
5. Eggert, T.: Star Note **424** (2000)
6. Rasulbaev, M. et al.: Appl. Radiat. Isotopes **66**(12), 1838 (2008)
7. Lasiuk, B., Whitten, C.A.: Star Note **360** (1998)

Hyperfine Interact (2012) 211:57–58
DOI 10.1007/s10751-012-0630-7

Creativity-innovation - the seed for frontier science

Johann Marton · Eberhard Widmann · Johann Zmeskal

Published online: 4 May 2012

The Symposium to celebrate Paul Kienle's 80th birthday titled: "Creativity and Innovation – the Seed for Frontier Science", highlights the broad and extremely successful scientific life of Paul Kienle, being an inventor and leader of frontier experiments on subatomic physics.

During his stay in Vienna as the director of the Stefan Meyer Institute (SMI), he was one of the founding fathers of this conference series Exotic Atoms and Related Topics. Therefore, with this Symposium - within EXA2011 - we also want to thank him for the time he spent in Vienna, always open to discuss ongoing and new physics projects, and for the many exciting and fruitful physics meetings we had.

The broad field of Paul Kienle's scientific work is reflected in the talks given by distinguished physicists in this afternoon, started with the Mößbauer effect and ended with an overview of the planned Facility for Antiproton and Ion Research (FAIR), to become one of the largest research facilities in the world, triggered by Paul Kienle. We have selected two contributions to be printed in these proceedings, representing the scientific life of Paul Kienle dealing with two quite different scientific fields.

J. Marton · E. Widmann · J. Zmeskal (✉)
Stefan-Meyer-Institute for Subatomic Physics, Austrian Academy of Sciences,
Boltzmanngasse 3, 1090 Vienna, Austria
e-mail: Johann.zmeskal@oeaw.ac.at

The programme started with a Laudatio given by Helmut Rauch, followed by talks covering the physics interest of Paul Kienle:

Walter F. Henning	A love affair with quantum structures: From Moessbauer effect to neutrino oscillations
Toshimitsu Yamazaki	Highlights on strangeness physics
Wolfram Weise	Theory of the antikaon-nucleon interaction in the age of SIDDHARTA
Walter Greiner	Giant nuclear systems and the decay of the vacuum insupercritical fields
Walter Kutschera	Super-heavy elements in nature
Günther Rosner	Facility for Antiproton and Ion Research (FAIR)
Laura Fabbietti	Paul Kienle's life in pictures

It was a great honour to celebrate Paul Kienle's birthday and we like to express our gratitude for his tremendous impact to modern nuclear physics, which was also extremely important for our Institute. We are confident that Paul Kienle will keep his enthusiasm in creating new methods and in triggering the work on new technologies to solve open problems in physics. Last but not least we are looking forward to a lot of exciting and interesting discussions and also for his advices to clarify open physics issues in our research fields.

Hyperfine Interact (2012) 211:59–67
DOI 10.1007/s10751-012-0629-0

Fundamental problems in physics – dedication to my friend Paul Kienle

Walter Greiner

Published online: 2 May 2012

I am happy to be here and honor Paul Kienle at his 80th birthday. He is an outstanding physicist. I met him first in 1964 when I came back from the University of Maryland (where I was Assistant Professor) to my alma mater Freiburg in Breisgau.

I heard from my teacher Hans Marschall and from Peter Brix that Paul Kienle was a young Extraordinarius at the Technische Hochschule in Darmstadt and that he applied the Mösbauer-effect to nuclear problems. Indeed I found the paper by P. Brix, S. Hüfner, P. Kienle and D. Quitman in which they measured the change of the nuclear mean square radius $< r^2 >$ of the first excited state of a nucleus relative to the ground-state (see Fig. 1).

I was very impressed and applied this right-away to test the collective model. Most of my calculations agreed with Paul Kienle's measurements. In my book with J.M. Eisenberg "Nuclear Theory I: Nuclear Models " I described Paul's work. This was—I think—Paul's first entrance into a widely spread textbook.

At this early time I had the choice of becoming Professor at various places; also in Darmstadt, but I decided to accept the chair in Frankfurt where I worked on various problems in Heavy Ion and Photonuclear Physics (Nuclear Molecules, Superheavy Elements, Coupling of Giant Resonances to Nuclear Surface vibrations etc.).

But the most fascinating work was that on Quantum Electrodynamics of Strong Fields (with B. Müller, J. Rafelski, J. Reinhardt, G. Soff, ...)—see Fig. 2. The decay of the vacuum in overcritical Coulomb fields is—in my mind—the most fundamental process in Quantum Electrodynamics. We predicted spontaneous e^+e^- - pair creation with the electron bound and the positron being emitted. Paul Kienle, always striving for fundamental physics, became immediately interested and—together with his collaborators developed the Double Orange e^+e^- - Spectrometer (Fig. 3). He observed the enormous increase of positrons produced in heavy ion collisions while approaching the supercritical charge (Fig. 4).

At first we expected in the positron spectrum a sharp line, which should be broadened because of the various sticking times of the heavy ions. It was only in recent years when Zagrebaev and I realized that the sticking times can be fixed due to the mass transfer or, also, due to the energy loss in deep inelastic encounters of heavy ions (Figs. 5 and 6).

W. Greiner (✉)
Frankfurt Institute for Advanced Studies (FIAS), Frankfurt, Germany
e-mail: greiner@fias.uni-frankfurt.de

Fig. 1 Isomer Shift on ^{151}Eu

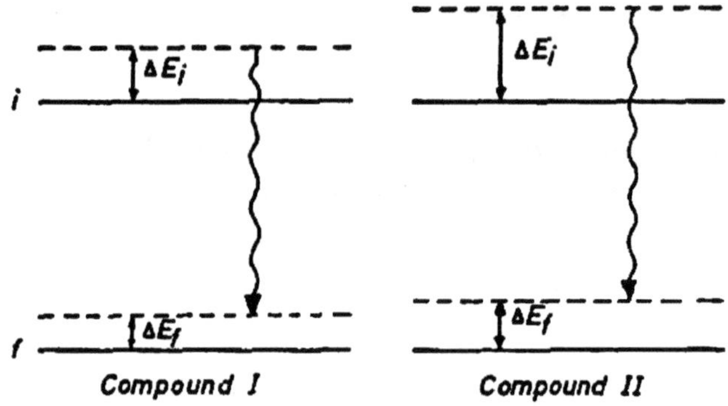

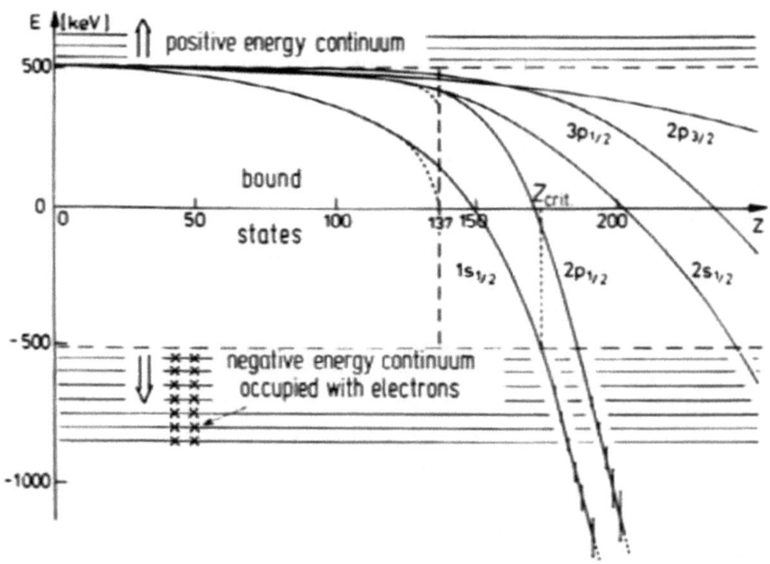

Fig. 2 The electron-positron-vacuum as a function of the proton charge Z of a central nucleus

Fig. 3 Double Orange e^+e^- Spectrometer

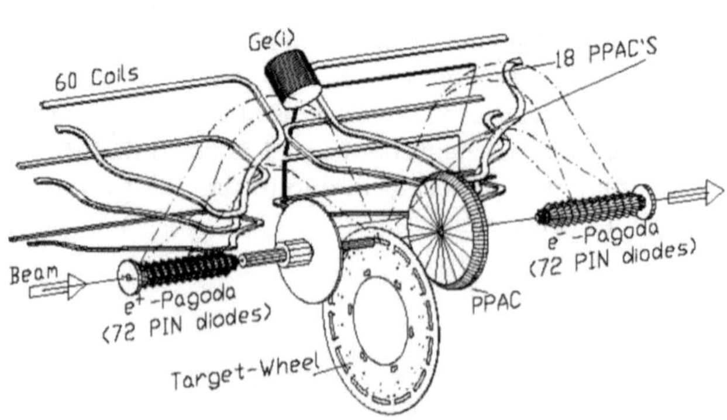

Fig. 4 Positrons from Uranium-atom collisions

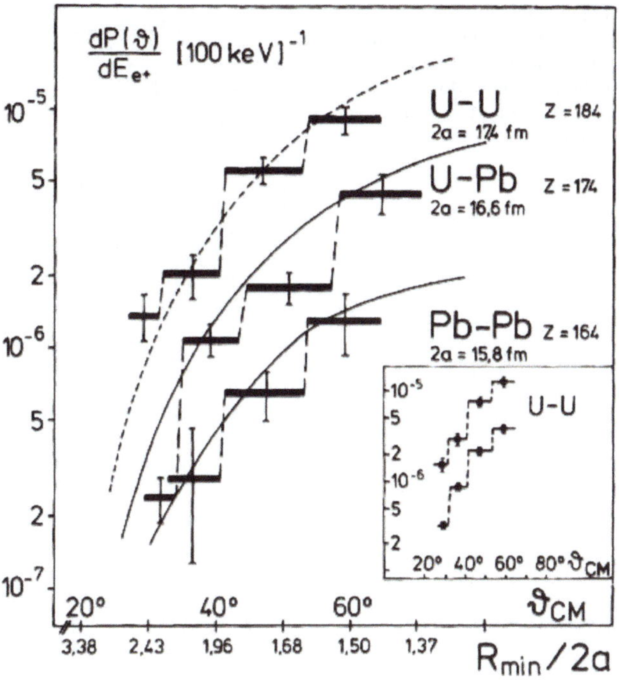

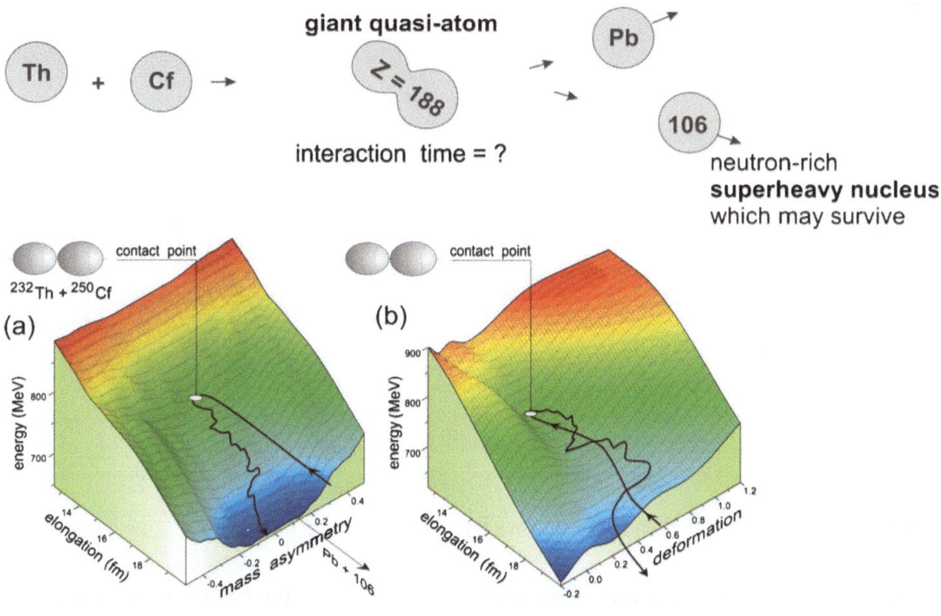

Fig. 5 The nuclei collide and exchange nucleons/deformations along the path. This "glues" them together

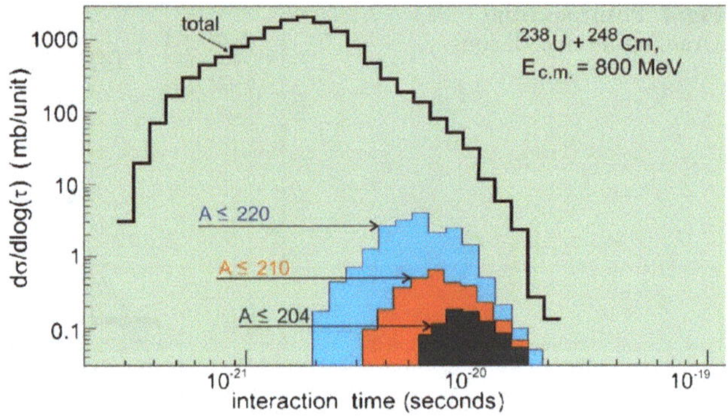

Fig. 6 Delay time distributions and possibility of spontaneous positron formation according to the theory of Zagrebaev and Greiner

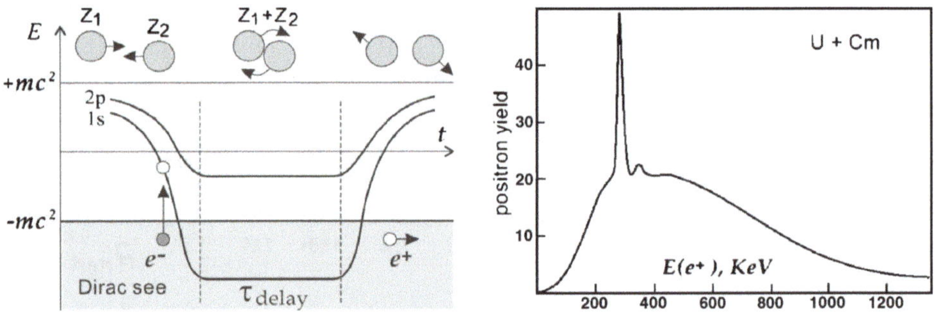

Fig. 7 Sharp time delay τ_{delay} of colliding nuclei leads to a narrow positron emission line

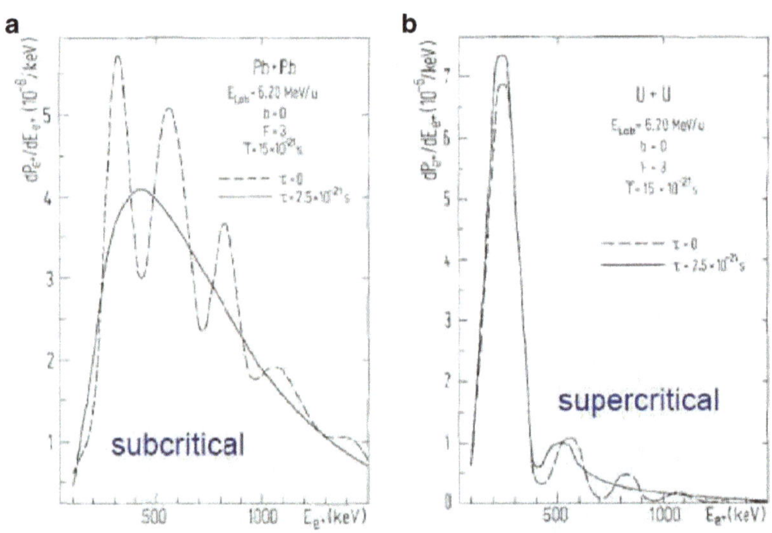

Fig. 8 Positron creation in time-delayed heavy ion collisions

We have developed a theory for calculating the connection between sticking, mass transfer and energy loss. Instead of a narrow line, as originally expected (Fig. 7) a huge line structure in the e+ -spectrum as shown in Fig. 8 is to be expected. For

Fig. 9 Long-living di-nuclear systems in U+U collisions. Large mass transfer (≈30 nucleons) is observed (Pb-like transfer products)

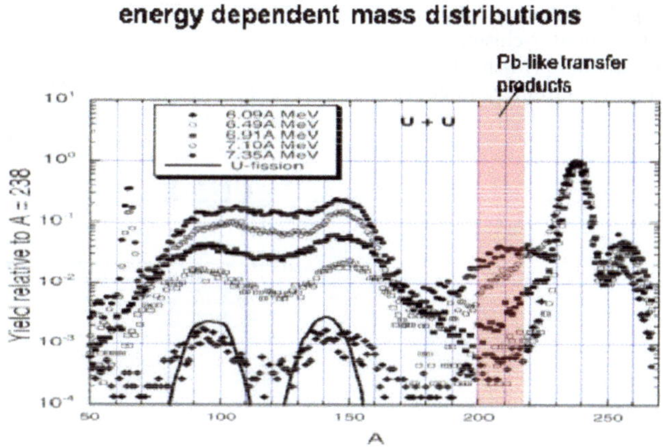

Fig. 10 Schematic illustration of positive and negative energy states of nucleons (Mishustin, Satarov, Auerbach)

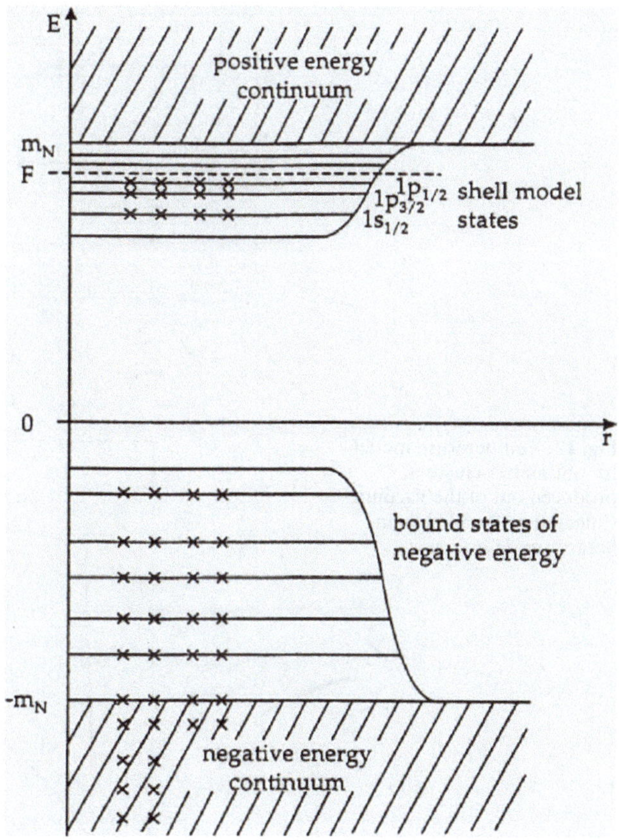

that purpose coincidence experiments (positron spectrum for large mass-transfer in e.g. U+U collisions) have to be performed. Sophie Heinz, a former student of Paul Kienle, is exploring and pushing this research and has already done some promising experiments carried out at Ganil (Fig. 9).

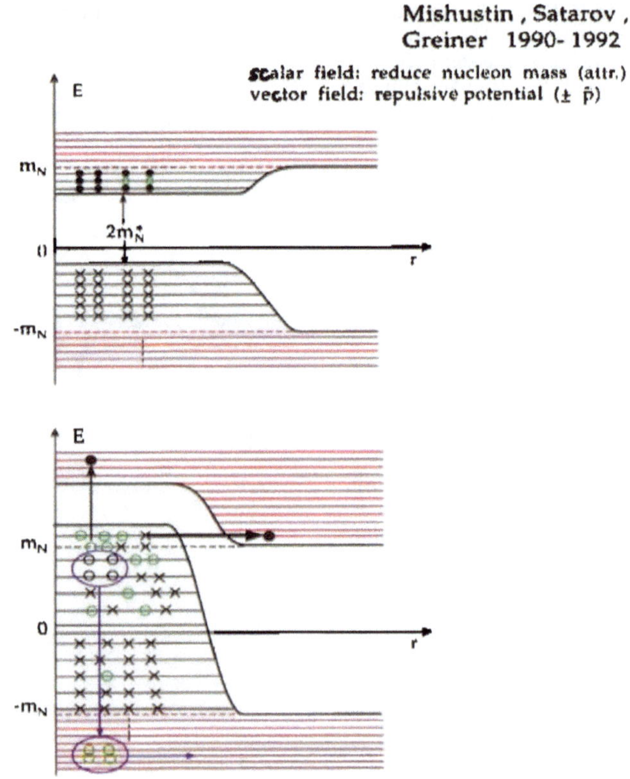

Fig. 11 Schematic illustration of antimatter-cluster production. (Mishustin, Satarov, Greiner, Stöcker)

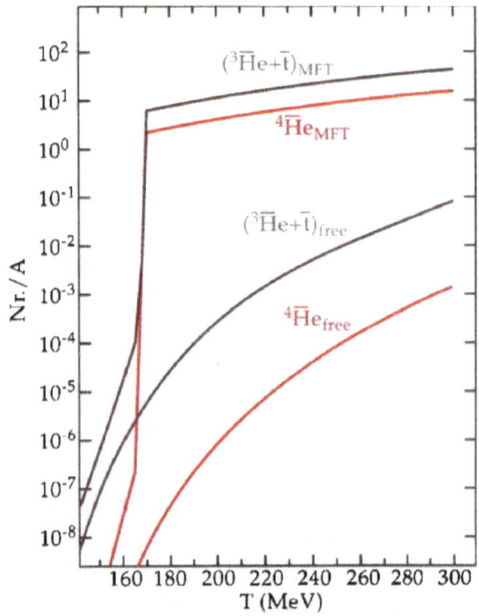

Fig. 12 Temperature model for antimatter clusters produced out of the vacuum. Calculations by Mishustin, Satarov and Greiner

Fig. 13 Nuclei and antinuclei yields in the experiment of STAR collaboration

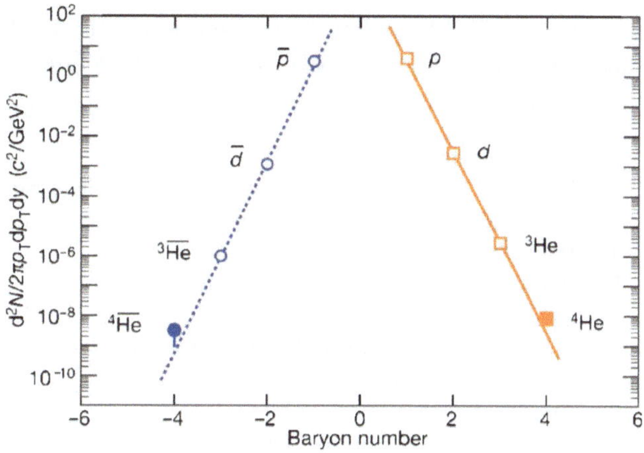

Not only the e^+e^- - vacuum is fundamental and interesting, but also the vacuum structure of the nuclear fields. The vacuum structure for the nucleons based on the Lagrangian according to the Relativistic Mean Field Theory shows fascinating new aspects. The positive and negative energy states of the nucleons are exhibited in Fig. 10. The upper well, emerging from the positive energy continuum, represents the nuclear shell model. The lower well, emerging out of the negative energy continuum, is new.

Under compression as e.g. in nuclear shock waves in a nucleus-nucleus collision this vacuum structure changes as exhibited in Fig. 11: The nuclear shell model vanishes completely. The lower well whose many levels are originally occupied by nucleons, increases tremendously and its levels are partly emptied due to the heat production (Rankine-Hugoniot) and spontaneous nuclear emission. Empty bound states in this well are obviously bound antinucleons. Hence it is now possible to produce directly antimatter clusters out of the vacuum. The conventional idea to produce antimatter clusters by anti nucleon emission into phase space followed by the capture of these anti nucleons into clusters is—intuitively quite clear—much less effective. A temperature model (Fig. 12) yields an increase of e.g. anti-helium production by 8 orders of magnitude.

Recent experiments at Brookhaven (Fig. 13) seem to indicate that anti-helium-production is indeed by a factor 6–8 larger than the extrapolation of ordinary anti-helium production.

At the FAIR-accelerator at GSI, where I expect much higher compression (the $\sim$ 40 GeV/nucleon for FAIR was chosen, because our various theoretical models yield higher compression in nucleus-nucleus encounters around this energy/nucleon). We are thus heaving the fantastic possibility to extend the Periodic System into the direction of antimatter.

I cannot describe it here, but I close with the remark that also strange matter clusters can be produced in a very similar way. Carsten Greiner, Horst Stöcker, Avraham Gal and others studied and predicted that strange matter clusters should exist.

Finally, I emphasize again, that Paul Kienle is a physicist of the extraordinary kind: He is interested in fundamentals, works hard and pushes the field forward. As a director at GSI he initiated the extension of the Unilac-GSI to higher energies. At the end I show a few pictures characterizing his time and activities at GSI.

Comments to the figures following their order from above left to the final right:

1. Paul is adored. He likes it!
2. Paul and the government authority (H. Reisenhuber)
3. Paul and the Gräfin zu Hessen
4. Paul likes exotics
5. A molecular state
6. Two bavarians
7. Paul with angular momentum. Fusion of two clusters.
8. The GSI supreme rulers
9. The FC-Bayern fan
10. Here the end of the test-phase of the injector device and its transfer to CERN are celebrated.

GSI-LBL CERN INJ

Hyperfine Interact (2012) 211:69–75
DOI 10.1007/s10751-012-0601-z

Toward cold and dense antikaonic nuclear clusters

Talk at symposium "Creativity-Innovation—the seed for frontier science" on the occasion of the 80th birthday of Professor Paul Kienle

Toshimitsu Yamazaki

Published online: 16 March 2012

Abstract Experimental search for cold and dense anti-kaonic nuclear cluster systems has been tried since 1998. Recently, an important indication for the most basic cluster K^-pp has been obtained from old data of DISTO on $p + p \rightarrow K^+ + p + \Lambda$ through $\Lambda(1405)$ production as a doorway. This success now triggers an extended search for a double kaonic cluster K^-K^-pp. Ultimately we have possibility to investigate multi-kaonic objects and kaonic strangelets in heavy-ion reactions. In this paper I trace my personal reminiscences of collaborative work with Paul Kienle during which I enjoyed the same feeling and excitement as Paul for more than a decade.

Keywords Dense nuclei · Kaonic nuclear cluster · Multi-kaonic nuclear clusters

1 Discovery of heavy pionic nuclei and partial restoration of chiral symmetry breaking

In early 1990's when I was struggling with my young colleagues on how to obtain a high-energy deuteron beam to produce deeply bound pionic states in heavy nuclei [1–3], Paul Kienle, then the director of GSI, was constructing a SIS18 synchrotron together with its associated experimental facilities. Since it was dedicated to heavy-ion physics, it would be out of question to expect a deuteron beam from the just born baby, but, surprisingly, the laboratory chef, Paul, proposed us to use the deuteron beam combined with the FRagment Separator (FRS) as a forward magnetic spectrometer !! After all, in the first run in 1996 we succeeded in observing the predicted deeply bound pionic states (1s and 2p) in ^{207}Pb [4]. Successively, we extended our

T. Yamazaki (✉)
Department of Physics, University of Tokyo, 7-3-1 Hongo, Bunkyo-ku, Tokyo 113-0033, Japan
e-mail: yamazaki@nucl.phys.s.u-tokyo.ac.jp

T. Yamazaki
Nishina Center, RIKEN, 2-1 Hirosawa, Wako-shi, Saitama-ken 351-0198, Japan

research to isotope shifts of pionic bound states in Sn isotopes, motivated by the idea of Paul to find experimental signature on chiral symmetry restoration in the nuclear medium [5, 6] based on the scenario of spontaneous chiral symmetry breaking of Nambu and Jona-Lasinio [7, 8].

A successful experiment toward this direction was carried out [9], and the isovector part of the πN interaction (represented by a parameter b_1) was shown to be significantly enhanced over the free nucleon case, indicating that

$$R_{\pi N}(\rho) = \frac{b_1(\rho)}{b_1(0)} = \frac{| < \bar{q}q >_0 |}{| < \bar{q}q >_\rho |} \propto \frac{1}{1 - \alpha\rho}, \tag{1}$$

where $\rho =$ the nuclear density, and $< \bar{q}q >_0$ and $< \bar{q}q >_\rho$ are (u, d) quark condensates at vacuum and at nuclear density ρ, respectively. The density dependence is expressed above with a parameter α, which was determined to be $\alpha = (0.36 \pm 0.08) \times \rho_0^{-1}$ [9]. In this way, Paul made essential contributions both experimentally and theoretically.

2 Kaonic nuclear clusters predicted: many surprises

In 1998, when we succeeded in finding the ground state of π^- in ^{205}Pb, Paul and myself decided to extend pionic nuclear spectroscopy to kaonic nuclear spectroscopy. By then, already interesting indications of strongly attractive $\bar{K}N$ interactions were known experimentally and theoretically, but how?

The key element was the $\Lambda(1405)$ resonance (abbreviated here as Λ^*), which was interpreted as the $I = 0$ K^--p bound state. Akaishi constructed complex $I = 0$ and $I = 1$ $\bar{K}N$ potentials by a $K^-p - \Sigma\pi$ coupled-channel procedure, and applied them to few-body nuclear problems by a variational method [10]. In this way, deeply bound $\bar{K}$ nuclei (called KNC after *Kbar Nuclear Cluster*) with moderate decay widths (because of the deep binding no decay to $\Sigma\pi$ final states is possible) were predicted. The most surprising and exciting consequence was an enormously large nuclear density; the strongly attractive force overcomes the short-range NN repulsion, producing exotic nuclei with average nuclear density as much as 3 times the normal density ρ_0. Here, for the first time, we showed a new possibility of nuclear systems that violate *the law of constant nuclear density*, which had been believed to be something like a fundamental law in nuclear physics. Comprehensive *unconstrained* many-body calculations were carried out [11, 12], which revealed more surprises.

3 Structure of K^-pp

Figure 1 shows various aspects of K^-pp, the most fundamental KNC. This was first predicted to be a strongly bound system [13], which possesses all the important features of KNC's [14–16]. Recent Faddeev calculations of this three-body system [17–19] also yielded a strongly bound K^-pp mass similar to the YA value [13].

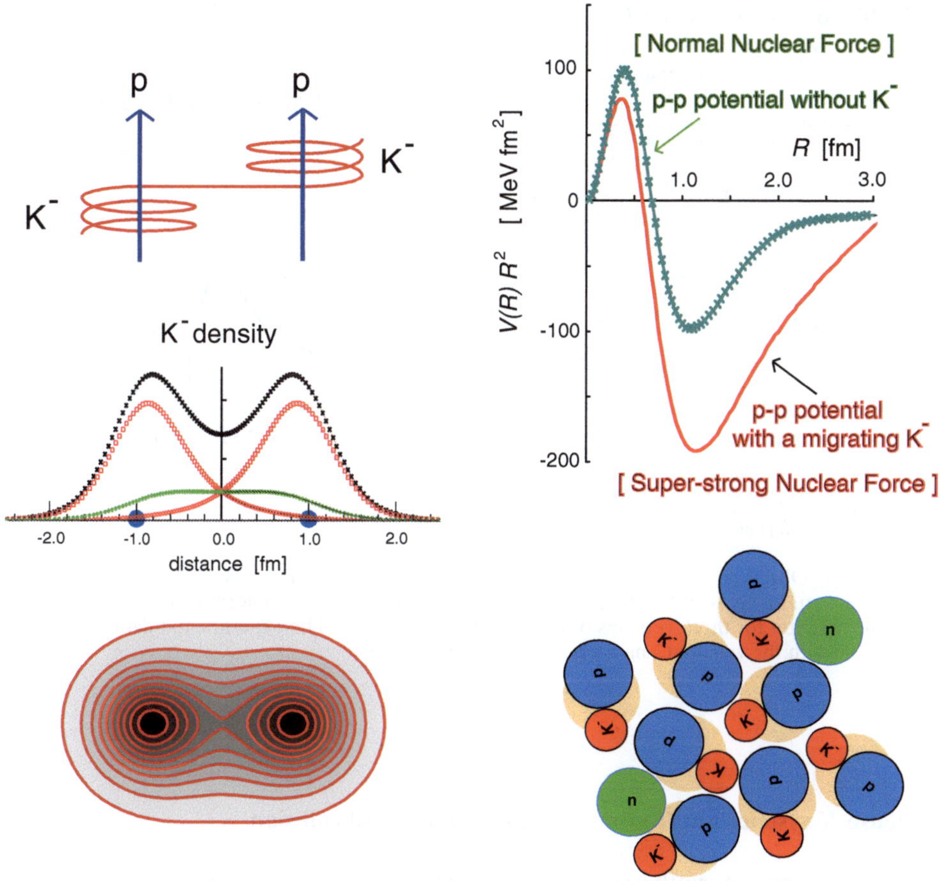

Fig. 1 (*Left*) Super-strong nuclear force, a new kind of covalent force induced by a real boson, K^-, between two protons, is revealed in K^-pp. (*Upper right*) The nucleon-nucleon potential (SSNF, *red curve*) in K^-pp as compared with the Yukawa–meson-mediated ordinary nuclear force (*green curve*). (*Lower right*) A cold and dense multi-baryon system with migrating $\bar{K}$ mesons. It is nearly equivalent to a multi-Λ^* system. From [14, 15]

The structure of K^-pp and its production by $p + p$ reactions is described in [14, 15]. The spatial distributions and correlations of K^-, p and p were studied, from which a molecule-type covalent bonding was recognized. Namely, a real boson, K^-, migrates between the two protons, forming $\Lambda^* - p \leftrightarrow p - \Lambda^*$, causes a very attractive exchange potential, as shown in Fig. 1. This is a new mechanism for nuclear binding, as Kazuhiko Nishijima called it *"kaonic origin of nuclear force"*, whereas the ordinary nuclear force is called *"pionic origin of nuclear force"*. This kaonic force is about 4 times as intense as the pionic force, and thus, we often call it *super-strong nuclear force (SSNF)*. This is the origin of the enormously dense nuclear systems. One can conceive such migrating $\bar{K}$'s in a multi-Λ^* system, as shown in Fig. 2 (lower right). This study justifies the molecular description of K^-pp by [20].

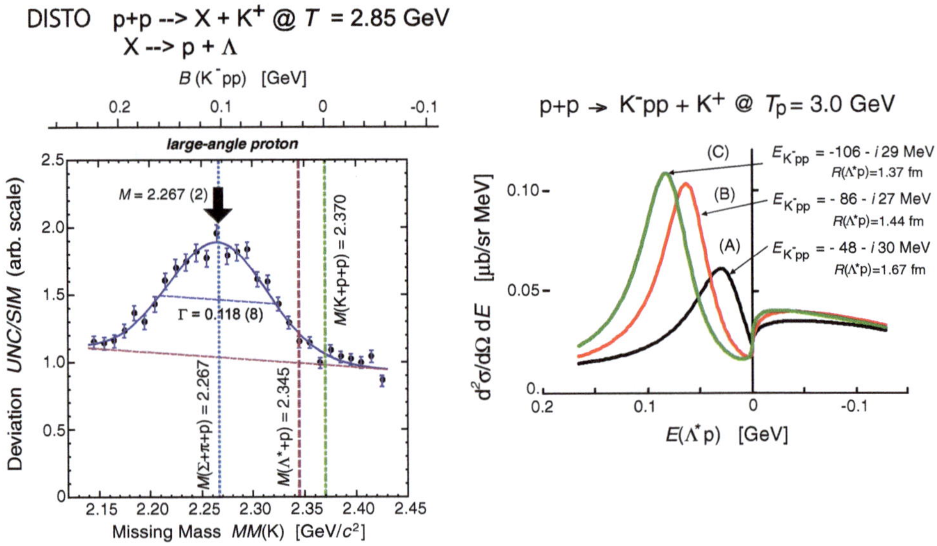

Fig. 2 (*Left*) Observed *DEVIATION* spectrum of the missing-mass of K^+, showing a dominant peak of K^-pp, in $pp \to K^+ + K^-pp$ [21]. (*Right*) Theoretical mass (energy) spectra of K^-pp in the same reaction for three versions of the $\bar{K}N$ interaction [15]. The experimental data seem to be compatible with the version C, namely, the 25% enhanced one

4 Dense $\bar{K}$ systems not from soft landing but by hard collision

For the purpose of synthesizing an exotic nuclear bound state, the traditional way was to use a reaction of low momentum transfer (*soft landing* and *large sticking*) of an exotic particle. Well known examples are seen in hypernuclear spectroscopy using (K^-, π^-) reactions and in pion-transfer reactions of $(d, {}^3\text{He})$ type. Thus, we tended to pursue nuclear reactions of low momentum transfer. During the course of designing possible experiments of easy experimental accesses using a proton beam as a projectile we came to the simplest $p + p$ reaction, but we thought it must be an impossible experiment because of a huge momentum transfer which would suppress sticking of K^- with nucleons. It was when Paul and myself were in a desperate mood that Akaishi found a totally new reaction mechanism, that is, *"large sticking of* $\Lambda^* +$ *p at high momentum transfer and short-range* $p + p$ *collisions"* [14, 15].

Akaishi clarified the following. i) A Λ^* is produced in a short range ($\sim \hbar/m_\rho \sim 0.7$ fm) and with high momentum transfer ($q \sim 1.6$ GeV/c) in a $p + p \to p + \Lambda^* + K^+$ reaction at 3 GeV. ii) This Λ^* and the participating p stick together at high probability, *if and only if the* $\Lambda^* - p$ *composite forms a dense bound state*. This is nothing but the predicted deeply bound K^-pp !!

Paul knew that there were lots of $p + p \to p + \Lambda + K^+$ reaction data at DISTO, in which a treasure might be hidden. So, we approached this group and we all agreed to form a collaboration group to make analyses of the data at 2.85 GeV. It was in September of 2006, five years ago from now, when we celebrated Paul's 75th birthday in Vienna.

5 New finding in DISTO and its implications

The first striking fact we observed was the proton angular distribution in $p + p \to p + \Lambda + K^+$ reactions was mostly forward/backward peaked, and the separated large-angle component revealed an intriguing peak in what we call *DEVIATION* spectra of $M(\Lambda p)$ and $\Delta M(K^+)$ [21], as shown in Fig. 2 (left). This peak, we call $X(2265)$, was found to be as strong as the $\Lambda(1405)$ production by using a neutral hyperon spectrum of $\Delta M(pK^+)$ [21]. This fact was understood as evidence for dense K^-pp production through the Λ^* doorway mechanism. Later, we confirmed the role of Λ^* in the formation of $X(2265)$ from its excitation function [22]. It is to be noted that this peak is consistent with an earlier observation in stopped-K^- reactions at FINUDA [23].

The peak, when interpreted as K^-pp, indicates: its mass $M_X = 2267 \pm 7$ MeV/c^2, width $\Gamma = 118$ MeV, and $B_K = 103$ MeV. The binding energy is deeper than the original prediction of 48 MeV. It is translated into an effective $\bar{K}N$ interaction, which is about 25 % enhanced over the original AY interaction [13, 15], as shown in Fig. 2 (right). It suggests various missing mechanisms [14]. One of them might be a density-dependent interaction arising from chiral symmetry restoration, similar to the case of the πN interaction, (1). This effect may become enormous, because an enhanced $R_{\bar{K}N}(\rho)$ would produce a denser and denser nuclear system, which in turn enhances the $\bar{K}N$ interaction furthermore.

6 Toward double- and multiple-$\bar{K}$ nuclear clusters

The next fundamental object is K^-K^-pp, which was predicted to exist with a large binding energy, 117 MeV [27], as shown in Fig. 3. How can we produce such an exotic object? Paul proposed to use a stopped $\bar{p}$ reaction in ^{3}He [24]. Another idea came out after finding the DISTO experiment. Namely, one can make use of double-Λ^* doorway in $p + p$ reactions to produce a high-density $\Lambda^* - \Lambda^* \to K^-K^-pp$. This process has recently been studied theoretically. Since the observed K^-pp bound state is much deeper than the original prediction, we assume the K^-K^-pp state to be also much deeper than the prediction and study the change of the spectrum shape.

The diagram of the proposed reaction and predicted $M(K^-K^-pp)$ distributions are presented in Fig. 4. The bound-state peak grows when the assumed bound-state becomes deeper from −50 to −200 MeV. Similar to the case of the K^-pp production, the K^-K^-pp production dominates, and its cross section increases with the density of the system. So, experiments of $p - p$ collisions with two K^+'s as a spectator are waited for.

One can directly produce those kaon clusters also in proton-induced and heavy-ion reactions on nuclear targets [27] (see Fig. 5). Some attempts were made by using the FOPI detector at GSI (Herrmann, private communication, [28]), but the incident energy of 2 GeV/N was too low to cause direct productions. We need the beam energy quite high to produce single-$\bar{K}$ and double-$\bar{K}$ systems in direct reactions (not through thermal equilibrium [29]): $T_p \sim$ 3GeV/N and $T_p \sim$ 7GeV/N, respectively. Thus, we can foresee a paradise of $\bar{K}$ nuclear physics at FAIR of Darmstadt (Germany) and at J-PARC of Tokai (Japan). In addition, it involves search for metastable strangelets.

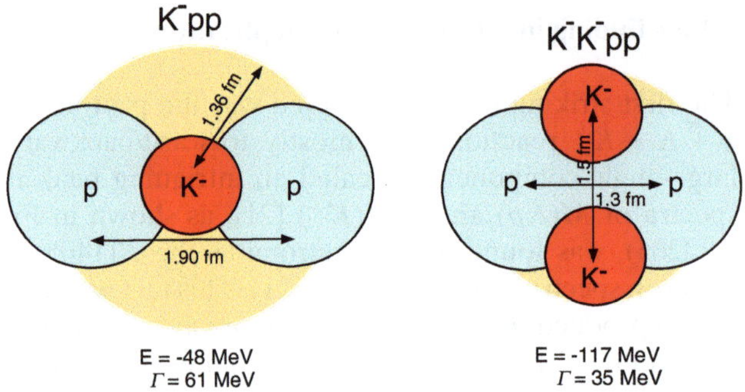

Fig. 3 Schematic structure diagrams for the calculated $K^- pp$ and $K^- K^- pp$ nuclei. The rms radius of K^- and the rms inter-nucleon and inter-$\bar{K}$ distances are shown. From [27]

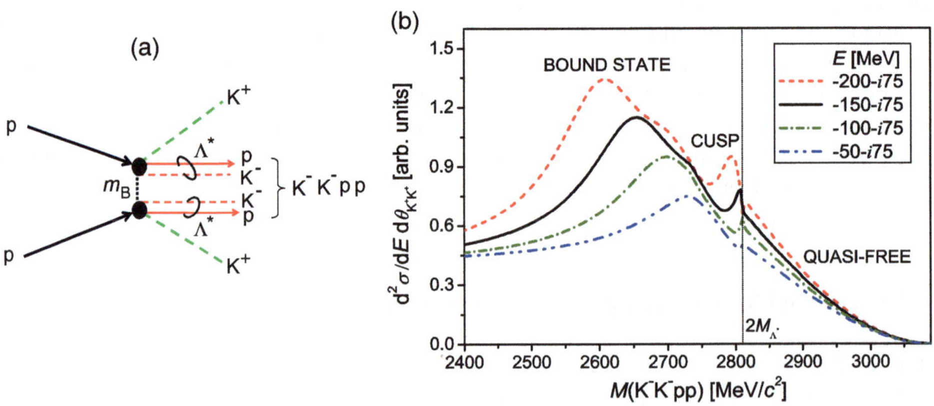

Fig. 4 (Color online) (a) Elementary process for $p + p \to K^+ + K^+ + K^- K^- pp$ via the production of two Λ^*'s as doorways with an intermediate boson of mass m_B. (b) Differential cross sections for various bound-state energies, E, of the $K^- K^- pp$ system for $T_p = 7.0$ GeV, $\Gamma = 150$ MeV, b (collision length) $= 0.3$ fm. From [26]

Fig. 5 Schematic picture of the formation of basic kaonic nuclear clusters, $K^- pp$ and $K^- K^- pp$ in heavy-ion reactions. From [25]

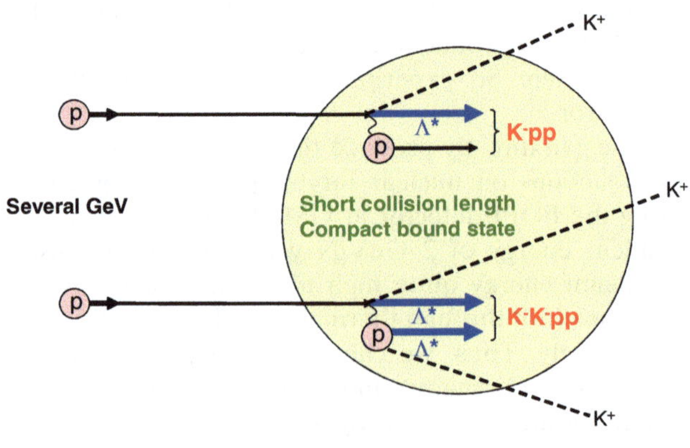

Acknowledgements The author acknowledges the enthusiatic collaborative work with Prof. Paul Kienle and Prof. Yoshinori Akaishi. This work is supported by Grant-in-Aid for Scientific Research of Monbu-Kagakusho of Japan.

References

1. Toki, H.,Yamazaki, T.: Phys. Lett. **B 213**, 129 (1988)
2. Toki, H., Hirenzaki, S., Hayano, R.S., Yamazaki, T.: Nucl. Phys. **A 501**, 653 (1989)
3. See details in Yamazaki, T., Hirenzaki, S., Hayano, R.S., Toki, H.: Phys. Reports (2012, in press)
4. Yamazaki, T., et al.: Z. Phys. **A355**, 219 (1996)
5. Kienle, P., Yamazaki, T.: Phys. Lett. **B 514**, 1-6 (2001)
6. Kienle, P., Yamazaki, T.: Prog. Part. Nucl. Phys. **52**, 85–132 (2004)
7. Nambu, Y., Jona-Lasinio, G.: Phys. Rev. **122**, 345 (1961)
8. Nambu, Y., Jona-Lasinio, G.: Phys. Rev. **124**, 246 (1961)
9. Suzuki, K., *et al.*, Phys. Rev. Lett. **92**, 072302-1 (2004)
10. Akaishi, Y., Yamazaki, T.: Phys. Rev. **C 65**, 044005 (2002)
11. Doté, A., Horiuchi, H., Akaishi, Y., Yamazaki, T.: Phys. Lett. **B 590**, 51 (2004)
12. Doté, A., Horiuchi, H., Akaishi, Y., Yamazaki, T.: Phys. Rev. **C 70**, 044313 (2004)
13. Yamazaki, T., Akaishi, Y.: Phys. Lett. **B 535**, 70 (2002)
14. Yamazaki, T., Akaishi, Y.: Proc. Jpn. Acad. Ser. **B 83**, 144 (2007)
15. Yamazaki, T., Akaishi, Y.: Phys. Rev. **C 76**, 045201 (2007)
16. Akaishi, Y., Yamazaki, T., Obu, M., Wada, M.: Nucl. Phys. **A 835**, 67–74 (2010)
17. Shevchenko, N.V., Gal, A., Mares, J.: Phys. Rev. Lett. **98**, 082301 (2007)
18. Shevchenko, N.V., Gal, A., Mares, J., Révai, J.: Phys. Rev. **C 76**, 044004 (2007)
19. Ikeda, Y., Sato, T.: Phys. Rev. **C 76**, 035203 (2007)
20. Faber, M., Ivanov, A.N., Kienle, P., Marton, J., Pitschmann, M.: Int. J. Mod. Phys. **E 20**, 1477–1490 (2011)
21. Yamazaki, T., *et al.*: Phys. Rev. Lett. **104**, 132502 (2010)
22. Kienle, P., *et al.*: arXiv-1102.0482v1
23. Agnello, M., *et al.*: Phys. Rev. Lett. **94**, 212303 (2005)
24. Kienle, P.: Int. J. Mod. Phys. **A 22**, 365–373 (2007)
25. Yamazaki, T., Akaishi, Y., Hassanvand, M.: Proc. Jpn. Acad. **B 87**, 362 (2011)
26. Hassanvand, M., Akaishi, Y., Yamazaki, T.: Phys. Rev. **C 84**, 15207 (2011)
27. Yamazaki, T., Doté, A., Akaishi, Y.: Phys. Lett. **B 587**, 167–174 (2004)
28. Herrmann, N.: J. Phys. G Nucl. Part. Phys. **37**, 094036 (2010)
29. Braun-Munzinger, P., Heppe, I., Stachel, J.: Phys. Lett. **B 465**, 15–20 (1999)

Author Index

MIX
Papier aus verantwortungsvollen Quellen
Paper from responsible sources
FSC® C105338

If you have any concerns about our products,
you can contact us on
ProductSafety@springernature.com

In case Publisher is established outside the EU,
the EU authorized representative is:
Springer Nature Customer Service Center GmbH
Europaplatz 3, 69115 Heidelberg, Germany

Printed by Libri Plureos GmbH
in Hamburg, Germany